3rd International Youth Conference on Interdisciplinary Problems of Nanotechnology, Biomedicine and Nanotoxicology (Nanobiotech 2015)

AA002732

IOP Conference Series: Materials Science and Engineering
Volume 98

Tambov, Russia
21 – 22 May 2015

Editors:

Magne Refsnes
Alexander Gusev
Anna Godymchuk
Anna Bogdan

ISBN: 978-1-5108-1635-0
ISSN: 1757-8981

Printed from e-media with permission by:

Curran Associates, Inc.
57 Morehouse Lane
Red Hook, NY 12571

Some format issues inherent in the e-media version may also appear in this print version.

Copyright© (2015) by the Institute of Physics
All rights reserved. The material featured in this book is subject to
IOP copyright protection, unless otherwise indicated.

Printed by Curran Associates, Inc. (2016)

For permission requests, please contact the Institute of Physics
at the address below.

Institute of Physics
Dirac House, Temple Back
Bristol BS1 6BE UK

Phone: 44 1 17 929 7481
Fax: 44 1 17 920 0979

techtracking@iop.org

Additional copies of this publication are available from:

Curran Associates, Inc.
57 Morehouse Lane
Red Hook, NY 12571 USA
Phone: 845-758-0400
Fax: 845-758-2634
Email: curran@proceedings.com
Web: www.proceedings.com

Table of contents

Volume 98

**3rd International Youth Conference on Interdisciplinary Problems of Nanotechnology, Biomedicine and Nanotoxicology (Nanobiotech 2015)
21–22 May 2015, Tambov, Russia**

**Accepted papers received: 14 October 2015
Published online: 3 November 2015**

Preface

Papers

IOP Conf. Series: Materials Science and Engineering **98** (2015) 011001 doi:10.1088/1757-899X/98/1/011001

Preface

The **3rd International Youth Conference "Interdisciplinary Problems of Nanotechnology, Biomedicine and Nanotoxicology"** (Nanobiotech2015) was held on 21-22 May 2015 in Tambov, Russia, and was jointly organized by Tambov Derzhavin State University (Russia), the Norwegian Institute of Public Health (Norway), the National University of Science and Technology MISiS (Russia), Tomsk Polytechnic University (Russia) and Tomsk State University.

The conference gathered experienced and young researchers, post-docs and students, working in the field of nanotechnologies, nanomedicine, nano(eco)toxicology and risk assessment of nanomaterials, in order to facilitate the aggregation and sharing of interests and results for better collaboration and visibility of activity.

The goal of Nanobiotech2015 was to bring researchers and practitioners together to share the latest knowledge on nanotechnology-specific risks to occupational and environmental health and assessing how to reduce these potential risks.

The main objective of the conference is to identify, systematize and solve current scientific problems in the sphere of nanobiotechnologies, nanomedicine and nanotoxicology, in order to join forces to determine prospective areas and compose working groups of interested co-workers for carrying out interdisciplinary research projects.

The topics of Nanobiotech2015 were: (1) Nanotechnologies in pharmaceutics and medicine; (2) Sources and mechanisms of nanoparticle release into the environment; (3) Ecological and biological effects of nanoparticles; (4) (Eco)toxicology of nanomaterials; (5) Methods for detection of nanoparticles in the environment and in biological objects; and (6) Physico-chemical properties of nanoparticles in the environment.

We want to thank the Organizing Committee, the universities and sponsors supporting the conference, and everyone who contributed to the organization of this meeting, for their contribution towards the conference and for their contributions to these proceedings.

The Organizing Committee of Nanobiotech2015

Editors:
Prof. Magne Refsnes
Norwagian Institue of Public Health, Norway
Dr. Alexander Gusev
Tambov Derzhavin State University, Russia
Dr. Anna Godymchuk
Tomsk Polytechnic University, Russia
Anna Bogdan
Tomsk Polytechnic University, Russia

Content from this work may be used under the terms of the Creative Commons Attribution 3.0 licence. Any further distribution of this work must maintain attribution to the author(s) and the title of the work, journal citation and DOI.
Published under licence by IOP Publishing Ltd

Peer review statement

All papers published in this volume of *IOP Conference Series: Materials Science and Engineering* have been peer reviewed through processes administered by the proceedings Editors. Reviews were conducted by expert referees to the professional and scientific standards expected of a proceedings journal published by IOP Publishing.

Content from this work may be used under the terms of the Creative Commons Attribution 3.0 licence. Any further distribution of this work must maintain attribution to the author(s) and the title of the work, journal citation and DOI.
Published under licence by IOP Publishing Ltd

Reproductive toxicity of carbon nanomaterials: a review

I Vasyukova[1], A Gusev[1,2] and A Tkachev[3]

[1]G.R. Derzhavin Tambov State University, 33, Internatcionalnaya st., Tambov, 392000, Russia
[2]National Research Technological University "MISIS", 4, Leninsky ave., Moscow, 119991, Russia
[3]Tambov State Technical University, 106, Sovetskaya st, Tambov, 392000, Russia

E-mail: vasyukovaia@gmail.ru

Abstract. In the current review, we assembled the experimental evidences of an association between carbon nanomaterials including carbon black, graphite nanoplatelets, graphene, single- and multi-walled carbon nanotubes, and fullerene exposure and adverse reproductive and developmental effects, *in vitro* and *in vivo* studies. It is shown that carbon nanomaterials reveal toxic effect on reproductive system and offspring development of the animals of various system groups to a certain degree depending on carbon crystal structure. Although this paper provides initial information about the potential male and female reproductive toxicity of carbon nanomaterials, further studies, using characterized nanoparticles, relevant routes of administration, and doses closely reflecting all the expected levels of exposure are needed.

1. Introduction

Carbon nanomaterials become more and more attractive and they gain wider popularity in the field of power accumulation, biomedicine, catalysis etc. It happens because of their ability to improve material properties significantly even being introduced in a small dosage due to their truly unique physical-chemical characteristics like high chemical stability, large specific surface, high flexibility and electric conductivity [1-4]. With their high biocompatibility, carbon nanomaterials show a great potential in the area of biomedicine as containers for addressed drug delivery as well as for bioimaging and tissue engineering purposes [5-11].

Potential toxicity of the carbon nanomaterials has recently attracted great attention because of their intensive industrial and medical utilization that brings not only obvious advantages but also probable human health hazard. At the same time, wide application of these materials will inevitably result in uncontrolled increase of their content in the environment that can't help causing alarm about safety of industrial waste, working places and surrounding objects because of their biological effect being insufficiently studied [12-15]. The results of investigation of carbon nanomaterials effect on reproductive functions of various systematic groups are currently contradicting and not systematized. In the present paper we present the experimental results of assessment of toxic effects on reproductive system of the animals displayed by carbon nanoparticles with various structures.

Content from this work may be used under the terms of the Creative Commons Attribution 3.0 licence. Any further distribution of this work must maintain attribution to the author(s) and the title of the work, journal citation and DOI.
Published under licence by IOP Publishing Ltd

2. Reproductive toxicity of carbon nanomaterials

2.1. Carbon Black
The results of the assessment of nanosized carbon black (CB) effect on the mice breed [16-18] show the absence of statistically significant difference in offspring quality from exposed females. However, in [19] malfunctions in gene expression of the offspring from mouse females exposed in pregnancy period were reported. The offspring of exposed mouse males have several indices of reproductive system functioning reduced, including an amount of sperm generated [20]. Results of the observations [21] of the influence on male reproductive system in experiments with mice of ICR line reveal negative effect of CB. Being administered intratracheally, the material at 0.1 mg/mouse dosage is shown to stimulate significant increase of testosterone level and force partial vacuolization of testicular tubules.

2.2. Graphite nanoplatelets, (GNPs)
Zanni et al. do not reveal any toxic effect of GNPs on reproductive functions of *Caenorhabditis elegans* [22].

2.3. Graphene
Toxicity assessment of graphene oxides (GO) at 0.5–100 mg/l dosage on model organisms *Caenorhabditis elegans* didn't show any toxic effect on reproductive system and offspring development [23-24].

In [25] the influence of nanostructured GO on reproductive system of mouse males of ICR line was in focus. GO were introduced through the tail vein in 6.25, 12.5 and 25 mg/kg dosage. According to the experimental results, secretion of gonadal hormones, testes and epididymis functioning as well as reproductive activity remains normal after the exposure. Among the offspring from unexposed females there was no significant difference revealed in the number, sex, weight and viability compared to the control group. Besides, authors give evidence to the absence of the toxical effect when GO are introduced intraperitoneally in higher dosage, namely up to 300 mg/kg. This fact lets them declare there is no or very small toxicity of GO for male reproductive system.

However, on exposing females, the GO negative effects reveal itself significantly both for female's organism and the offspring. In [26] the effect of GO on the females before and after breeding as well as on the offspring development was studied at GO dose of 6.25, 12.5, and 25 mg/kg. The dependence of toxic effect on fetuses upon pregnancy stage was detected. So, if females are exposed to GO before breeding or at the early pregnancy stages the offspring will be healthy with high probability. However, females' exposure in 6.25, 12.5 and 25 mg/kg dosage at the later stages of pregnancy leads to miscarriages and anomalies of fetuses, while the exposure at the maximal concentration led to death of 85% females. The results obtained give evidence to extremely high toxic effect of GO on females at late pregnancy stages. In [27] was shown that oral GO exposure of females in lactation period at 0.5 and 0.05 mg/ml dosage leads to the offspring defects consisting in blood biochemical indices altering, microstructure of gastrointestinal tract abnormalities, and general development inhibition. These results confirm the ability of GO to cause directly or indirectly offspring development abnormalities.

2.4. Single-walled carbon nanotubes
Studies of single-walled carbon nanotubes (SWCNT) biological effect on embryonic development of model species of *Danio rerio* (zebrafish) reveal significant embryonic development inhibition induced by more than 120 mg/l exposure. DWCNT exposure also leads to embryonic development inhibition, but at greater concentrations (more than 240 mg/l). The data obtained point to probable negative effects of SWCNT on flora and fauna when penetrated into the water environment [28].

SWCNT exposure of *Drosophila melanogaster* females to 0, 0.005%, 0.01%, 0.05%, 0.1%, 0.5% mass / volume concentrations over 14 days has no effect on reproductive system of females as well as on offspring's quantity and quality [29].

Investigation results of CNT embryotoxicity on mammals [30] prove the negative effect of non-functionalized and oxidized SWCNT from 100 ng/mouse to 30 μg/mouse dosage on mouse embryos development. The injurious effect on test objects of oxidized SWCNT is stronger than of non-functionalized ones, since the part of miscarriages and fetuses malfunctions within the former group was higher than that within the latter one. Similar results were obtained by authors on higher dosage investigation [31]. They introduced oxidized SWCNT at 3.3-1000 μg/kg dosage in pregnant females intravenously. When nanomaterial was administered at high concentrations the number of miscarriage was dramatically increased. Groups exposed to small dosage had apparent defects of embryonic development. There was an evidence of oxidative stress in placenta and embryo, but it didn't succeed in registering SWCNT there. These data are also confirmed by [29]. It was shown that oral SWCNT administration at 10 mg/kg dosage in pregnant females of CD-1 line during organogenesis significantly increased the number of fetuses resorption as well as led to morphological and skeletal anomalies development. In similar studies devoted to reproductive toxicity assessment when SWCNTs were introduced intraperitoneally at 1 and 10 mg/kg dosage in pregnant females the absence of toxic effect on maternal organism was observed. However, decrease in the number of offspring was detected in the group taking SWCNT at 10 mg/kg dosage. Such indices as lactation period, offspring viability and sex distribution didn't show any deviation. [32].

In [33] the study of ability to penetrate breast milk and distribution in offspring's tissue of tritium labeled shorted SWCNT-COOH was carried out on the model of lactating outbreeding rats. The material was introduced at 0.8 mg/kg dosage through one-time intravenous injection. As a result, authors establish SWCNTs penetration into rats breast milk which estimated peak concentration was approximately 0.12 μg/l. Quite uniform distribution of SWCNT in tissue and organs of pups including brain, kidneys, skeletal muscles etc. was detected. Maximal concentration was observed in blood, cardiac muscle and spleen, while the minimal one was in lungs.

As for toxic effect of SWCNT on male reproductive system, the result of mammalian spermatogonial chromosome aberration test also revealed the ability of SWCNT to increase the rate of teratosperm [34]. This fact indicates the toxicity of SWCNT for reproduction and development.

2.5. Multi-walled carbon nanotubes

Fluorescently labeled multi-walled carbon nanotubes (MWCNT) spread to the whole blastoderm after one-time injection in *Danio rerio* embryos. But in 96 hours after MWCNT introduction embryonic cells were completely cleaned from foreign substance by immune response generation. Their further development didn't have any anomalies, however, the next generation of *Danio rerio* from MWCNT exposed ones had less viability indices. It may be evidence to delayed negative MWCNT effect to the animals [35].

In order to assess the effect on female reproductive system of the mammals MWCNT were administered in rats through stomach pump at 40, 200, and 1000 mg/(kg·day). All animals had survived to the end of experiment. The results show that repeated MWCNT gavage at the studied dosage cause minimal toxic effect on female reproductive system. Thymus weight reduction was only detected in the group with high MWCNT dosage [36].

Assessment of MWCNT influence on mammal's embryonic development hasn't yielded unambiguous results yet. In [37, 38] MWCNT influence on impregnation successfulness was investigated, mouse females of the line C57Bl/6J were exposed. Nanomaterial was administrated intratracheally at the dosage of 67 μg per animal. As a result, inflammatory reactions in lung and liver of exposed females were detected, and insufficient delay of pregnancy time took place, but there was no effect on the offspring's quality. On the 125th day of the life of male offspring the analysis of their sperm genesis per day was performed. According to it, there was no deviation from control values. As authors think, some pregnancy period delay might be related to inflammatory processes in the female's lung. Similar results are also obtained by intraperitoneal MWCNT introduction in mouse females of NMRI line at the dosage of 1 and 10 mg per mouse. Negative effect of MWCNT on female organism and pregnancy process was not detected, but there was certain rise in anxiety state among the offspring [39].

In order to assess the effect on the rat's embryos MWCNT were administered into stomach at the dosage of 40, 200, and 1000 mg/kg per day over 13 days, beginning with 6th day of pregnancy. No anomalies of embryonic development were detected. On the base of the results obtained authors made up decision that even high-dosage oral administration of MWCNT can be considered safe for embryo development [36].

However, the results [40] prove the ability of MWCNT to cross placental barrier and alter embryonic development being introduced at various pregnancy stages. Oxidized MWCNT with radioactive labels on the surface were introduced by intravenous injections in pregnant females of laboratory mice at 20 mg/kg dosage. The probability of spontaneous expulsion of a fetus and offspring development anomalies were assessed. In that work, MWCNT placenta crossing was established by means of transmission electron microscopy and paper chromatography. Significant increase of resorption number and external anomalies of embryos development was also mentioned. Similar results were also obtained by authors [41]. In their investigations, they introduced MWCNT intraperitoneally and intratracheally in pregnant mouse females of ICR line one time on the 9th day of pregnancy at 2, 3, 4, and 5 mg/kg. All fetuses were detached from womb on 18th pregnancy day to assess external and skeletal development anomalies. When MWCNT were introduced intraperitoneally sufficient development defects were observed at all concentrations. Intratracheal administration resulted in embryonic development anomalies at 4 and 5 mg/kg dosages only. At the same time, the higher is MWCNT dosage, the greater is the number of breed with fetuses having skeletal anomalies. It leaves MWCNT embryotoxicity question open-ended.

As for toxic effect of MWCNT on male reproductive system, there is little information about it now. Scientific team from China [42] studied MWCNT influence on male reproductive function. They have shown that repeated intravenous injections of water-soluble MWCNT in mice males can result in reversible testis damage without affecting fertility. According to obtained experimental data, MWCNT accumulated in testes of exposed laboratory mice of BALB/c line leading to oxidative stress and reducing gonadal epithelium thickness on 15th day, but all defects were regenerated on 60th and 90th day. Quantity, quality, and consistence of the sperm, as well as levels of three basic gonadal hormones have no significant damage over all the experiment. Fertility of the animals under investigation remained within the normal limits, pregnancy period and successfulness of the birth process in control and experimental groups had no difference. This work confirms the probability of crossing hemato-testicular barrier by MWCNT, that can't help causing concerns about offspring quality.

In [43] dose-dependent decrease in testosterone level and fertilizing capacity was revealed for orally administered MWCNT at 0.3, 3, and 30 mg/kg. At the same time, no anomalies of development of the offspring from exposed males were detected.

In vitro assessment of the influence of MWCNT at 10-200 μg/ml concentrations on sperm didn't reveal any cytotoxic effect on spermatozoa. Moreover, when nanotube concentration was not so high (10-40 μg/ml) spermatozoa activization was detected, the number of active cells during their viability period increased in some cases by more than 20% [44].

2.6. Fullerene

In [45-47] an assessment of fullerene influence on embryonic development of zebrafish was performed. It is shown that fullerenes induce embryonic cell necrosis and lead to the increase in embryonic mortality. Fullerene exposure also effect negatively on embryonic development of mammals that is confirmed by *in vitro* and *in vivo* investigations on laboratory mice [48].

In [49] the ability of fullerenes to ease the consequences of diabetes treatment due to their antioxidant properties has been reported. It is shown that daily oral fullerene administration in diabetic male rats of Vistar line over 5 weeks results in reduction of diabetic-induced oxidative stress and diabetic treatment consequences such as testicular dysfunction and spermatogenesis disruption. The authors also noted the absence of toxic effect on male reproductive system of healthy animals.

3. Conclusion

So, carbon nanomaterials reveal toxic effect on reproductive system and offspring development of the animals of various system groups to a certain degree depending on carbon crystal structure. However, the reported results are insufficient to establish reliable nanomaterial's reproductive toxicity gradation based upon its carbon allotropic modification.

Currently, one can consider CNT and CB to be studied most carefully from the point of view of reproductive toxicity. There are experimental evidence to their ability to cross the hemato-testicular [42] and placental [19, 40] barriers, as well as breast milk penetration [33].

Graphene was first synthesized quite recently; therefore, their biological effect is studied to the least degree among modifications presented. Nevertheless, there are experimental evidences of graphene toxic effect on female reproductive system and offspring development.

In regard to fullerene and graphite nanoplatelets, the data obtained to the present time doesn't allow their reproductive and embryonic toxicity assessment.

Main toxicity mechanism of carbon nanomaterials as well as of many other types of nanoparticles is considered to be the oxidative stress [50-52]. Their effect, for example, in liver can be reduced to the minimum. However, testes don't have such protection mechanisms [53]. This is the cause of male reproductive system being more sensitive to external adverse impacts [54, 55] than female's one. Consequently, it is affected by anthropogenic environmental pollution much more often [56], as confirmed by most of carbon nanomaterial toxicity research. It should be noted that negative consequences for the offspring reveal itself if parent animals have been exposed to any allotropic modification of the carbon.

References

[1] Yeh M, Lin L, Chang L, Leu Y, Cheng W, Lin J and Ho K 2014 *Chem Phys Chem* **15** 1175

[2] Eatemadi A, Daraee H, Karimkhanloo H, Kouhi M, Zarghami N, Akbarzadeh A, Abasi M, Hanifehpour Y and Joo S 2014 *Nanoscale Research Letters* **9** 393

[3] Ganesh E 2013 *Int. J. of Innovative Technology and Exploring Engineering* **2**(4) 311-320

[4] Avouris P, Chen Z and Perebeinos V 2007 *Nature Nanotechnology* **2**(10) 605-615

[5] Yang K, Feng L, Shi X, Liu Z 2013 *Chem. Soc. Rev.* **42** 530

[6] He H, Pham-Huy L, Dramou P, Xiao D, Zuo P, Pham-Huy C 2013 *BioMed Research International* 1-12

[7] Singh B, Baburao C, Pispati V, Muthy N, Prassana S and Rathode B 2012 *Int. J. of Research in Pharmacy and Chemistry* **2**(2) 523

[8] Liu Z, Robinson J, Tabakman S, Yang K and Dai H 2011 *Materials Today* **14**(7-8) 316-323

[9] Lamberti M, Pedata P, Sannolo N, Porto S, De Rosa A and Caraglia M 2015 *Int. J. Immunopathol. Pharmacol.* **28**(1) 4

[10] Sun X, Liu Z, Welsher K, Robinson J, Goodwin A, Zaric S and Dai H 2008 *Nano Res.* **1** 203-212

[11] Wong B, Yoong S, Jagusiak A, Panczyk T, Ho H, Ang W and Pastorin G 2013 *Advanced Drug Delivery Rev.* **65** 1964

[12] Ostrowski A, Martin T, Conti J, Hurt I and Harthorn B 2009 *J. Nanopart. Res.* **11** 251-257

[13] Wani M, Hashim M, Nabi F and Malik M 2011 *Adv. in Phys. Chem.* **2011** 15

[14] Guo X and Mei N 2014 *J. of Food and Drug Analysis* **22** 105.

[15] Bour A, Mouchet F, Silvestre J, Gauthier L and Pinelli E 2015 *J. of Hazardous Materials* **283** 764

[16] Yoshida S, Hiyoshi K, Oshio S, Takano H, Takeda K and Ichinose T 2010 *Fertility and Sterility* **93**(5) 1695

[17] Jackson P, Vogel U, Wallin H and Hougaard K 2011 *Basic & Clinical Pharmacology & Toxicology* **109** 434
[18] Boisen A, Shipley T, Jackson P, Wallin H, Nellemann C, Vogel U, Yauk C and Hougaard K 2013 *Reproductive Toxicology* **41** 45
[19] Jackson P, Hougaard K, Vogel U, Wu D, Casavant, L, Williams A, Wade M, Yauk C, Wallin H and Halappanavar S 2012 *Mutation Research/Genetic Toxicology and Environmental Mutagenesis* **745** 73
[20] Kyjovska Z, Boisen A, Jackson P, Wallin H, Vogel U and Hougaard K 2013 *Reproductive Toxicology* **36** 88
[21] Yoshida S, Hiyoshi K, Ichinose T, Takano H, Oshio S, Sugawara I, Takeda K and Shibamoto T 2009 *Int. J. Androl.* **32**(4) 337
[22] Zanni E, De Bellis G, Bracciale M, Broggi A, Santarelli M, Sarto M, Palleschi C and Uccelletti D 2012 *Nano Lett.* **12** 2740
[23] Zhang W, Wang C, Li Z, Lu Z, Li Y, Yin J, Zhou Y, Gao X, Fang Y, Nie G and Zhao Y 2012 *Advanced Materials* **24** (39) 5391
[24] Wu Q, Yin L, Li X, Tang M, Zhang T and Wang D 2013 *Nanoscale* **5** 9934
[25] Liang S, Xu S, Zhang D, He J and Chu M 2015 *Nanotoxicology* **9** 92
[26] Xu S, Zhang Z and Chu M 2015 *Biomaterials* **54** 188
[27] Fu C, Liu T, Li L, Liu H, Liang Q and Meng X 2015 *Biomaterials* **40** 23
[28] Cheng J, Flahaut E and Cheng S 2007 *Environ. Toxicol. Chem.* **26** 708
[29] Philbrook N, Walker V, Afrooz A, Saleh N and Winn L 2011 *Reproductive Toxicology* **32** 442
[30] Pietroiusti A, Massimiani M, Fenoglio I, Colonna M, Valentini F, Palleschi G, Camaioni A, Magrini A, Siracusa G, Bergamaschi A, Sgambato A and Campagnolo L 2011 *ACS Nano* **5**(6) 4624
[31] Pietroiusti A, Campagnolo L and Fadeel B 2013 *Small* **9**(9-10) 1557
[32] Ivani S, Karimi I, Tabatabaei S, Syedmoradi L 2014 *Toxicology and Industrial Health* 1-9
[33] Aldobaev V, Eremenko L, Mazanova A, Biketova D, Kovalyova E, Kvacheva L, Badun G, Muradyan V, and Maslikov A 2013 *Toksikologicheskiy vestnik (in Russia)* **2**(119) 26-30
[34] Zhao Z, Liu M, Jia X, Wang H, Liu Z, Zhang J, Sun L and Zhang M 2014 *Nano LIFE* **4**(3) 1441009
[35] Cheng J, Chan C, Veca L, Poon W, Chan P, Qu L, Sun Y and Cheng S 2009 *Toxicology and Applied Pharmacology* **235** 216
[36] Lim J, Kim S, Shin I, Park N, Moon C, Kang S, Kim S, Park S and Kim J 2011 *Birth Defects Res B Dev Reprod Toxicol.* 92(1) 69
[37] Hougaard K, Jackson P, Kyjovska Z, Birkedal R, De Temmerman P, Brunelli A, Verleysen E, Madsen A, Saber A, Pojana G, Mast J, Marcomini A, Jensen K, Wallin H, Szarek J, Mortensen A and Vogel U 2013 *Reproductive Toxicology* **41** 86
[38] Hougaard K, Jackson P, Jensen K, Saber A, Vogel U and Wallin H 2010 *Reproductive Toxicology* **30**(2) 239
[39] Ivani S, Karimi I, Tabatabaei S and Reza S 2012 *J. Toxicol. Sci.* **37**(6) 1191
[40] Qi W, Bi J, Zhang X, Wang J, Wang J, Liu P, Li Z and Wu W 2014 *Scientific Reports* **4** 4352
[41] Fujitani T, Ohyama K, Hirose A, Nishimura T, Nakae D and Ogata A 2012 *J. Toxicol. Sci.* **37**(1) 81
[42] Bai Y, Zhang Y, Zhang J, Mu Q, Zhang W, Butch E, Snyder S and Yan B 2010 *Nature Nanotechnology* **5** 683
[43] Vasyukova I, Gribanovskii S, Gusev A, Ubogov A, Khaliullin T, Fatkhutdinova L and Tkachev A 2015 *Nanotechnologies in Russia* **10**(5-6) 458
[44] Ivanov L, Kramar M, Chernykh V, Kovalenko S, Kartel N and Gubin Y 2009 *Poverhnost* **1**(16) 314
[45] Usenko C, Harper S and Tanguay R 2007 *Carbon* **45** 1891
[46] Zhu X, Zhu L, Li Y, Duan Z, Chen W, Alvarez P 2007 *Environ. Toxicol. Chem.* **26**(5) 976

[47] Usenko C, Harper S and Tanguay R 2008 *Toxicology and Applied Pharmacology* **229** 44
[48] Tsuchiya T, Oguri I, Yamakoshi Y and Miyata N 1996 *FEBS Lett.* **393**(1) 139
[49] Bal R, Türk G, Tuzcu M, Yilmaz O, Ozercan I, Kuloglu T, Gür S, Nedzvetsky V, Tykhomyrov A, Andrievsky G, Baydas G and Naziroglu M 2011 *Toxicology* **282** 69-81
[50] Hougaard K and Campagnolo L 2012 *Adverse Effects of Engineered Nanomaterials* ed B Fadeel, A Pietroiusti and A Shvedova (UK: Elsevier) 225
[51] Aitken R, Smith T, Jobling M, Baker M and De Iuliis G 2014 *Asian J Androl.* **16**(1) 31
[52] Aschberger K, Johnston H, Stone V, Aitken R, Hankin S, Peters S, Tran C and Christensen F 2010 *Critical Reviews in Toxicology* **40**(9) 759
[53] Agarwal A, Virk G, Ong C and du Plessis S 2014 *World J. Mens Health* **32**(1) 1
[54] Saleh R and Agarwal A 2002 *J. Androl.* **23** 737
[55] Lan Z and Yang W 2012 *Nanomedicine* **7**(4) 579
[56] Bonde J 2010 *Asian J. of Andrology* **12** 152

IOP Conf. Series: Materials Science and Engineering **98** (2015) 012002 doi:10.1088/1757-899X/98/1/012002

Ag electrode modified with polyhexamethylene biguanide stabilized silver nanoparticles: a new type of SERS substrates for detection of enzymatically generated thiocholine

A A Tepanov, N L Nechaeva, T A Prokopkina, A A Kudrinskiy, I N Kurochkin, G V Lisichkin
Lomonosov Moscow State University, 1, Leninskiye Gory, Moscow, 119991, Russia

Email: atepanov@inbox.ru

Abstract. The detection of thiocholine is one of the most widespread techniques for estimation of the cholinesterase activity – acetylcholinesterase and butyrylcholinesterase. Both cholinesterases can be inhibited by organophosphates and carbamates and accordingly can be considered for estimation of these pollutants in the environment. In the current work, SERS spectroscopy was applied for the thiocholine detection. The Ag electrodes modified with silver nanoparticles stabilized by polyhexamethylene biguanide were for the first time suggested as SERS-substrates for that purpose. Such electrodes can be applicable for SERS detection of submicromolar concentrations of thiocholine.

1. Introduction

Human cholinesterases have important physiological functions: acetylcholinesterase (AChE) is implicated in neural signal transmission catalyzing hydrolysis of neuromediator acetylcholine [1, 2], butyrylcholinesterase (BChE) acts as a stoichiometric acceptor of organophosphates [3, 4]. There are numerous chemicals and drugs that interact with the cholinergic nervous system: nerve agents, prophylactic antidotes, myorelaxants and other therapeutic drugs. Since thiocholine is a hydrolysis product of human cholinesterases, its accumulation can represent enzymatic status in organism. Abnormal low values of cholinesterase activity may be caused by some anemias [5], dermatomyositis [6], organophosphorous intoxication [7, 8], and drug use [9]. Additionally, abnormal high levels of cholinesterase activity may indicate the nephrotic syndrome [9]. Therefore the determination of thiocholine in blood with high accuracy is of great importance for clinical diagnostics.

At the present time, the electrochemical methods [10-13] are commonly used for the thiocholine detection due to low limit of detection (LOD) [11, 14]. In particular, MnO_2-modified screen-printed graphite electrodes allow detection of 60 nM of thiocholine [15, 16]. Moreover, electrochemical biosensors fabricated with layer-by-layer technology also show very low LOD – about 130 nM [17]. Nevertheless, the further development of detection techniques is a significant problem. In this regard, SERS spectroscopy is very promising in view of its ability to detect single molecules [18, 19].

This article reports the application of SERS spectroscopy for the thiocholine detection. We suggest the Ag electrodes modified with silver nanoparticles stabilized by polyhexamethylene biguanide (PHMB) as SERS substrates for determination of enzymatically generated thiocholine. It was found that the Raman signal intensity of thiocholine applied onto the surface of Ag electrode modified with PHMB-stabilized silver nanoparticles is, at least, 10 times higher in comparison with one applied onto pure Ag electrode surface. At the same time, PHMB don't significantly influence on the Raman signal intensity of thiocholine. Ag electrodes modified with PHMB-stabilized nanoparticles can be applied for the quantitative analysis of thiocholine. Preliminary results indicate that limit of detection of thiocholine by this technique is about 200 nM.

Content from this work may be used under the terms of the Creative Commons Attribution 3.0 licence. Any further distribution of this work must maintain attribution to the author(s) and the title of the work, journal citation and DOI.
Published under licence by IOP Publishing Ltd

2. Experimental

2.1. Reagents and materials

For the preparation of silver nanoparticles, silver nitrate $AgNO_3$ (99%, Reakhim, Russia), sodium borohydride $NaBH_4$ (99%, Acros Organics) and polyhexamethylene biguanide hydrochloride (PHMB, 20% wt. solution in water, Arch) were used. All reagents were applied as purchased. The structure of PHMB is shown in figure 1. For the generation of thiocholine, butyrylthiocholine iodide (Sigma, USA), butyrylcholinesterase from equine serum (BChE, EC 3.1.1.8) with activity 264 U/mg (Sigma, USA), 5,5'-dithiobis(2-nitrobenzoic acid) (DTNB) (Sigma, USA), and phosphate buffered saline (PBS) 0.1 M, pH 7.0 were used. Choline bromide (Sigma, USA) was used as internal standard.

For substrate preparation, silver screen-printed electrodes were applied. All solutions were made with milliQ water (18.2 MΩ*cm).

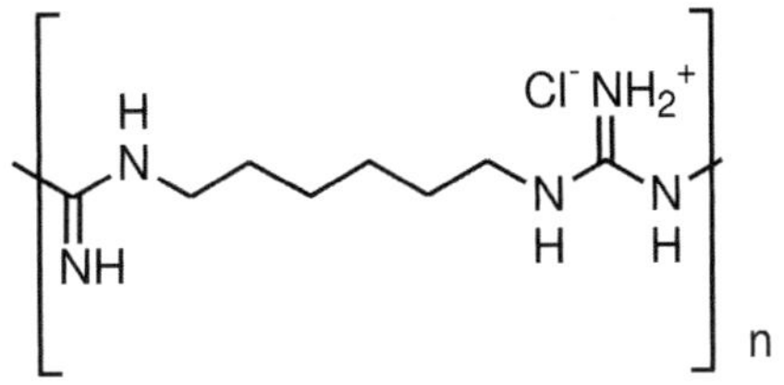

n = 8-20

Figure 1. Structure of polyhexamethylene biguanide (PHMB).

2.2. Preparation of silver nanoparticles

Silver nitrate was reduced in aqueous solution with sodium borohydride under vigorous stirring [20] in the presence of PHMB. The concentrations of reagents were chosen to obtain stable colloids with concentration of silver nanoparticles of 50 μg/ml. Sodium borohydride was added in 2-fold molar excess compared to the silver nitrate. The silver/PHMB molar ratio was 1:1.

2.3. Immobilization of silver nanoparticles on Ag electrodes

Silver electrodes were immersed into colloidal solutions of the preliminarily prepared nanoparticles in the presence of 0.08 M KCl and were kept for 12 hours. Then as-prepared substrates were rinsed thoroughly with water and dried.

2.4. Synthesis of thiocholine

Thiocholine solution was obtained by the enzymatic hydrolysis of butyrylthiocholine iodide. For that purpose, 22.6 mg of butyrylthiocholine iodide were dissolved in 600 μl of 0.1 M PBS buffer (pH 7.0). Then 400 μl of 0.1 mg/ml BChE solution in the same buffer were added. The mixture was incubated for 1 h.

The Ellman technique [21] was applied to determine the thiocholine concentration. Preliminarily obtained thiocholine solution was diluted in 40 times with PBS, 10 μl of it introduced in the cell of microplate spectrophotometer xMark (BioRad, USA) with 180 μl of PBS followed by adding 10 μl of 5 mM solution of DTNB. Then the absorbance was measured at 405 nm. The thiocholine concentration was calculated by the equation $c = A/(\varepsilon_{405}d)$, where A is the absorbance, ε_{405} is the absorption coefficient of 5-thio-2-nitrobenzoate ione (TNB) at 405 nm (13.3 ml/μmol*cm), and d is the optical path equal to 0.6 cm.

2.5. SERS measurements of thiocholine

Each sample for Raman spectroscopy was prepared by applying of a drop of an aqueous solution of thiocholine on Ag electrode and drying it at 60°C.

2.6. Characterization

The UV-Vis spectra of nanoparticles were recorded using UV-visible spectrometer Specord UV-Vis (Carl Zeiss, Germany) in the 300-700 nm range. The spectrum of pure water was used as a background.

The size distribution of silver nanoparticles was measured with Zetasizer Nano ZS (Malvern Instruments Ltd., UK) equipped with a 633 nm He-Ne laser (power 4 mW).

The measurements of ζ-potential were carried out by applying an electric field to the cell with sol of nanoparticles using technique based on the laser Doppler anemometry.

TEM images were taken using a Leo 912 AB Omega microscope (Leo Ltd., Germany). Each sample under study was prepared by drying a drop of an aqueous solution of nanoparticles on a copper grid coated with a carbon film. The nanoparticle size distributions were calculated using ImageJ 1.48v program.

XPS experiments were carried out on LAS-3000 (Riber, France) microscope with an XPS chamber equipped with an X-ray anode source (AlKα, 1486.6 eV) and an OPX-150 hemispherical energy analyzer.

SEM images of substrates were obtained with a Carl Zeiss Supra 40 microscope (accelerating voltage 10 kV; 30 μm aperture) with extreme resolution 3-4 nm.

SERS spectra were measured in a BWTech InnoRam Raman spectrometer equipped with a diode near IR 785 nm laser (350 mW). The laser spot of the diode near IR laser was a circle with diameter 100 μm. The spectra were analyzed at a resolution of 4 cm^{-1} within 64-3011 cm^{-1}.

3. Results and discussion

3.1. Preparation of silver nanoparticles

The reduction of $AgNO_3$ in aqueous solution with $NaBH_4$ in the presence of polyhexamethylene biguanide hydrochloride yields PHMB-capped silver spherical nanoparticles with typical sizes between 5-20 nm as determined by TEM (figure 2). The aqueous colloidal dispersions of these particles have the characteristic surface plasmon resonance [22] at 397±2 nm in the UV-Vis absorbance spectrum (figure 3).

The ζ-potential value for PHMB-stabilized silver nanoparticles based on dynamic light scattering (DLS) measurements is +39.1 (±1.5) mV. The obtained sols of Ag NPs were stable for a long time. A typical microdiffraction pattern of silver particles (figure 2D) clearly indicates that nanoparticles consist of crystalline silver.

PHMB-stabilized nanoparticles are less stable in the presence of KCl. Addition of potassium chloride decreases the electrical double layer thickness and can break the binding between the stabilizer and the nanoparticle surface due to specific sorption of chloride ions. As a result, the colloidal solution becomes unstable and coagulates in the presence of excess of KCl. The minimal concentration of potassium chloride corresponding to the coagulation threshold of PHMB-stabilized nanoparticles is about 0.11 M.

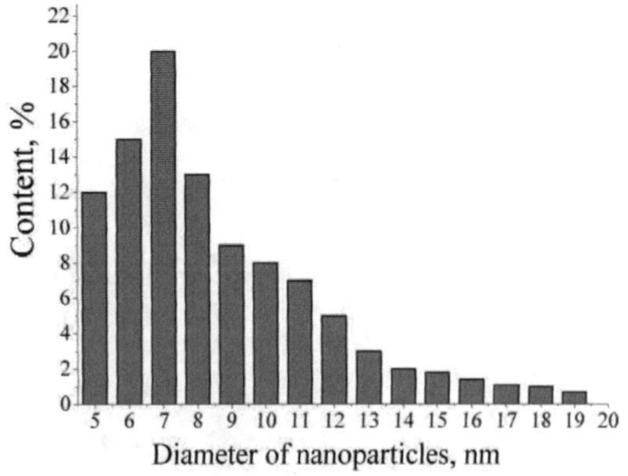

A

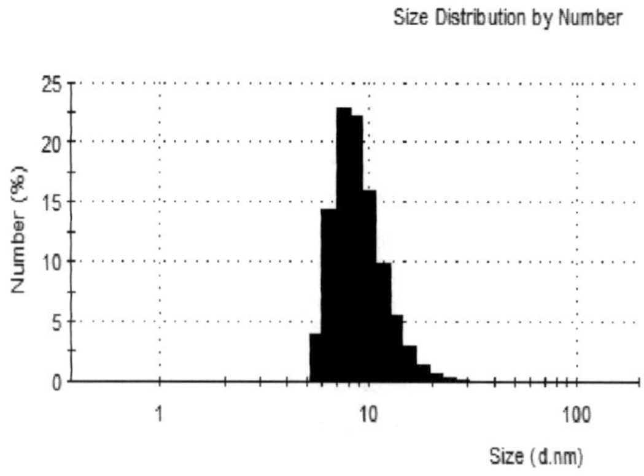

B

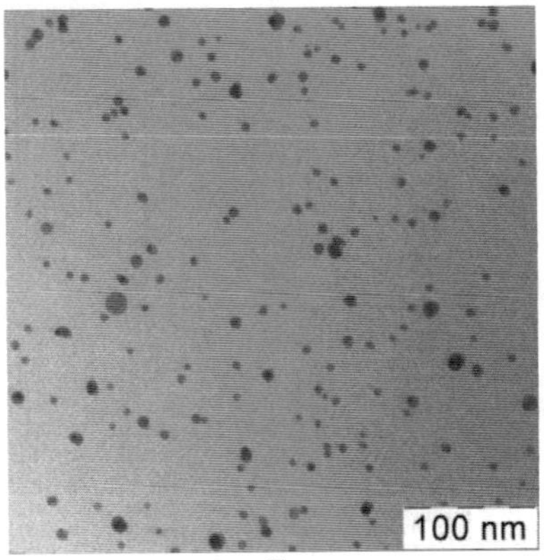

C

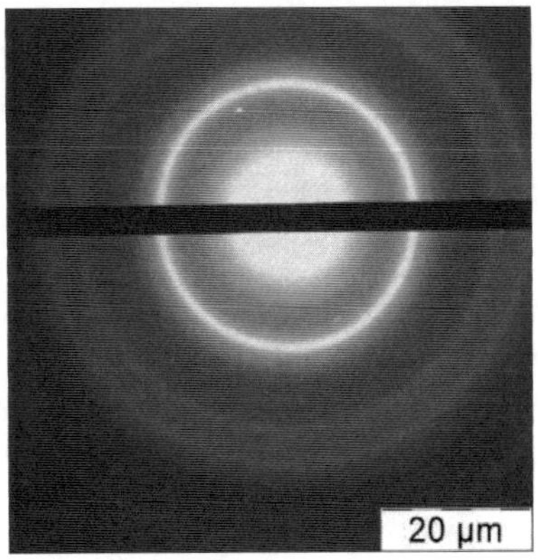

D

Figure 2. The size distributions based on TEM (A), DLS (B), TEM image (C) and microdiffraction pattern (D) of PHMB-stabilized silver nanoparticles.

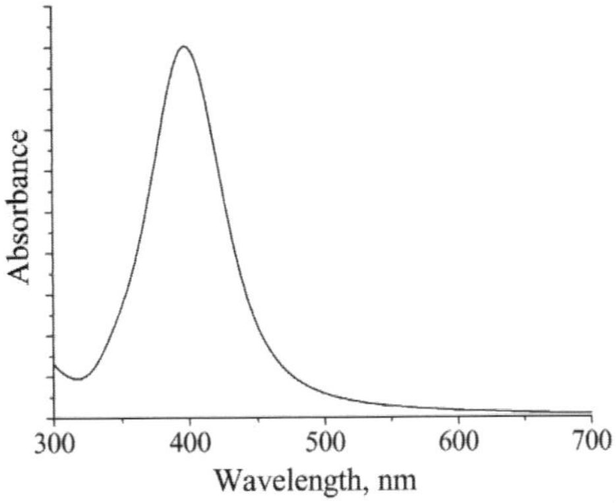

Figure 3. The absorption spectrum of PHMB-stabilized silver nanoparticles (mass concentration 50 μg/ml, dilution 10 times).

Nanoparticle composition was investigated by XPS. Figure 4 depicts XPS-spectra of silver nanoparticles. There are two bands with energies of 368.1 eV (3d 5/2) and 374.0 eV (3d 3/2) corresponding to crystalline silver [23] in the Ag 3d line (figure 4A).

The carbon atoms in PHMB have different oxidation states which correspond to aliphatic and imine groups. The carbon 1s line is shown as a superposition of two peaks (figure 4B). The most intensive band with binding energy of 285.1 eV apparently related to aliphatic carbon atoms whereas the second band (binding energy 288.0 eV) corresponds to imine carbon atoms [24]. Because of overlapping of the N 1s line with the satellite of Ag 3d photoelectron line the interpretation of this spectral range is complicated. Hence it is possible to make an only qualitative conclusion about the band with energy 400-405 eV that corresponds to the presence of nitrogen (figure 4C).

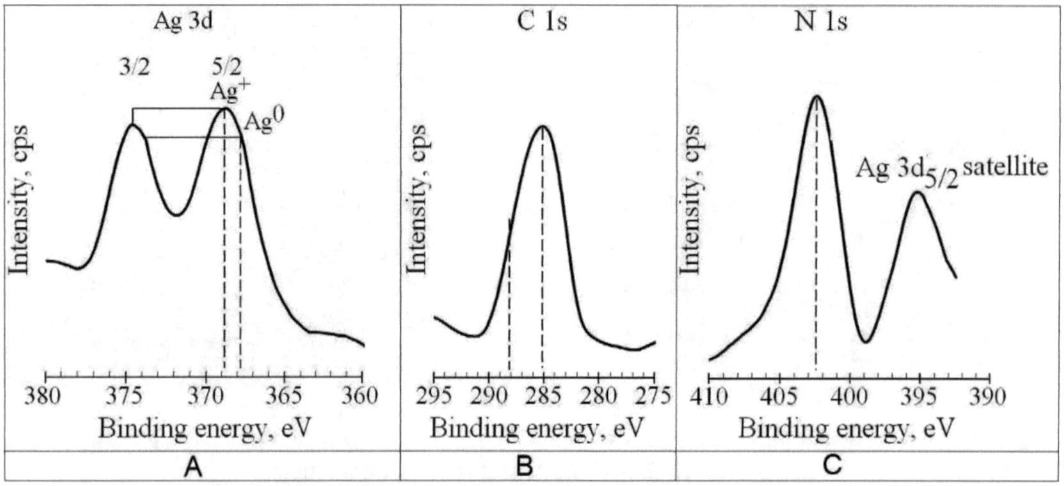

Figure 4. XPS-spectra of PHMB-stabilized nanoparticles: A) Ag 3d; B) C 1s; C) N 1s.

3.2. Immobilization of silver nanoparticles onto silver electrode surface

In order to successfully immobilize nanoparticles onto silver electrode surface it is necessary to break binding between stabilizing molecules and nanoparticle surface. Addition of potassium chloride to modifying sol during immobilization decreases aggregative stability of nanoparticles – the ζ-potential value of PHMB-stabilized particles decreases from +39.1 (±1.5) mV to 17 (±1.5) mV while adding 0.08 M KCl. The KCl-assisted immobilization of nanoparticles was carried out at concentration of chloride ions of 0.08 M that is 1.4 times smaller then the coagulation threshold for PHMB-stabilized nanoparticles to prevent formation of large conglomerates. As a result, friable aggregate films of silver nanoparticles were formed on the silver electrode surface (figure 5A). Molecules of PHMB apparently can act as linkers for nanoparticles during immobilization.

Addition of potassium chloride into colloidal solutions of nanoparticles do not influence on the oxidation state of nanoparticle surface. Figure 5B illustrates Ag 3d XPS-spectrum of PHMB-stabilized nanoparticles immobilized onto Ag electrode surface in the presence of 0.08 M KCl. The spectrum is identical to the same for PHMB-stabilized nanoparticles in colloidal solution (figure 4).

3.3. Synthesis of thiocholine

For the thiocholine solution preparation, butyrylthiocholine iodide was used according to the Scheme 1.

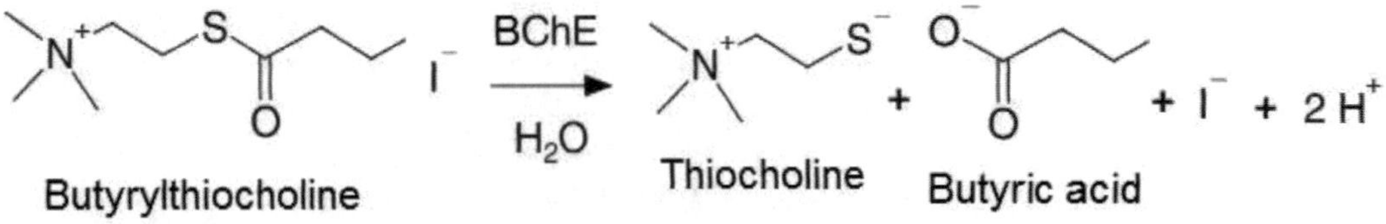

Scheme 1. Formation of thiocholine by hydrolysis of butyrylthiocholine.

A B

Figure 5. Typical SEM image of Ag electrode modified with PHMB-stabilized silver nanoparticles (A), Ag 3d XPS-spectrum of PHMB-stabilized nanoparticles immobilized onto Ag electrode surface in the presence of 0.08 M KCl for 12 hours (B).

The reaction yields two products, thiocholine and butyric acid, the latter does not interfere thiocholine SERS detection. Concentration of obtained thiocholine solution was determined by Ellman's assay, which represents a color reaction (scheme 2). Colored product of this reaction is 5-thio-2-nitrobenzoate ione (TNB). The reaction which is shown in Scheme 2 is rapid and stoichiometric, with the addition of one mole of thiol releasing one mole of TNB. The conversion of butyrylthiocholine is about 100%.

Concentration of thiocholine stock solution was equal to 0.12 M.

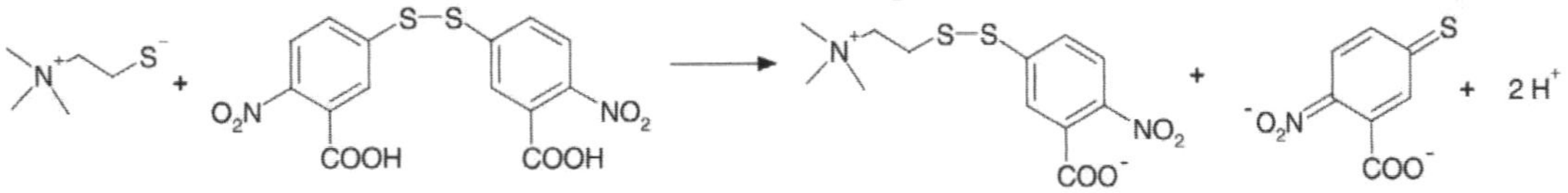

Thiocholine DTNB 5-Thio-2-nitrobenzoate (TNB^-, yellow)

Scheme 2. Formation of colored reaction product.

3.4. SERS measurements of thiocholine

The mixture of products of the enzymatic hydrolysis of butyrylthiocholine containing thiocholine and butyric acid was applied onto silver electrode surface, dried at 60°C and then was detected by SERS. It was found that the Raman signal intensity of the mixture applied to Ag electrode modified with PHMB-stabilized silver nanoparticles is, at least, 10 times higher in comparison with one applied to the pure Ag electrode surface (figure 6A).

The Raman signal intensity of PHMB solution applied on the electrode surface is less than intensity of signal of detected mixture applied onto PHMB-treated electrode surface at wavelength excitation of 785 nm (figure 6B). Concentration of PHMB solution was equal to PHMB concentration in the nanoparticles colloidal solution. The treatments with PHMB solution and with 40 μM thiocholine solution were carried out under the same conditions. It is apparent that the presence of PHMB on the electrode surface also promotes the enhancement of thiocholine signal but this effect is small compared to enhancement due to the nanoparticle presence.

Thus, PHMB molecules don't complicate the thiocholine detection. Therefore the Raman signal enhancement of thiocholine solution applied onto Ag electrode surface modified with PHMB-stabilized nanoparticles is related mainly to the presence of immobilized nanoparticles. The observed increasing of Raman signal intensity could be employed for semi-quantitative analysis of thiocholine.

IOP Conf. Series: Materials Science and Engineering **98** (2015) 012002 doi:10.1088/1757-899X/98/1/012002

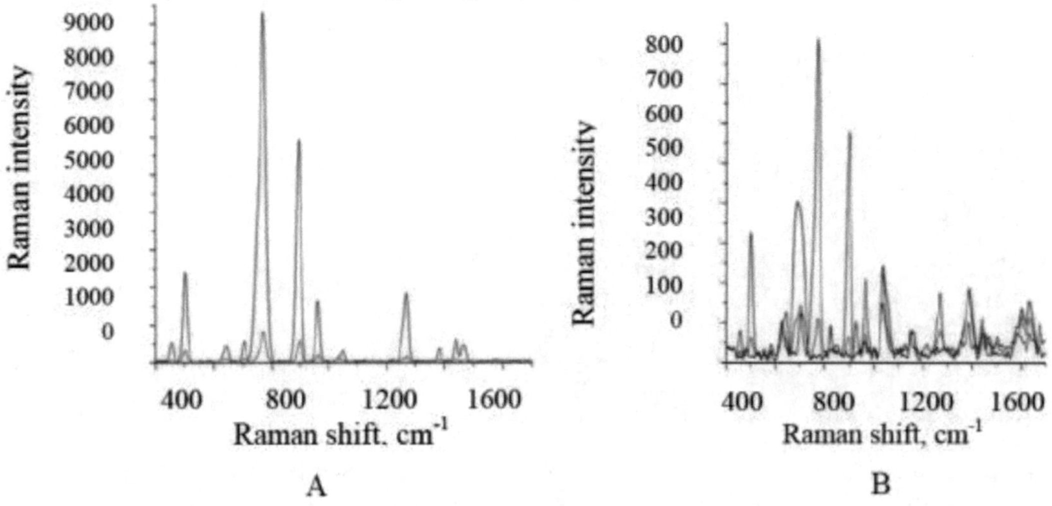

Figure 6. The SERS spectrum (red line) of 40 μM solution of thiocholine applied onto Ag electrode surface and (blue line) onto Ag electrode surface modified with PHMB-stabilized nanoparticles (A); the Raman spectrum (blue line) of 40 μM thiocholine solution applied onto pure Ag electrode surface compared to the spectra (red line) of 40 μM thiocholine solution applied onto Ag electrode surface treated with PHMB and (black line) of PHMB solution applied onto Ag electrode surface (B).

The as-prepared SERS substrates allow detection of mixtures containing submicromolar concentrations of thiocholine. In view of complexity of SERS method the internal standard is needed for the quantitative analysis. In our work, choline was applied as internal standard because its characteristic band at 714 cm^{-1} doesn't interfere with thiocholine bands. The SERS spectra were normalized on the choline signal intensity at 714 cm^{-1}. The intensity of bands at 770 cm^{-1} and 1267 cm^{-1} corresponding to thiocholine vibrations increases symbatically with thiocholine concentration. Therefore, Ag electrodes modified with PHMB-stabilized nanoparticles can be applied for the quantitative analysis of thiocholine.

4. Conclusion

In this work we have presented a new substrate for the SERS detection of thiocholine. For that purpose Ag electrodes were modified with PHMB-stabilized silver nanoparticles. It was found that the Raman signal intensity of thiocholine solution applied onto Ag electrode surface modified with PHMB-stabilized silver nanoparticles is, at least, 10 times higher in comparison with one applied onto pure Ag electrode surface. At the same time, PHMB don't contribute significantly on the thiocholine Raman signal intensity. The achieved enhancement factor is appropriate for thiocholine detection. Ag electrodes modified with PHMB-stabilized nanoparticles can be applied for the quantitative analysis of thiocholine.

References

[1] Rosenberry T 2006 *Acetylcholinesterase* (Alton M Advances in Enzymology and Related Areas of Molecular Biology) **43** 103
[2] Marrs T 2007 *Toxicology of organophosphate nerve agents, chemical warfare agents* (Marrs T, Maynard R, and Sidell F Toxicology and Treatment. New York: Wiley) 191
[3] Bajgar J 1992 *British J. of Industrial Medicine*, **49** 648
[4] Wilson B and Henderson J 1992 *Reviews of Environ. Contamination and Toxicology*, **128** 55
[5] Bey T, Sullivan J Jr, and Walter F 2001 *Organophosphate and carbamate insecticides. Clinical environmental health and toxic exposures* (Philadelphia: Lippincott Williams & Wilkins) 1046
[6] Reigart J and Roberts J 1999 *Recognition and management of pesticide poisoning* (Washington, DC, U.S. Environmental Protection Agency)

[7] Zejli H, Hidalgo-Hidalgo de Cisneros J, Naranjo-Rodriguez I, Liu B, Temsamani K, and Marty J 2008 *Talanta*, **77**(1) 217
[8] Zhang L, Zhang A, Du D, and Lin Y 2012 *Nanoscale*, **4**(15) 4674
[9] Clark R 2002 *Insecticides: organic phosphorus compounds and carbamates* (Goldfrank L, Flomenbaum N, Lewin N, Howland M, Hoffman R, and Nelson L Goldfrank's Toxicological Emergencies. New York: McGraw-Hill Professional 7th edn) 1346
[10] Tian Y, Ye S, Shi X, Jing L, Liang C, and Xian Y 2011 *The Analyst*, **136**(23) 5084
[11] Arduini F, Cassisi A, Amine A, Ricci F, Moscone D, and Palleschi G 2009 *J. Electroanalytical Chemistry*, **626**(1-2) 66
[12] Parsajoo C and Kauffmann J-M 2013 *Talanta*, **109** 116
[13] Liu G, Riechers S, Mellen M, and Lin Y 2005 *Electrochemistry Communications*, **7**(11) 1163
[14] Sgobbi L, Razzino C, and Rosset I 2013 *Electrochimica Acta*, **112** 500
[15] Eremenko A, Dontsova E, Nazarov A, and Kurochkin I 2014 *Thick film thiol sensors for cholinesterases assay* (Toxicological Problems. Sofia: Military Publishing House) 82
[16] Eremenko A, Dontsova E, Nazarov A, Evtushenko E, Amitonov S, Savilov S, Martynova L, Lunin V, and Kurochkin I 2012 *Electroanalysis*, **24**(3) 573
[17] Kurochkin I, Sigolaeva L, Eremenko A, Dontsova E, Gromova M, Rudakova E, and Makhaeva G 2014 *Layer-by-layer electrochemical biosensors for blood esterases assay* (Toxicological Problems. Sofia: Military Publishing House) 51
[18] Nie S and Emory S 1997 *Science*, **275**(5303) 1102
[19] Blackie E, Le Ru E, and Etchegoin P 2009 *J. Am. Chem. Soc.*, **131**(40) 14466
[20] Evanoff D Jr and Chumanov G 2005 *Chem. Phys. Chem.*, **6** 1221
[21] Ellman G, Courtney K, Andres V Jr, and Featherstone R 1961 *Biochemical Pharmacology*, **7** 88
[22] Moores A and Goettmann F 2006 *New J. Chem.*, **30** 1121
[23] Kumar A, Joshi H, Pasricha R, Mandale A B, and Sastry M 2003 *J. Colloid Interface Sci.*, **264** 396
[24] Beamson G and Briggs D 1992 *High Resolution XPS of Organic Polymers: the Scienta ESCA300 Database (*New-York J. Wiley & Sons)

IOP Conf. Series: Materials Science and Engineering **98** (2015) 012003 doi:10.1088/1757-899X/98/1/012003

Extremely low level of Ag nanoparticle excretion from mice brain in *in vivo* experiments

A Antsiferova[1,3,6], Yu Buzulukov[1,3], V Demin[1,3], P Kashkarov[1,3,4], M Kovalchuk[1,5], E Petritskaya[2]

[1]National Research Centre "Kurchatov Institute", 1, Academica Kurchatova sq., Moscow, Russia
[2]M.F. Vladimirsky Moscow Regional Research Clinical Institute, 61/2, Schepkina st., Moscow, Russia
[3]Moscow Institute of Physics and Technologies, Institutskii lane, Dolgoprudny, Moscow Region, Russia
[4]Lomonosov Moscow State University, GSP-1, Leninskie Gory, Moscow, Russia
[5]Saint Petersburg State University, 7-9, Universitetskaya emb., St. Petersburg, Russia

Email: antsiferova_aa@nrcki.ru

Abstract. Silver nanoparticle accumulation in mice organs as well as the excretion processes from them were studied. The investigation included a one-time oral administration of silver nanoparticles and a series of prolonged oral administrations of the same nanoparticles to study the long-term impact of the nanoparticles. In these experiments, the mice had been fed with colloid silver and in these prolonged experiments, administrations lasted for 2 months. The nanoparticle administration was then cancelled for one month. The elemental composition of tissue samples was studied by Nuclear Physical technique, which allowed us to obtain the masses of the key element, namely silver. It was demonstrated that silver concentrations in tissues were redistributed with time. The main result of this work was the discovery of extremely low level of silver nanoparticle excretion from mice brain (just 6% per month) following the cancellation of NP administration. However, the rates of excretion from blood and liver appeared to be rather high (about 80% per month). Thus, the accumulation effect of silver nanoparticles in the mice brain was observed, which is of great practical importance. It changes the approach to the toxicity assessment of silver nanoparticles as a result of the prolonged injection of colloidal silver.

1. Introduction

Due to the increased practical use of nanoparticles (NPs) in various areas of industry, including the food industry, medicine, pharmacy and others, their interaction with living organisms and recycling processes are of significant interest for applied science as well as for fundamental one. Thus, silver NPs are widely used as food supplements, antibiotics, antiviral agents and hygiene agents [1-10]. However, the data on the influence of silver on mammal organisms either demonstrates a pronounced cytotoxic effect, in particular in nanoform [11-22], which is likely due to the high penetrability of NPs. Another problem is the pollution of the environment from the NPs released from consumption goods. In this, case NPs eventually enter to the food chain reaching human organisms as a result. Having in mind both the reasons, it seems to be significant to study their natural ways of exposure.

Content from this work may be used under the terms of the Creative Commons Attribution 3.0 licence. Any further distribution of this work must maintain attribution to the author(s) and the title of the work, journal citation and DOI.
Published under licence by IOP Publishing Ltd

IOP Conf. Series: Materials Science and Engineering **98** (2015) 012003 doi:10.1088/1757-899X/98/1/012003

The main objective of this work is to study the biokinetics of silver NPs in organs and the biological fluids of laboratory animals, i.e. their biodistribution in the organisms as a result of natural ways of injection, penetration through various histohematogenous barriers of the organism and NP following transport via blood. The bio-objects under investigation in this study are blood, the liver and the brain. The answers to the important questions concerning these organs could be beneficial in the formation of the basis of toxic and hygienic standards, imposing certain restrictions on the use of silver NPs as consumer goods.

It has been already proven that silver injected into an organism in nanoform is able to penetrate through the blood-brain barrier [23, 24], perhaps into the cells of the neural system. At least by additional measurements of the Fe isotope activity in blood and the whole brain, it was fairly confirmed that the major fraction of silver of the same type as described in the present paper concentrates not in the blood vessels of brain but in the cellular or perhaps intracellular medium of brain [23]. It makes the problem of silver nanosafety extremely topical and important. Therefore, in the present research we especially focused on the ability of silver NPs to penetrate into the brain.

However, the ability to penetrate different tissues strongly depends on experimental conditions such as the route of ingestion, NP size, surface functionalization [25], which designate the specific pathway of cellular uptake. Significant variety of these characteristics as well as the application of methods with diverse sensitivity and accuracy leads to controversial experimental data. In order to obtain unanimous meaning about such aspects as biokintetics and biodistribution, as well as toxicity assessment, experiments are supposed to be conducted with metrologically assured high-precision methods designed for NP content measurements and advance characterization of the key NPs.

2. Methodological basis of the work

Table 1. Characteristics of techniques for NP tracing

#	Method	Advantages	Disadvantages
1.	Electron Microscopy	Visualization of measurement results	Low representativeness (microsections and microsamples), complexity of sample preparation, impossibility of biophilic nanoparticle analysis
2.	Optical Spectroscopy and Mass-Spectrometry	Relatively high accessibility	Relatively low representativeness (small and microsamples), difficulties in converting of solid samples into liquids, destructive techniques, low accuracy, impossibility of biophilic nanoparticle analysis
3.	Nuclear-Physical methods	Possibility of nondestructive macrosample analysis (whole tissues or their parts), possibility of biophilic nanoparticle analysis (Zn, Fe, Se etc.), high sensitivity (up to 10-11 g) and metrological accuracy (~1-10%)	Relatively low accessibility

There are a few techniques by which it is possible to estimate the concentration of NPs in complex biological tissues. Nuclear-Physical methods demonstrate a number of advantages in comparison with other techniques: high accuracy, representativeness, integrity, possibility of concentrations measurements of biophilic NPs. Table 1 demonstrates the comparative analysis of different techniques designed to trace nanoparticles in complex biological matrices. There are two major types of Nuclear-

IOP Conf. Series: Materials Science and Engineering **98** (2015) 012003 doi:10.1088/1757-899X/98/1/012003

Physical technique: Radiolabelling technique and Neutron Activation Analysis. The first method implies administration of radiolabelled nanoparticles and tracing their transport via blood to different organs and tissues with time. In the second one, nonradioactive nanoparticles are administrated into animal organisms and then after irradiation of the tissues containing unknown amounts of NPs, elemental analysis of the tissues is conducted [26].

In the present work Instrumental Neutron Activation Analysis (INAA) to study biokinetics of silver NPs in biological tissues was applied. INAA is held in conjunction with gamma spectrometry which allows to obtain information about radioactive isotopes in the measured probes, and the levels of their activity are converted into concentrations of the key elements.

There are two possible pathways which can be applied for obtaining values of masses and concentrations of the key element. In absolute measurements, calculations are performed using different known nuclear-physical constants and measured values of activities. Otherwise, in more clear and reliable approach activities of measured samples are compared with activity of reference material containing the known amount of the key element, which is irradiated simultaneously with the measured samples at the same conditions.

3. Experiment

1. Materials

In the present work, biokinetics of silver nanoparticles stabilized with polyvinylpyrrolidone (PVP) performed by colloidal solution of food supplement "Argovit-C" with initial concentration of 10 mg/ml were investigated. Size and aggregation kinetics were studied with the use of Dynamical Light Scattering (DLS). Solutions of the food supplement were diluted down to 0.5 mg/ml and sonicated for 15 minutes in an ultrasonic bath then sizes of NPs were measured by DLS technique. DLS data showed that the NPs were not aggregated and the average size of NPs was about 34 nm, which is in a good agreement with manufacturer's data. The histogram of size distribution by number of particles is demonstrated in figure 1.

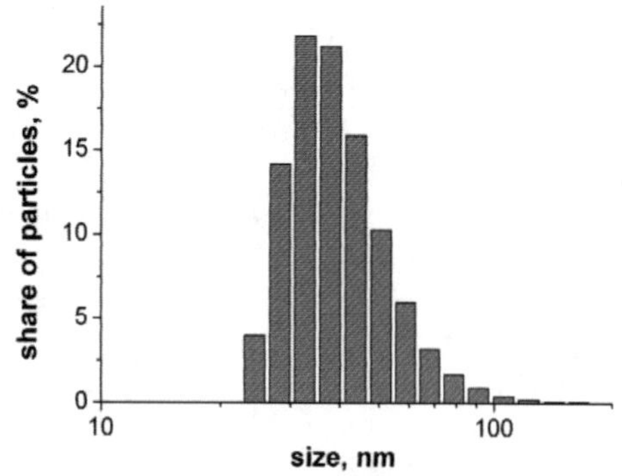

Figure 1. Size distribution of silver NPs "Argovit-C" obtained by DLS.

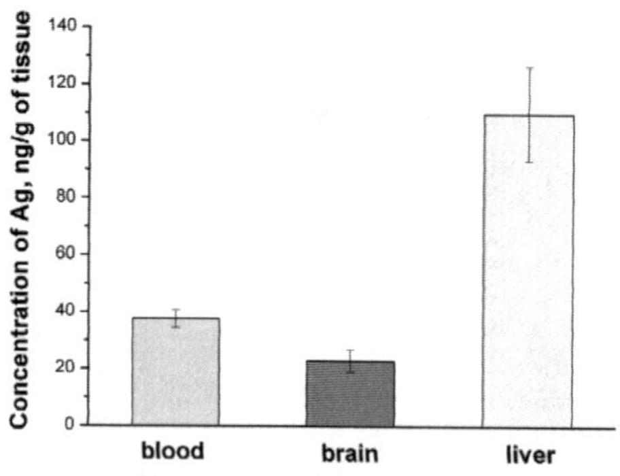

Figure 2. Ag concentrations in different organs after once-time exposure to NPs.

2. Scheme of the experiments

Diluted with distilled water solutions of “Argovit-C” were administrated orally into white mice SHK digestive tract. The mice were divided into 6 groups:

-group 1: control group, 5 mice, 1 day of maintaining;

-group 2: control group, 2 mice, 2 months of maintaining;

-group 3: control group, 2 mice, 3 months of maintaining;

-group 4: once-time oral administration of 100 micrograms of silver NPs, 5 mice, 1 day of maintaining;

-group 5: repeated oral administration of silver NPs (100 microgram per day), 6 mice, 2 months of maintaining.

-group 6: repeated oral administration of silver NPs (100 microgram per day) for 2 months and administration of distilled water for 1 following month after cancelling of silver injection, 6 mice, 3 months of maintaining.

After the end of each experiment the mice were decapitated and a number of the following organs were taken, weighted and then dried for 24 hours at 70ºC (liver, brain, blood).

2. Instrumental Neutron Activation Analysis (INAA)

Simultaneously with the control and experimental samples, the reference samples containing known amount of silver (100 micrograms) were prepared. The reference samples were irradiated together with the experimental ones. Preparation of the reference samples was necessary for the determination of amounts of silver in experimental samples based on comparison of their activities.

All the samples were put into aluminum containers and irradiated in vertical cannels VEC-9 and VEC-10 of a nuclear research reactor IR-8 at the National Research Centre “Kurchatov Institute” with thermal neutron flux approximately $3{\cdot}10^{12}$ $c^{-1}cm^{-2}$ for 12 hours. As a result, experimental samples and reference materials were labeled with ^{110m}Ag radioactive isotope with half-life around 250 days. It is obvious that during such the irradiation many the elements in the samples are activated possessing the radioisotopes formed this way. After decay of high energy isotopes with short half-life samples were analyzed by measuring of ^{110m}Ag activities in each sample using gamma spectroscopic device.

4. Results

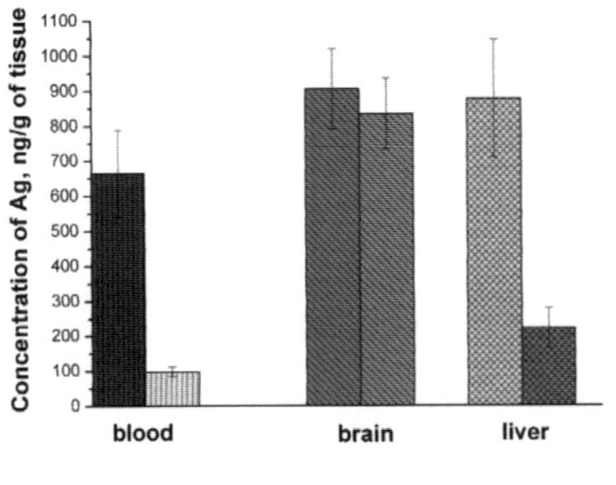

Figure 3. Ag concentrations in different organs after long-time exposure to NPs. Left bar is related to group 5 and the right bar is to group 6.

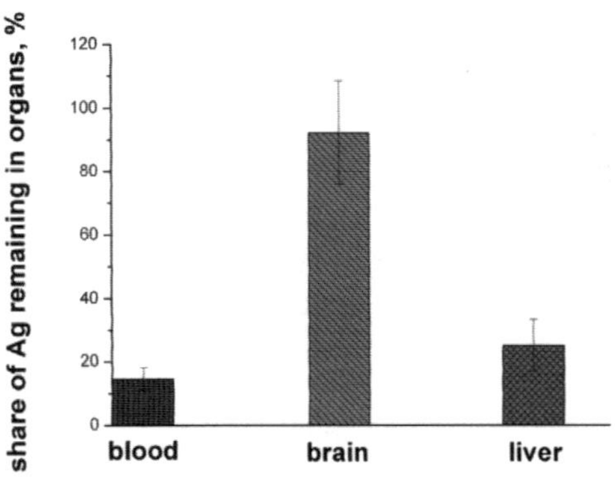

Figure 4. Fractions of silver remaining in mice organs after 1 month of washing up.

The amounts of silver in the control samples were lower than the sensitivity of the method and neglected. Results of the single administration and 2 months of exposure to NPs are demonstrated in figure 2 and 3 (left bars) respectively. The result obtained for group 6 is demonstrated in figure 3 (right bars). It can be seen that the highest concentrations related to a one-time exposure are observed in liver and blood, and in 2 months of NPs administration obvious redistribution of silver is shown. After long-time exposure the highest concentrations of silver were found in brain and liver which may be explained by accumulation of silver in brain and filtering properties of liver while blood seemed to be quite self-cleansing. After washing silver up during one month concentrations of silver in all organs reduced.

In order to estimate probable accumulation effects, average concentrations of silver in corresponding mice organs from group 5 and group 6 were compared. The obtained fractions of the rest silver in different tissues are presented in figure 4. It is seen that the rates of excretion of silver NPs from blood and liver after 1 month of washing NPs up are rather high (around 80% and 75% respectively) and the level of excretion of silver NPs from brain is extremely low (6%) which might be explained by the active functional properties of a blood-brain barrier such as mechanisms of endocytosis and exocytosis by which a NP may be taken up or extracted. Thus, no accumulation effects in liver and blood were observed. Presumably, further intake of pure water without any silver NPs and protection from other contaminating sources would bring to reducing of the amounts of silver in these organs to neglecting values. Though an extremely low level of silver excretion from brain was demonstrated, this asserts the assumption about possible accumulation of nanosilver in brain made in the previous work [23].

5. Conclusions

According to the results obtained, relatively high levels of excretion from liver and blood indicate no significant hazardous effect of silver NPs on these organs - NPs do not tend to accumulate there. On the other hand, low level of silver excretion from brain causing growing accumulation of the element in this organ might imply risks for humans using products containing nanosilver. The reason for such unexpected blood-brain barrier behavior in relation to Ag NPs is not understood and is the subject for further research. The condition of silver, which penetrates into brain is not yet a finally resolved question. Are the silver NPs conserved in the nanoform or solved to the ionic state when reaching the blood-brain barrier? In spite of that, silver is widely believed to be a non-metabolizable element and is hardly oxidized. We did not reach a unanimous agreement in the literature concerning the transformations of Ag NPs in living organism. Nevertheless, the discovery and confirmation of the fact of the accumulation of silver NPs in brain is of great practical importance, which can modify existing approaches of nanosilver toxicity assessment during the long-time exposure to organisms. Therefore, there is a need for further studies to determine the features of extended kinetics of silver accumulation in brain and the actual structural forms of silver in different tissues after possible transformations in living organism. Moreover, it is not known if the concentrations of silver measured in the brain (up to 1200 ng/g of tissue) can influence the cognitive functions of animals.
All questions mentioned above are the important subjects for the following research.

It is also should be noted that the use of Nuclear-Physical methods described in this work is highly suitable for tracing NPs in complex heterogeneous macroscopic samples such as the biological organs and tissues. At the same time due to its high sensitivity, selectivity and metrological accuracy this technique can be successfully applied for different puposes – elemental analysis of samples such as food products, drugs, soils and other inorganic complex materials and their certification. Thus, the use of INAA and Radiolabelling technique can help to reveal the unexpected data of different phenomena and promises to be one of the future top techniques.

Acknowledgments

This work was financially supported by Ministry of Education and Science of the Russian Federation (grant # RFMEFI57514X0072).

References

[1] Suppi S, Kasements K, Ivask A, Kunnis-Beres K, Shihtmae M, Kurvet I, Aruoja V and Kahru A *Journal of Hazardous Materials* 2015 **286** 75
[2] Bondarenko O, Juganson K, Ivask A, Kasemets K, Mortimer M and Kahru A *Archives of Toxicology* 2013 **87**(7) 1181
[3] Hu R, Li S, Kong F, Hou R, Guan X and Guo F *Geneticas and Molecular Research* 2015 **13**(3) 7022
[4] Ramar M, Manikandan B, Marimuthu P, Raman T, Mahalingam A, Subramanian P, Karthick S and Munusamy A *Spectrochimica Acta* 2015 **140** 223
[5] Kujda M, Oćwieja M, Adamczyk A, Bochenska O, Bras G, Kozik A, Bielanska E and Barbasz J *Journal of Nanoscience and Nanotechnology* 2015 **15**(5) 3574
[6] Sheehy K, Casey A, Murphy A and Chambers G Journal *of Colloidal and Interface Science* 2015 **443** 56
[7] Sohrabnezhad S, Pourahmad A, Mehdipour Moghaddam M and Sadeghi A *Spectrochimica Acta* 2015 **136**(PC) 1728
[8] Shameli K, Akhmad M, Jaffar Al-Mula E, Shabanzadeh P and Bagheri S *Research on Chemical Intermediates* 2015 **41**(1) 251
[9] Ortego L, Gonzalo-Asensio J, Dolores V and Concepcion G *Journal of Inorganic Biochemistry* 2015 **146** 19
[10] Salem W, Leitner D, Zingl F, Schratter G, Prassl R, Goessler W, Reidl J and Schild *International Journal of Medical Microbiology* 2015 **305**(1) 85
[11] Prabhu S and Poulose E *International Nano Letters* 2012 2:32
[12] Xu L, Shao A, Zhao Y, Wang ZH, Zhang C, Sun Y, Deng Y and Lee Chou L *Journal of Nanoscience and Nanotechnology* 2015 **15**(6) 4215
[13] Wu J, Yu C, Tan Y, Hou Z, Li M, Shao F and Lu X *Environmental Research* 2015 **138** 67
[14] AshaRani P, Low Kah Mun G, Hande MP and Valiyaveettil S *ACSNano* 2009 **3**(2) 279
[15] Bar-Ilan O, Albercht R, Fako VE and Furgeson DY *Small* 2009 **5**(16) 1897
[16] Gwinn M and Vallyathan V *Environmental Health Perspectives* 2006 **114**(12) 1818
[17] Roco M *Environmental Science and Technology* 2005 **39**(5) 106A
[18] Liu H, Yang H, Lin B, Zhang W, Tian L, Zhang H and Xi ZH *Toxicology Research* 2015 **4**(2) 486
[19] Simard J, Vallieres F, de Liz R, Lavastre V and Girard D *Journal of Biological Chemistry* 2015 **290**(9) 5926
[20] Zhang X, Choi YJ, Han JW, Kim E, Park JH, Gurunathan S and Kim JH *International Journal of Nanomedicine* 2015 **10** 1335
[21] Heyrdnejad M, Samani R and Aghaeivanda S *Biological Trace Element Research* 2015, 6p
[22] Korani M, Ghazizadeh E, Korani S, Hami Z and Mohammadi-Bardbori A *European Journal of Nanomedicine* 2015 **7**(1) 51
[23].Antsiferova A, Buzulukov Yu, Demin V, Demin V, Rogatkin D, Petritskaya E, Abaeva L and P. Kashkarov *Nanotechnologies in Russia* 2015 **10**(1-2) 100
[24] Borm P and Kreyling W *Journal of Nanoscience and Nanotechnology* 2004 **4** (5) 521
[25] Khlebtsov N and Dykman L *Chemical Society Reviews* 2011 **40** 1647.
[26] Demin V, Antsiferova A, Buzulukov Yu, Demin V and Soloviev V *Medical Radiology and Radiation Safety* 2015 **60**(2) 60

IOP Conf. Series: Materials Science and Engineering **98** (2015) 012004 doi:10.1088/1757-899X/98/1/012004

Effect of platinum nanoparticles on morphological parameters of spring wheat seedlings in a substrate-plant system

T Astafurova, A Zotikova, Yu Morgalev, G Verkhoturova, V Postovalova, S Kulizhskiy, S Mikhailova
Tomsk State University, 36, Lenina ave., Tomsk, 634050, Russia, tel./fax: +7(3822) 534519

E-mail: yu.morgalev@gmail.com

Abstract. When wheat is cultivated in the media contaminated with platinum nanoparticles, the change in the morphological and physiological indexes of wheat seedlings depends on the physico-chemical parameters of the germination substrate. The changes become less pronounced with the decreasing bioaccessability of the nanomaterial in the following order: water suspension – luvisols – phaeozems. Contamination with nanoparticles affects the height parameters and activates the mechanisms protecting the plant from stress. When using wheat seedlings as test organisms for biotesting the environmental safety of NPs, it is advisable to use the following parameters: weight of roots, weight of aerial part, leaf area, and flavonoid content.

1. Introduction

With nanomaterial production on the steady increase, nanoparticles (NPs) are being released into the environment at different stages of their life cycle, which makes it necessary to research possible consequences of their interaction with elements of the biota and ecosystems. Recent research has shown that NPs that penetrate into biological tissue cause various bioeffects, both positive and negative [1-3]. Interacting with various cell structures, NPs can be catalysts in different reactions producing not only growth promoters but also inhibitors [4,5]. The effect of NPs on plants may differ from that of the salts of the same metals. Even in very low dosage, they can change physiological and biochemical processes, while maintaining prolonged action [6]. The discovered bioactivity of technogenous NPs confirms the necessity to research their migration and accumulation in plants growing on the territories subject to contamination with finely-dispersed materials.

The aforementioned papers contain little data on the research into platinum NPs phytotoxicity [7-9]. We know, however, that they show a strong regenerative ability related to their high antioxidant activity as well as unique catalytic propertie [10].

Phytotoxicity is difficult to assess, because toxicity does not only depend on the NP physical nature, production method, size and structure, but also on features of the biological models and, quite possibly, their growing technique.

The purpose of this paper was to compare the effect of platinum NPs 4 nm in size on the morpho-physiological traits of wheat seedlings, when growing them in water culture as well as on different soil types.

2. Materials and methods

Content from this work may be used under the terms of the Creative Commons Attribution 3.0 licence. Any further distribution of this work must maintain attribution to the author(s) and the title of the work, journal citation and DOI.
Published under licence by IOP Publishing Ltd

The object of the study was water and soil cultures of 10-day plants of soft spring wheat (Triticum aestivum L.), variety Novosibirskaya - 29, which were grown in a laboratory setting in a temperature chamber at a constant temperature of 22 °C and light intensity of 150 W/m2 (photoperiod 12/12 hours). One set of experiments compared water and soil cultures. In water cultures, we grew the plants in a fine medium containing platinum NPs with the initial concentration of 10 mg/l and used distilled water as a control. In soil cultures, we grew the seedlings in containers with 200 g of Garant soil containing, per 100 grams of soil, 30-50 mg of nitrogen, 70 mg of phosphorus, 80-100 mg of potassium, pH 6.5, micronutrients, and humic growth promoters. Before seeding, the substrate was irrigated by a suspension of NPs until their concentration reached 10 mg/kg of soil. Another set of experiments used two types of soils, luvisol and phaeozem, through which we had poured a 10 mg/l suspension of platinum NPs in a filtration column. Luvisol retained 13.4 % (C = 1 mg/kg) of the platinum NPs introduced and phaeozem, 22.5 % (1.34 mg/kg). Uncontaminated soil served as a control.

Platinum nanoparticles (nPt) were obtained by laser ablation in distilled water from high-purity (99.97%) platinum bars [Morgalev et al., 2012]. The characteristics of NPs were verified by TEM (Phillips CM-12, France), dynamic light scattering (Zetasizer Nano ZS, USA), and BET (TriStar 3000, USA). Particle size Δ50=5 nm and specific surface S =36 m2/g.

Platinum nanoparticles bioaccumulation in the plants from the substrates was determined by ICP-MS (ELAN DRC-e, USA) in homogenates from the plant roots and aerial parts (leaves + stem). The morphometric changes were evaluated by the length of the root system, height of the seedlings and weight of the organs. The content of chlorophylls and flavonoids in the leaves was measured by a Dualex-4 sensor (France). The nitrogen index was calculated as a ratio of the sum of chlorophylls to that of flavonoids.

The agrochemical soil composition was studied according to the following GOSTs: pH of the salt extract, GOST 26483-85; labile phosphorus (P2 O5) and labile potassium (K2O) fractions, GOST R 54650-11; nitrate nitrogen, GOST 26951-86; exchangeable ammonium, GOST 26489-85; humus, GOST 26213-91; exchangeable calcium and exchangeable magnesium, GOST 26487-85; hydrolytic acidity, GOST 26212-91.

Statistica 7 software package was used for statistical data processing. All the indicators were analyzed for normality of distribution. The significance threshold was set at $p<0.05$. The tables and figures show the mean data and SEM.

3. Results and discussion

The nPt introduced into the cultivation medium triggered changes in the height and weight parameters of the leaf and root system in wheat seedlings when those were cultivated in a water dispersion medium and Garant substrate. In water cultures, the length and weight of the root increased significantly, which indicates that platinum NPs actively interfere with the complicated mechanism of plant growth regulation. Soil cultures were superior to their water counterparts in height and weight parameters and the effects of NP introduction were less prominent (see Table 1).

One of the reasons for such differences is the NP lifetime in the nanosized state, which depends on the properties of the dispersion medium [11]. In water solutions with natural stabilizers, it ranges from several hours to 40 days and sometimes even exceeds a year [12, 13]. As to soil, there are very little data showing that NP inactivation in soils may be long-term as indicated for ultradispersed metal powders: they gradually oxidize in soils and serve as micronutrients for plants during their growth and development [14].

Among other things, NP aggregation may reduce their bioaccessibility, which must affect the bioaccumulation processes in the plant tissues. Our studies have shown that platinum NPs are massively accumulated in the root system of wheat seedlings at their early ontogenetic stages, if they are grown in water culture (see Table 2).

Table 1. Morphometric parameters of wheat seedlings grown in an aqueous dispersion medium or soil comprising nPt a concentration of 10 mg /l and 10 mg / kg.

Variant	Root		Aboveground part	
	Length (cm)	Weight (mg)	Height (cm)	Weight (mg)
		Water culture		
Control	6.67 ± 2.80	31±3	16.33 ± 0.85	93±2
Experiment	9.22 ± 0.72*	38±1*	17.93 ± 0,87	100±2*
		Soil culture		
Control	13.78 ± 1.98	57±1	24.4 ± 3.11	133±3
Experiment	14.12 ± 2.34	59±2	25.01 ± 2.49	145±5*

Note: the asterisk hereinafter marks significant differences of parameters as compared to the control with the confidence figure of $p<0.05$.

Table 2. nPt content in water and soil cultures of wheat .

Variant	nPt content (mkg/g wet weight)			
	Water culture		Soil culture	
	Root	Aboveground part	Root	Aboveground part
Control	0.0012 ± 0.0001	0.00062 ± 0.00010	0.0016 ± 0.0005	≤ 0.0001
Experiment	42.96 ± 8.69	0.44 ± 0.08	0.81 ± 1.16	0.10 ± 0.02

Water culture is an effective tool to evaluate not only the adaptative reserves of a growing organism but also the unique role played by the tillering node in the initial growth and development of plants. Located at the stem-root junction, it functions as a powerful physiological barrier to substance transport, which, along with other possible mechanisms, helps restrict the access of NPs into the vegetative part of plants at the early stage of their development. Therefore, fewer NPs entered the aerial part, yet even in this case, the differences between test and control samples were tremendous. It is noteworthy that NPs can not only migrate through xylem vessels from the root to leaves but also move in the opposite direction with the help of a phloem, which is known to transport photosynthates from leaves to where they are consumed [13].

In soil cultures, on the contrary, the nPt accumulation is much lower; nonetheless, the difference between the test and control samples of the root system was still considerable (more than 50 times). Crystalline nanostructures have much higher surface energy than macroforms, so their dissolution in the soil moisture and transition to the germination zone must be significantly easier [15]. In our research, however, the leading role in reducing NP bioaccumulation seems to belong to their adsorption by soil elements, since both the oxidative modification and dissolution are negligible for nPt in this time interval.

The other set of experiments studied the influence of the soil type on the display of biological effects of nanomaterials. We used typical soils of the Siberian region: luvisol and phaeozem.

The phaeozem had somewhat higher fertility level vs. luvisol due to a greater content of humus (6.4±0.6 vs. 5.4±0.5)%, labile phosphorus (175±35 vs.75±15) mg/kg, labile potassium (155±23 vs.100±15) mg/kg, exchangeable calcium and magnesium (20.4±1.5 vs.15.5±1.2 and 2.7±0.2 vs.4.9±0.59 respectively) mmol/100g. Both the soil types are favorable for grain farming.

Soil contamination with nPt led to a significant ($p<0.05$) increase in the content of exchangeable ammonium (from 4.5±0.7 to 9.0±1.4) mg/kg in the luvisol and (from 15.1±3.0 to 4.7±1.4) mg/kg in the phaeozem and hydrolytic acidity (from 4.9±0.59 to 6.66±0.8) mmol/100g in the luvisol and (from 5.01±0.6 to 6.8±0.8) mmol/100g in the phaeozem. At the same time, the content of nitrate nitrogen dropped (from 16.6±3.3 to 10.5±2.1) mg/kg in the luvisol and (from 15.1±3.0 to 4.7±1.4) mg/kg in the phaeozem. Thus, we can see the effect of nPt on the agrochemical properties of the soils under study and, therefore, on their fertility level.

When cultivated in the phaeozem, the control sample had a greater dry weight of its roots and aerial part as well as root length and leaf area than in the luvisol, which corresponds to the agrochemical evaluation (see Table 3).

However, although the agrochemical composition of both the soil types changed after nPt contamination, the height parameters and vegetative weight of the plants changed significantly only in the luvisol, and these are important indicators of cell division and elongation intensity.

Table 3. nPt impact on morphological parameters of 10-day wheat seedlings grown in different soils.

	Luvisols		Phaeozems	
Parameters	Control	Experiment	Control	Experiment
Leaf area (cm^2)	4.15±0.49	5.14±0.24*	6.57±0.54	5.20±0.65
Dry above-ground mass (mg)	34.5 ± 0.2	47.6±0.2*	53.5±0.0	49.1±0.4
Length of root (cm)	15.9±1.2	16.5±1.8	20.9±2.6	18.7±1.6
Dry weight of roots (mg)	25.1 ±0.2	26.2±0.0	30.2±0.2	27.2±0.3

Under nPt contamination of the luvisol, both the leaf area and the overall aerial weight of the wheat seedlings increased significantly, whereas the root system remained unchanged (see Table 3). According to the literature, the stimulating effect may be due to the increased activity of redox enzymes and photochemmical activity in chloroplasts as well as activating ATP synthesis and resynthesis as one of the major factors of metabolisn regulation [16].

The absence of pronounced effects of the phaeozem contamination may be due to NP adsorption by soil elements. As noted above, this soil retained more nanomaterial than the luvisol did, when we poured the suspension through it. This may have been caused by different content of exchangeable divalent cations promoting NP adsorption by pore walls. A more acidic reaction of the medium in the luvisol could have contributed as well, judging by the pH of the salt extract (5.1±0.1 vs. 5.3±0.1). Greater content of humus in the phaeozem could also have enhanced the protective function guarding the plant roots from NP penetration.

This fact requires a cautious attitude to evaluating the factor of NP bioaccumulation by plants from the soil. Its calculation as a ratio of the marker element concentration in a plant to that in a soil may be misleading, since the bioaccessability of NPs may be considerably lower due to adsorption and aggregation processes in the soil.

When exposed to heavy metals, plants usually accumulate secondary phenolic substances, which form complexes with heavy metals, thus inactivating them. Besides, flavonoids (vegetable polyphenols) have antioxidant properties and can neutralize free radicals produced by oxidative stress. NPs of some metals can induce active forms of oxygen, which cause significant damage to cell structures [17]. On the other hand, increasing content of flavonoids does not only indicate unfavorable processes going on but also the activation of mechanisms preventing destructive changes.

The experiments have shown that soil contamination with nPt leads to variation of flavonoids in seedlings grown in both the luvisol and phaeozem soils (see Figure 1). Especially striking is the flavonoid content increase in wheat in the luvisol: 40% after the introduction of nPt. Most probably, the growing amount of flavonoids in wheat seedlings is due to the defense mechanisms activated as a response to nPt impact. The efficiency of this mechanism reveals itself in the stable content of chlorophylls when exposed to nPt, since their amount is heavily influenced by the changes in the microelement soil composition.

The nitrogen index calculated as a ratio of chlorophylls to flavonoids was changing downwards, which indicated that the growth and development processes were intensified in the seedlings and they required additional nitrogen nutrition.

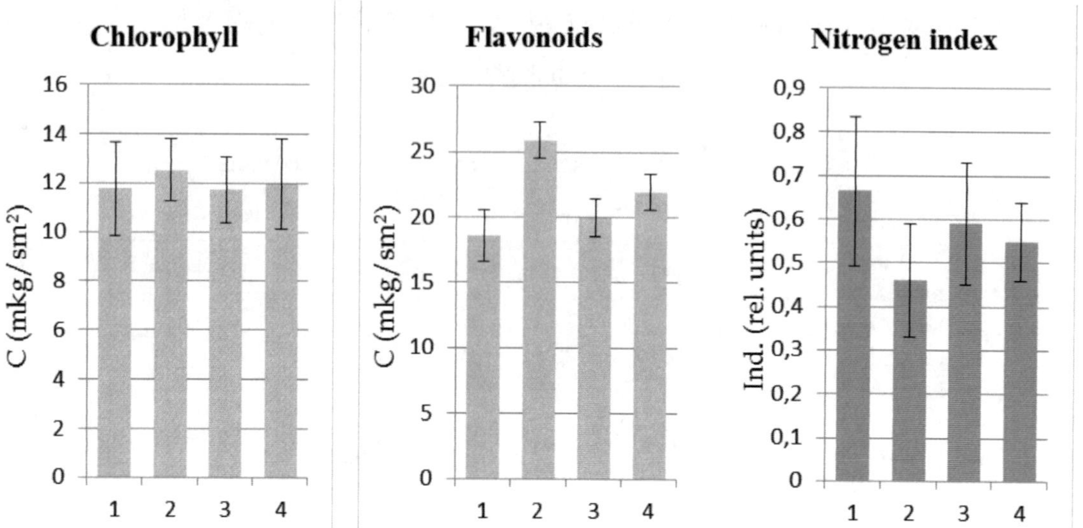

Figure 1. nPt impact on biochemical parameters of 10-day wheat seedlings grown in luvisol and phaeozem soils. Control and test samples, luvisol (1, 2); control and test samples, phaeozem (3, 4).

4. Conclusion

The research findings show that nPt penetrate plants better when they are cultivated in water media as compared to soil substrates. Pt nanoparticles change the agrochemical composition of both the luvisol and phaeozem soils. The morphological, physiological and biochemical changes in the plants under study, when introducing nPt into the cultivation medium, become less pronounced depending on the type of substrate and bioaccessibility of the nPt introduced in the following order: water suspension – luvisol – phaeozem. Here, nPt affect the height parameters and leave the chlorophyll content virtually unchanged, while activating protective mechanisms, which is confirmed by the increase in flavonoids.

When using wheat seedlings as test organisms for biotesting the environmental safety of NPs, it is advisable to use the following parameters: weight of roots, weight of aerial part, leaf area, and total flavonoid content.

Acknowledgments

The results obtained in the framework of the state task № 37.901.2014/K Ministry of education and science of Russia.

Work was conducted with the application of the Tomsk regional common use centre technical equipment acquired thanks to a grant of the Russian Ministry of the Agreement No.14.594.21.0001 (RFMEFI59414X0001).

References

[1] Adili A, Crowe S, Beaux M 2008 *Nanotoxicology* **2** 1–8
[2] Ma X, Geiser-Lee J, Deng Y and Kolmakov A *Sci. Total Environ* 2010 **408** 3053–61
[3] Salama H. *J. Biotechnology* 2012 **3** 190–7

[4] Prasad M 2003 *Plant physiology* **50** 764–80
[5] Lam C, James J, McClyskey R *Crit. Rev. Toxicology* 2006 **36** 189–217
[6] Churilov G and Ampleeva L 2010 (Rjazan: Rjazan State University Press) p 148
[7] Morgalev Yu, Astafurova T, Borovikova G, Zotikova A, Zaytseva T, Postovalova V, Verkhoturova G and Morgaleva T *Nanotechnic* 2012 **3** 81–6
[8] Zotikova A, Astafurova T, Mikhaylova C and Bender O. 2013 *Conf. Factors of plant resistance to extreme natural and man-made environment* (Tomsk) 94 –7
[9] Astafurova T, Morgalev Yu, Borovikova G, Zotikova A, Zaytseva T, Postovalova V, Verkhaturova G and Morgaleva T 2013 *Plant Physiology and Genetics* **45** 544–9
[10] Ershov B *Rus.Chen.J.* 2001 **45** 20–30
[11] Morgalev S, Morgaleva T, Morgalev Yu and Gosteva I 2015 *Adv. Mat. Res.* **1085** 424-30
[12] Svetlichnyi Vand Lapin I 2013 *Russian Physics Journal* **56** 581-7
[13] Wang Zhenyu and Xie Xiaoyan, Zhao Jian et al. 2012 *Environ. Sci. Technol.* **46** 4434–41
[14] Panitchkin L and Raikova A 2008 *Nanotechnology in agriculture* **3** 79–81
[15] Montaño M 2014 *Environmental Chemistry* **4** 351–66
[16] Davronov K, Usmanov R and Kutchkarov K 2008 *Agricultural biology* **1** 65–9
[17] Markovic Z, Todorovic-Markovic B, Kleut D, Nikolic N, Vranjes-Djuric S, Misirkic M, Vucicevic L, Janjetovic K, Isakovic A, Harhaji L, Babic-Stojic B, Dramicanin M and Trajkovic V 2007 *Biomaterials* **28** 5437–48

Effect of zink oxyde nanoparticles on the test function of water organisms of different trophic levels

Yu Morgalev, T Morgaleva, I Gosteva, S Morgalev, S Kulizhskiy, T Astafurova

Centre "Biotest-Nano", Tomsk State University, 36, Lenina ave., Tomsk, 634050, Russia, tel./fax: +7(3822) 534519

E-mail: yu.morgalev@gmail.com

Abstract. The toxicity of zinc oxide nanoparticles (nZnO) with particle size Δ_{50} = 20 nm was evaluated according to the degree of toxicity of the aqueous disperse system (DS) with biological testing methods using a set of test organisms representing the major trophic levels.We observed the influence of the concentration degree of nZnO on toxic effects level on the fluorescence of the bacterial biosensor "Ekolyum-13", the chemotactic response of ciliates *Paramecium caudatum*, the rate of growth of unicellular algae *Chlorella vulgaris Bayer*, mortality of entomostracans *Daphnia magna Straus* and fish *Danio rerio.* The detected values of L(E)C_{50} are: for biosensor "Ekolyum-13" - 0.30 mg/L, for ciliates *Paramecium caudatum* - 0.14 mg/L, for *Chlorella vulgaris Bayer* - 0.17 mg/L and for *Daphnia magna Straus* - 0.52 mg/L. No toxicity of nZnO was detected in relation to fish *Danio rerio*, L(E)C_{50} > 100 mg/L. In assessing the maximum effect of nZnO according to GHS and EU Directive 93/67/ EEC, it is assigned to dangerous substances with a high degree of toxicity "Acute toxicity 1".

1. Introduction

Development of nanotechnology, growing production volumes of fine disperse materials (FDM), the exponential growth in the number of products containing engineered nanoparticles (NP), on the one hand, opens up new possibilities of using FDM in biomedicine, pharmacology, food production, in addressing environmental and agricultural issues. On the other hand, the unique properties of FDM may carry a potential risk to the environment.

The need to determine the safety of FDM, approved by regulatory documents of Russia and OECD, ISO, EU and the Environmental Protection Agency [1, 2], arises from the large volumes of production and a wide range of application of FDM.

Despite the fact that currently the volume of industrial production of various NP is more than 100,000 tons a year, studies on the safety of nanotechnologies and nanoproducts are scattered, fragmentary, incomplete and significantly far behind the FDM production rate growth and expansion of their scope of usage. To date, most of the published papers are devoted to the study of biological effects of the most popular metal oxides NP. According to Bondarenco et. al. [3] among metal oxides NP, the NP of zinc oxide (nZnO) hold the third place in terms of volume of production (nSiO2 - 55,000 tons/year, nTiO2 - 3000 t/year, nZnO - 550 tons/year). The wide range of applications of nZnO in the field of cosmetology, medicine, materials science [4, 5, 6, 7] is due to its anti-bacterial, surfactant and catalytic properties [8, 9]. The presence of biological effects when NP contact with representatives of the biota makes the research on the biological effects of FDM, identifying risks to the environment and ensuring biosafety as a necessary component in the development of

Content from this work may be used under the terms of the Creative Commons Attribution 3.0 licence. Any further distribution of this work must maintain attribution to the author(s) and the title of the work, journal citation and DOI.
Published under licence by IOP Publishing Ltd

nanotechnology a priority. This necessity is determined by the growth of consumer goods on the basis of GMR, including those on the basis of nZnO, which will inevitably lead to their release into the environment and as a result in aquatic ecosystems - the ultimate receiving environment.

It is an acute problem that there is still no single unified analytical system that could provide a reliable risk assessment of NP for wildlife objects.

As a result of earlier (in 2008-2013) target programs and research studies we have developed a biological model and a series of biotests for ecotoxicity assessment [10]. An essential component of the algorithm for estimating toxicity we developed is the expanded core set of test organisms (representatives of the main trophic levels), which allows us to study not only of the FDM biocidal action, but also changes in the test - functions of living aquatic organisms.

2. Materials and methods

NZnO dispersed system (DS) with an initial concentration of 50 mg/l was obtained by laser ablation [11]. According to BET analysis and transmission electron microscopy the average particle size was $\Delta_{50} = 20$ nm.

NZnO toxicity study was conducted using the test organisms representing the main trophic levels of ecosystems: luminescent bacteria on the basis of genetically engineered strains of *Eshcerichia coli* (biosensor "Ekolyum-13") was used for studying biolyuminescence change [10]; unicelled animals *Paramecium caudatum* - for the chemotactic response [10]; unicellular algae – for the optical density of the test culture algae *Chlorella vulgaris Beijer* [10]; entomostracans *Daphnia magna Straus* - in terms of mortality rate [10]; *Danio rerio* fish – also in terms of mortality rate [12]. The research design is similar to the described in [13, 14].

nZnO DS was formed according to the method we developed [15]. The study used nZnO DS at concentrations of 0.0001, 0.001, 0.01, 0.1, 1.0, 5.0 and 10.0 mg/l. Right before biotesting nZnO DS was sonicated with ultra sound (30 W/L) for 5 minutes for redispersion. Evaluation of toxicity of the analysed nZnO DS was carried out in accordance with the criteria prescribed in the certified techniques [16].

Statistical data processing was executed in Excel 2010. Probit analysis was used to calculate the values of $L(E)C_{10}$ (NOEC), $L(E)C_{20}$ and $L(E)C_{50}$. Work was conducted with the application of metrologically gauged instruments: Autolumat LB953 (Germany), «Shimadzu» RF-5301 PC spectrofluorimeter (Japan), IPS-03 photoelectric colorimeter (Russia), Biotester-2 concentration meter (Russia) and diffractometer «Shimadzu» XRD 6000 (Japan).

3. Results

Studies revealed that introduction of a low concentration of nZnO particles ($S \leq 0.001$ mg/l) resulted in suppression of the luminescence of the biosensor within acceptable values not exceeding 20 %. The toxicity index of nZnO DS for the tested range of concentrations was calculated according to the degree of luminescence decrease (figure 1). nZnO DS with particle concentration of 0.01 mg/l had a toxic effect on the biosensor and the toxicity index was slightly higher than the acceptable level ($I=25.25\pm3.60$ %). Further increase of concentration of nZnO particles amplifies DS toxicity: when $C=0.1$ mg/l ($I=41.36\pm3.03$%) and $C=1.0$ mg/l ($I=51.72\pm3.78$ %) the suppression of the biosensor luminescence is doubled. Further increase of nZnO concentration to 5 mg/l causes luminescence inhibition up to 59.84 ± 0.46%. Almost complete suppression of biosensor luminescence was observed with nZnO DS concentration of 10 mg/l ($I = 93.75 \pm 6.48$%).

When exposing *Paramecium caudatum* in nZnO DS in concentrations ranging from 0.0001 mg/l to 0.1 mg/l the index of toxicity did not exceed the acceptable range ($I=22.5\pm1.4$%, $I=1.1\pm0.1$%, $I=9.4\pm2.6$% and $I=9.6\pm0.8$%) (Figure 1). nZnO DS concentrations with 1.0, 5.0, and 10.0 mg/l for 30 minutes reduced the yield of *Paramecium caudatum* in the test medium to almost complete suppression of positive chemotaxis ($I=93.5\pm0.5$%; $I=94.4\pm1.0$% and $I=85.0 \pm0.9$%).

When cultivating *Chlorella v. B.* in nZnO DS with concentrations of 0.1, 0.01, 0.001 and 0.0001 mg/l the toxicity index did not exceed the acceptable range (figure 1). nZnO DS concentration of 1.0

mg/l had more pronounced toxic effect on the growth of chlorella cells. The toxicity index at that significantly exceeded the acceptable range (I=63.5±0.73 %). When chlorella was exposed to nZnO DS concentration of 5.0 mg/l and 10.0 mg/l it almost completely inhibited the growth of *Chlorella v.B.* The toxicity index was I=97.3±0.56 % and I=98.4±0.89 % respectively (figure 1).

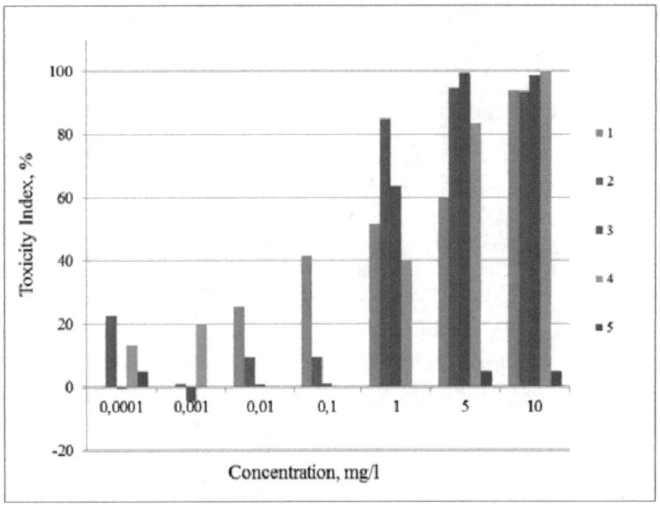

Figure 1. Index of the nZnO DS toxicity with a particle size of 20 nm for various test organisms (1 - «Ekolyum-13», 2 - *Paramecium caudatum*, 3 - *Chlorella vulgaris Beijer*, 4 - *Daphnia magna Straus*, 5 - *Danio rerio*)

When Daphnia was cultivated in DS nZnO with concentrations of 0.1 mg/l, 0.01 mg/l, 0.001 mg/l, 0.0001 mg/l for 48 hours, the mortality of these crustacean changed within the range of acceptable values from 0 to 20 % (figure 1). nZnO DS with concentration of 1.0 mg/l caused a high mortality rate of *Daphnia magna Straus* (I=40.0±4.1 %), which is rated as not providing acute toxic effects, but having a fairly pronounced biological effects. Further increase of the concentration of nZnO DS (C=5 mg/l) resulted in an increase in mortality of crustaceans (I=83.3±2.4 %). nZnO DS with concentration of 10 mg/l caused 100 % mortality of *Daphnia magna Straus* (figure 1).

NZnO DS (20 nm) in the test concentrations from 0.0001 mg/l to 10.0 mg/l did not cause mortality in fish.

Ecotoxicological parameters NOEC ($L(E)C_{10}$), ($L(E)C_{20}$ and $L(E)C_{50}$of nZnO DS have been defined for each test organism. When formulating the expert opinion on ecotoxicity of the test sample we took into account the principle of assessment according to the most pronounced reactions reflected in the regulations [16, 17].

Observations proved that nZnO DS does not have biological effects on the biosensor with the concentration of up to 0.028 mg/l (NOEC). Toxic effects began to show with the increase in concentrations of up to 0.063 mg/l ($L(E)C_{20}$); $L(E)C_{50}$ = 0.30 mg/l.

NZnO DS concentrations ranging less than 0.002 mg/l (NOEC) caused no biological effects in infusoria *Paramecium caudatum* The toxic effect was observed starting from concentrations of 0,008 mg/l ($L(E)C_{20}$); $L(E)C_{50}$ = 0.14 mg/l.

NZnO DS concentrations ranging less 0.013 mg/l (NOEC) caused no biological effects in the algae *Chlorella vulgaris Bayer* The toxic effect was observed starting from concentrations of 0.059 mg/l ($L(E)C_{30}$); $L(E)C_{50}$ = 0.17 mg/l.

NZnO DS concentrations ranging less 0.007 mg/l (NOEC) caused no biological effects in *Daphnia magna Straus* nZnO DS toxicity manifested itself starting with a concentration of 0.031 mg/l ($L(E)C_{20}$); $L(E)C_{50}$ = 0.518 mg/l.

DS nZnO proved to be not toxic for fish *Danio rerio*; $L(E)C_{50}$> 100 mg/l.

Basing on the data of ecotoxicological assessment conducted according to the criteria of regulatory documents [16, 17] we carried out an integrated assessment of nZnO toxicity. The distribution of test organisms according to the "toxicity zones" (figure 2) can be represented more clearly using the scheme proposed by Bondarenko et. al. [3].

In contrast to these authors, the figure presents the results of the experiments we conducted according to a unified algorithm using a core set of test organisms, which was looking into both biocidal properties of FDM and their impact on test functions of living organisms.

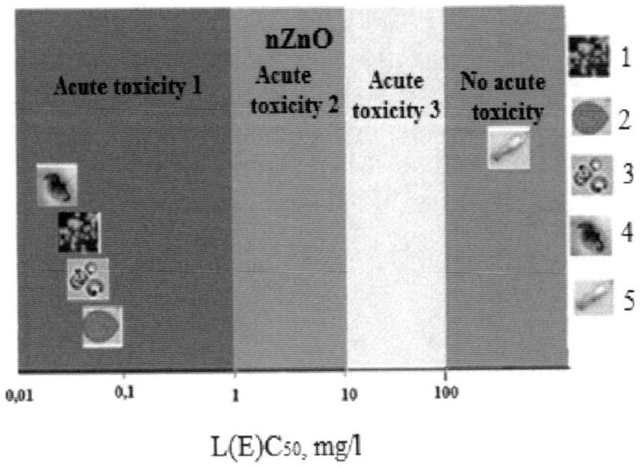

Figure 2. Classification of nZnO toxicity: 1 - «Ekolyum-13», 2 - *Paramecium caudatum*, 3 - *Chlorella vulgaris Beijer*, 4 - *Daphnia magna Straus*, 5 - *Danio rerio*

4. Discussion

The lack of a single unified analytical system, that provides reliable risk assessment of FDM to wildlife, does not allow us to properly compare the obtained results with the data available in the literature, most of which were obtained for 2, in a few cases for 3, test organisms. The authors of publications that attempt at presenting an integrated assessment of FDM toxicity, as a rule, do not take into account the size of particles, the specificity of NP DS preparation method and NP behaviour in cultured and natural media with varying degrees of mineralization, which significantly affect the assessment of toxicity of NP under study in the natural medium.

Bondarenko O. et. al. [3] notes in his review that the greatest interest lies in studying the toxicity of the three most popular types of NP: nAg, nCuO and nZnO. The number of publications devoted to the study of nZnO toxicity in recent years has grown up to 4,000.

Our results are mostly consistent with the published data on the damaging effects of nZnO on certain elements of ecosystem.

Pakrashia S. et. al. point out the damaging influence of nZnO on aquatic types of bacteria (*Vibrio fischeri*) depending on its concentration [18]. The toxic effect of nZnO sized 50-70 nm on ciliates *Tetrahymena thermophila*, according to Mortimer et. al. [19], may be associated with the transition of NP into the ionic form and nZnO aggregation.

Pendashte H. et. al. [20] investigated the effect of nZnO sized 20 nm on the change of *Chlorella v.* and *Scenedesmus d.* cell density. The inhibitory effect on *Chlorella v.* cell growth manifested itself as early as 24 hours after the start of the experiment: the number of cells decreased with the increasing of nZnO concentrations. The most significant decrease in cell growth was noted at a concentration of 1.0 mg/l. The authors found that the degree of toxicity of nZnO is different for species: nZnO was more toxic for *Chlorella v.*

Lopes S. et. al. [21] estimated the toxic effect of nZnO with a particle sizes of 30 nm and 80-100 nm on the survival rate of *Daphnia magna Straus* The values of $L(E)C_{50}$ for nZnO were 1.02±0.24 mg/l and 1.10±0.05 mg/l, respectively, which is a sign of acute toxicity of the tested NP. The present study did not indicate dependence between toxicity and the initial particle size. The authors suggest that during the experiment, due to nZnO sedimentation, Daphnia were exposed not to individual particles, but agglomerates, which were likely to modify the toxicity. Authors believe that nZnO can act as an additional source of Zn ions being released into the environment, causing changes in daphnia reproductive function. Dependence of toxicity on the NP size was established by Liu et. al. [22] in their research, which studied the effect of nZnO (30 nm, 50 nm) on the survival and reproduction of *Daphnia magna Straus* and the development of *Danio rerio* fish embryos. The authors found that nZnO sized 30 nm is more toxic than nZnO with 50 nm particle size. Researchers believe that nZnO (30 nm), unlike nZnO (50 nm), easily penetrates into the lymphatic, digestive and blood circulatory systems, thereby causing a greater toxic effect. Danio rerio, exposed to nZnO sized 30 nm and 50 nm, produced less embryos with the increase of concentration from 1.0 mg/l to 25.0 mg/l. Furthermore, not a single embryo appeared after applying nZnO of the same sizes in concentrations ranging from 50.0 mg/l to 100.0 mg/l [22].

The conducted research allowed us to determine the "limiting elements" of the ecosystem stability to the nZnO environment contamination. The degree of NP effect depends on the type of a test organism and the concentration of NP. The most sensitive to nZnO are the protozoaires *Paramecium caudatum* ($L(E)C_{50}$ = 0.14 mg/l) and unicellular algae *Chlorella vulgaris Bayer* ($L(E)C_{50}$ = 0.17 mg/l).

5. Conclusion

It has been determined that nZnO, obtained by laser ablation, with a particle size of $\Delta 50 = 20$ nm, belongs to dangerous substances with a high degree of toxicity "Acute toxicity 1".

It has also been found that the acceptable/safe level of the tested NP does not exceed the concentration of 0.001 mg/l. Higher concentrations of nZnO put at risk some representatives of the trophic chain, which causes damage to the integrity of the natural biocenosis and its ability to recover.

The increasing of concentrations of nZnO to 5.0 mg/l leads to significant toxic damage to test organisms (excluding fish).

Acknowledgments

The results obtained in the framework of the state task № 37.901.2014/K Ministry of education and science of Russia.
Work was conducted with the application of the Tomsk regional common use centr technical equipment acquired thanks to a grant of the Russian Ministry of the Agreement No.14.594.21.0001 (RFMEFI59414X0001).

References

[1] The concept of toxicological studies, risk assessment methodology, identification and quantification of nanomaterials 2007 *Approved by the Resolution of the Chief State Inspector of the Russian Federation October 31*
[2] OECD 2007 Current Developments in Delegations And Other International Organisations on The Safety of Manufactured Nanomaterials Tour de Table 3rd Meeting of the Working Party on Manufactured Nanomaterials
[3] Bondarenko O, Juganson K, Ivask A, Kasemets K, Mortimer M, Kahru A *Arch Toxicology* **87** 1181–1200
[4] Soundery N, Zhang Y 2008 *Recent Patents on Biomedical Engineering* **1** 34-42
[5] Chiu W, Khiew P, Cloke M, Isa D, Lim H, Tan T, Huang N, Radiman S, Abd-Shukor R, Hamid M, Chia C 2010 *J. Phys. Chem. C* **114** 8212–8218
[6] Ahn K, Kwon K et. al. 2011 *Biosensors & Bioelectronics* **28 1** 378–385
[7] Ovissipour M, Rasco B and Sablani S 2013 *J Fisheries Livest Prod* http://dx.doi.org/10.4172/2332-2608.1000e106
[8] Dastjerdi R, Montazer M 2010 *Colloid Surf B* **79** 5–18
[9] Song W, Zhang J, Guo J, Zhang J, Ding F, Li L, Sun Z 2010 *Toxicol Lett* **119** (3) 389–397
[10] Morgalev Yu, Hoch N, Morgaleva T, Dunaevsky G,. Morgalev S. 2010 *Methodological Guide* Tomsk
[11] Svetlichnyi V, Lapin I 2013 *Russian Physics Journal* **56** (5) 581-587

[12] The safety assessment of nanomaterials in vitro and in model systems in vivo: Guidelines. M. 2009 Federal Center of Hygiene and Epidemiology

[13] Gosteva I. Morgalev Yu, Morgaleva T, Morgalev S 2015*: submitted to Journal of* Advanced *Materials Research*

[14] Morgaleva T, Morgalev Yu, Gosteva I, Morgalev S 2015: *submitted to Journal of Advanced Materials Research*

[15] Morgalev S, Morgaleva T, Morgalev Yu, Gosteva I 2015 *Journal of Advanced Materials Research* **1085** 424-430

[16] Globally Harmonized System of Classification and Labelling of Chemicals (GHS) The third revised edition 2009 UN

[17] Technical Guidance Documenton Risk Assessment in support of Commission Directive 93/67/EEC on Risk Assessment for new notified substances 1996 Part II, Environmental Risk Assessment Luxembourg

[18] Pakrashia S, Dalaia S, Prathna T, Trivedia S, Mynenia R, Raichurb A, Chandrasekarana N, Mukherjee A 2013 *Journal Aquatic Toxicology* **132-133** 34– 45

[19] Mortimer M, Kasemets K, Kahru A 2010 *Journal Toxicology* **10** 182-189

[20] Pendashte H, Shariati F, Keshavarz A and Ramzanpour Z 2013 *World Journal of Fish and Marine Science* **5** (5) 563-570

[21] Lopes S, Ribeiro F, Wojnarowicz J, Łojkowski W, Jurkschat K, Crossley A, Soares A and. Loureiro S 2014 *Environmental Toxicology And Chemistry* **33** (1)

[22] Liu J, Fan D, Wang L, Shi L, Ding J, Chen Y, Shen SH 2014 *Environment Protection Engineering* **40** (1) 139-149

Evaluation of the toxicity of superfine materials to change the physiological functions of aquatic organisms of different trophic levels

S Morgalev, T Morgaleva, I Gosteva, Yu Morgalev

Centre "Biotest-Nano", Tomsk State University, 36, Lenina ave., Tomsk, 634050, Russia, tel./fax: +7(3822) 534519

E-mail: 3030@sibmail.com

Abstract. We assessed ecological and biological effects caused by the physical and chemical properties of nanomaterials on the basis of the laboratory researches into water test-organisms of different trophic levels. We studied the physiological functions of water organisms on adding into the environment superfine materials of various chemical nature and structural characteristics: metallic nanoparticles of nikel (nNi), argentum (nAg), platinum (nPt), aurum (nAu), binary NPs (powder of titanium dioxide - $nTiO_2$, aluminum oxide - nAl_2O_3, zink oxide - nZnO, silicon nitride - nSi_3N_4, silicon carbide (nSiC) and carbon nanotubes (BT-50, MCD-material). We observed the dependence of developing the complex of unfavourable biological effects in water plants and entomostracans' organisms on the physical and chemical properties of superfine materials. We determined the values of NOEC, $L(E)C_{20}$ and $L(E)C_{50}$ for aquatic organisms of various regular groups. We found out the most vulnerable elements of the communities' trophic structure and the possibility of a breakdown in the water ecosystem food pyramid.

1. Introduction

In the last two decades nanotechnology development has resulted in a rapid increase in the diversity of engineered nanoparticles (NPs) and nanomaterials (NMs), many of which differ from natural NPs radically in their physical and chemical characteristics including element composition, higher dispersion, inner structure and complicated spatial organization.

Unlike natural NPs engineered nanoparticles that are used increasingly in various life activity spheres may be potentially hazardous for the environment. Despite a growing number of publications on ecotoxicology and ecological effects of a direct or indirect NPs' impact on a living organism, including water organisms, the issue of NPs bio toxicity still remains urgent and is connected with studying both the mechanism of their toxic effect development and their circulation in nature. In connection with this the necessity of complex research is being discussed; such research will make it possible to assess not only biological activity of various NMs, ways of distribution and building NPs into food chains, but also long-term effects of their influence on the ecosystem as a whole.

In the last decade researchers have paid a lot of attention to the studies on NPs ecotoxicity. However, there are still no guidelines for a quantitative assessment of the effects of superfine materials (SFM), and the issue of applying the methodologies at hand to a new class of substances remains open. In connection with this studying SFM influence on the functions of bio-objects, in particular the physiological functions of water organisms, is of greater interest rather than studying SFM biocidal

Content from this work may be used under the terms of the Creative Commons Attribution 3.0 licence. Any further distribution of this work must maintain attribution to the author(s) and the title of the work, journal citation and DOI.
Published under licence by IOP Publishing Ltd

effects. The most sensitive and appropriate indicators of assessing SFM toxicity are, undoubtedly, the methods of bio-testing that imply creating test-system models (laboratory and field) of NPs' impact on the specially chosen test organisms and the assessment of eco toxicity based on changes in the informative test-parameters. The behaviour, condition and activity of various systematic groups allow stating the toxicity of the environment regardless of the fact what are the substances and what is their combination that cause it.

The literature data at hand concerning the influence of NPs of metals and their oxides on the growth and development of water organisms are limited, controversial and difficult to be compared by dose, NPs size and the species of water plants and animals [1, 2, 3, 4, 5, 6, 7, 8]. The issues of the physiological functions and behavior of organisms in the water ecosystems contaminated with SFM need investigating.

The paper is aimed at revealing the peculiarities of various SFM influence on the physiological functions of water organisms at different trophic levels through experimental modelling.

2. Materials and methods

We studied the functional properties of different NPs: argentum (nAg, with particle size Δ_{50}=2-5 nm and specific surface S_{sp} =60-80 m^2/g), aurum (nAu, Δ_{50}=2-5 nm, S_{sp} = 40 m^2/g), nikel (nNi, Δ_{50}=3-5 nm, S_{sp} = 80-100 m^2/g), platinum (nPt, Δ_{50}=3-5 nm, S_{sp}= 36 m^2/g), zink oxide (nZnO, Δ_{50}=12 nm, S_{sp} = 47 m^2/g), aluminium oxide (α-form) (α-nAl_2O_3, Δ_{50}=7nm and 70 nm, S_{sp} =112-122 m^2/g and 20-25 m^2/g), aluminium oxide (elecrtocorundum) (nAl_2O_3-corundum, Δ_{50}=70 nm, S_{sp} = 45 m^2/g), titanium dioxide (nTiO_2, Δ_{50}=5 nm, 50 nm and 90 nm, S_{sp} = 119 m^2/g, 35 m^2/g and 6 m^2/g), silicon carbide (nSiC, Δ_{50}=95 nm, S_{sp} = 15,7 m^2/g), silicon nitride (nSi_3N_4, Δ_{50}=80 nm, S_{sp}= 11,5 m^2/g) and carbon nanotubes (BT-50 diameter of 4 nm, «Baytubes®»), and MCD-material, diameter of 4 nm, «Sigma Gr.», RF). We used the tested substances in macro form as a control. Nanoparticles with sizes less than 15 nm were prepared by a procedure [9], nTiO_2, Δ_{50}=50 nm obtained by electric explosion («SibTermoChim» Ltd, RF), larger NPs made with the help of the pneumo circuital method («Mipor» Ltd, RF).

SFM influence on the physiological functions was assessed by the change in the test-responces characteristic of water organisms: the chemotactic response of ciliates *Paramecium caudatum* [10], the growth of unicellular algae *Chlorella vulgaris Beijer* [11], the indicators of the trophic activity of entomostracans *Daphnia magna Straus* [10]. We studied SFM effect on the mortality of *Daphnia magna Straus* [12].

The characteristics of the tested SFM were verified by TEM («Phillips CM-12», France), by method of dynamic light scattering with the help of the disperse particles' sizes analyser «Zetasizer Nano ZS» (USA), by method of BET («TriStar 3000», USA) to measure S_{sp}, and the method of X-ray structure analysis of the phase composition and structural characteristics (X-ray diffractometer «Shimadzu XRD-6000», Japan). The disperse systems (DS) SFM were made in accordance with our methodology [13] including the stages of ultrasonic dispersing (30 W/l) during 5 minutes. The toxicity of DS SFM samples were studied in the concentration C=1.0 mg/l corresponding their possible stable concentration in natural environment, which was grounded experimentally in our previous works [13]. The toxicity of the analysed DS SFM was assessed in accordance with the criteria of the certified methods [10, 11, 12].

3. Results and discussion

One of the most important factors that determine organisms' test-responses is the concentration (C) of the substance added to the test-system for analysis, which should correlate with the value of the concentration possible in the natural environment. The previous researches showed that in creating DS SFM test-system the increase in the concentration compared to the initial one (C>5 mg/l) results in the increase in the aggregation speed (V_{ag}) of NPs and the sufficient decrease in their final C [13, 14]. «Self-purification» of the hydrosphere from SFM due to forming conglomerates accompanied by sedimentation and elimination from water environment solves the problem of assessing effects of large C of NPs (except for short-term effects of emergency discharges).

We stated that in the in-situ conditions SFM concentration in a created DS becomes stable within the interval of several milligrams per a litre in two days after the addition of SFM to water environment. We indicated the following regulations:

– the increase in the initial C results in the increase in V_{ag} of NPs in DS SFM;

– the most rapid growth of V_{ag} of NPs occurs with C >5 mg/l, the ratio "C – Stability of DS SFM " is optimal within the range 1-5 mg/l;

– V_{ag} of NPs grows in the row «Distilled water», «medium of Lozino-Lozinsky» - growth medium (GM) of ciliates, «Drinking water» - GM of fish and daphnids, 0.9 % NaCl and 3.0 % NaCl – GM of luminescent bacteria [13].

Our researches showed that test-organisms' response on SFM in C=1.0 mg/l is ambiguous. The nature of changes in the physiological functions and behaviour depends on the physical and chemical properties of SFM, test-organism type and the test-functions under study. The addition of SFM into GM in most cases did not cause the death of test-organisms but caused serious failures of their physiological functions and behaviour: the negative chemotaxis of ciliates *Paramecium caudatum*, changing mortality/survival in *Daphnia magna Straus,* the failure of the trophic activity of entomostracans *Daphnia magna Straus* and the change in the growth speed of the cellular weight of *Chlorella vulgaris Beijer.* We stated that some SFM kinds have the inhibiting effect on the growth speed the alga *Chlorella vulgaris Beijer,* partially or fully inhibiting its growth (figure 1).

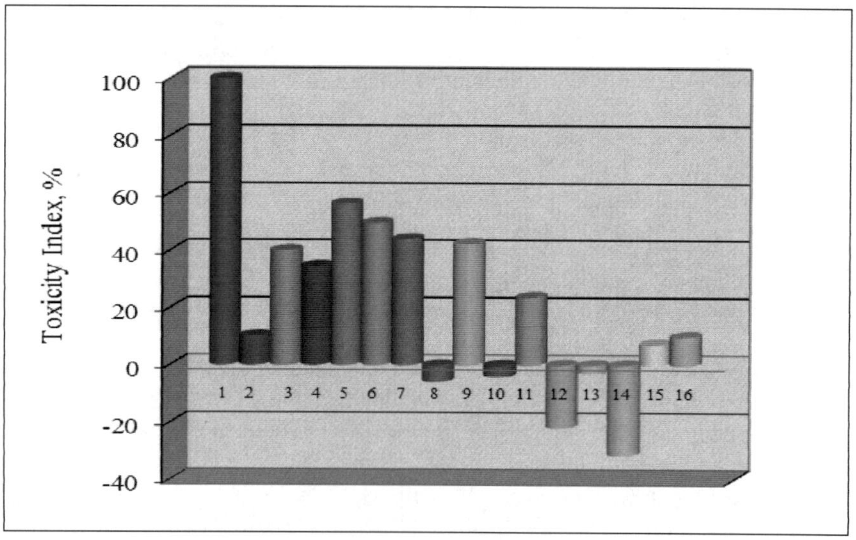

Figure 1. The toxicity index of DS SFM and macro form of the substances with various chemical composition and dispersion rate determined in accordance with the change in the weight growth of unicellular algae *Chlorella vulgaris Beijer* (C=1.0 mg/l): 1 - nAg, Δ_{50} = 2-5 nm; 2 - nPt, Δ_{50} = 3-5 nm; 3 - nAu, Δ_{50} = 2-5 nm; *4 - nNi,* Δ_{50} *= 3-5 nm; 5 - nZnO,* Δ_{50} *= 12 nm; 6 - α-nAl*$_2$*O*$_3$*,* Δ_{50} *= 7 nm; 7 - α-nAl*$_2$*O*$_3$*,* Δ_{50} = 70 nm; 8 - nAl_2O_3- corundum, Δ_{50} = 70 nm; 9 - nTiO_2, Δ_{50} = 5 nm; 10 - nTiO_2, Δ_{50} = 50 nm; 11 - nTiO_2, Δ_{50} = 90 nm; 12 - TiO_2, Δ_{50} = 10 mkm; 13 - nSi_3N_4, Δ_{50} = 80 nm; 14 - nSiC, Δ_{50} = 95 nm; 15 - BT-50; 16 - MCD- material

For example, nAg inhibits algae growth completely (100 %), while nAu, nNi, nZnO and α-nAl_2O_3, – only partially: nAu - by 40.0 %, nNi - by 34.5 %, nZnO - by 56.2 %, α-Al_2O_3 - by 49.2 %

(Δ_{50}=7 nm) and 43.9 % (Δ_{50}=70 nm), $nTiO_2$ (Δ_{50}=5 nm) – by 42.3 % and $nTiO_2$ (Δ_{50}=90 nm) – by 23.7 %. Some SFM kinds such as nSiC, (Δ_{50}=95 nm) stimulate the unicellular algae growth (by 31.3 %). We found the group of SFM that hardly ever change the chlorella growth speed, for example NPs nPt, nAl_2O_3-corundum (Δ_{50}=70 nm), $nTiO_2$ (Δ_{50}=50 nm), Si_3N_4 (Δ_{50}=80 nm) and carbon nanotubes (BT-50 and MCD-material).

The substances in macro form (Δ_{50}>100 nm, C=1.0 mg/l) do not have a sufficient influence on the chlorella growth speed, i.e. they are biologically inert in relation to unicellular algae.

The addition of SFM into GM of ciliates *Paramecium caudatum* (C=1.0 mg/l) causes changes in the chemotactic response in 7 out of12 testes DS SFM. Only 4 types of DS SFM has a sufficient effect on the chemotaxis of ciliates: nNi (Δ_{50}=3-5 nm) and nAg (Δ_{50}=2-5 nm) completely (100 %) inhibited ciliates' release into the environment with the above-mentioned NPs, nZnO (Δ_{50}=12 nm) – by 85.1 %, $nTiO_2$ (Δ_{50}=5 nm) - by 51.3 % (figure 2). A less sufficient inhibition of the chemotactic response was observed after the addition of NPs nAl_2O_3 (Δ_{50}=70 nm), $nTiO_2$ (Δ_{50}=50 nm) and MCD-material into GM. The indicated NPs inhibit ciliates' release by 28.0 %, 18.0 % and 27.5 % correspondingly. The macro form of the substances under study does not cause any changes in the chemotactic response of *Paramecium caudatum*.

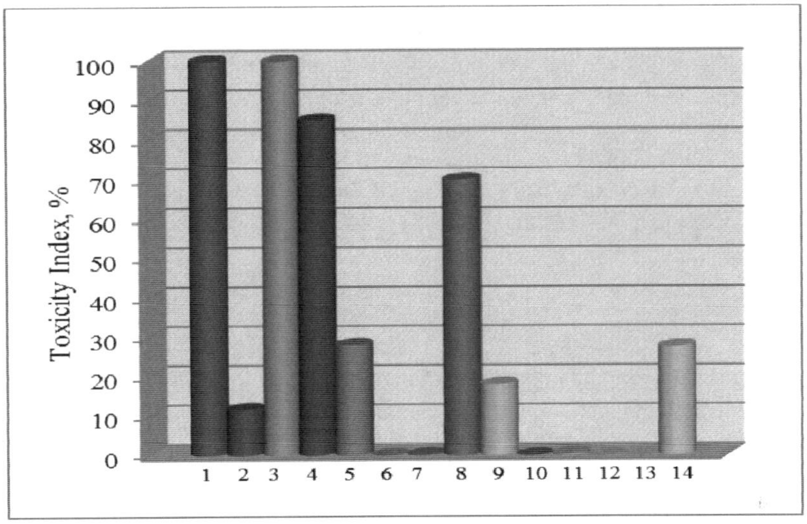

Figure 2. The toxicity index of nanomaterials and macro form of the substances with various chemical composition and dispersion rate determined in accordance with the change in the chemotactic response of ciliates *Paramecium caudatum* (C=1.0 mg/l): 1 - nAg, Δ_{50}=2-5 nm; 2 - nPt, Δ_{50}=3-5 nm; 3 - nNi, Δ_{50}=3-5 nm; 4 - nZnO, Δ_{50}=12 nm; 5 - α-Al_2O_3, Δ_{50}=70 nm; 6 - α-Al_2O_3, Δ_{50}=4 mkm; 7 - nAl_2O_3- corundum, Δ_{50}=70 nm; 8 - $nTiO_2$, Δ_{50}=5 nm; 9 - $nTiO_2$, Δ_{50}=50 nm; 10 - TiO_2, Δ_{50}=10 mkm; 11 - nSi_3N_4, Δ_{50}=80 nm; 12 - nSiC, Δ_{50}=95 nm; 13 - BT-50; 14 - MCD-material

The addition of SFM into the composition of GM of entomostracans results in the mortality of *Daphnia magna Straus* only in 2 out of 14 tested DS (figure 3): in DS nAg (Δ_{50}=2-5 nm) and nZnO (Δ_{50}=12 nm) (100 % and 40 % correspondingly).

A number of researches have shown that it is impossible to correctly assess the medium quality with the help of the principle, widely used in ecotoxicology, that takes into account the characteristics of 2 test-organisms, especially in the cases when one of the criteria is the mortality of entomostracans *Daphnia magna Straus* [10, 12, 15].

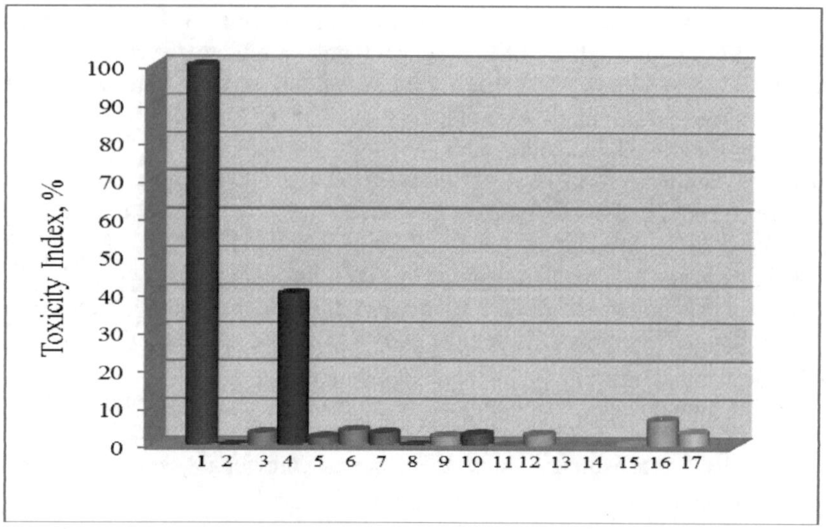

Figure 3. The toxicity index of NMs with various chemical composition and dispersion rate determined in accordance with the indicator of mortality of entomostracans *Daphnia Magna Straus* (C=1.0 mg/l): 1 - nAg, Δ_{50}=2-5 nm; 2 - nPt, Δ_{50}=3-5 nm; 3 - nNi, Δ_{50}=3-5 nm; 4 - nZnO, Δ_{50}=12 nm; 5 - α-Al_2O_3, Δ_{50}=7 nm; 6 - α-nAl_2O_3, Δ_{50}=70 nm; 7 - α-Al_2O_3, Δ_{50}=4 mkm; 8 - nAl_2O_3- corundum, Δ_{50}=70 nm; 9 - nTiO_2, Δ_{50}=5 nm; 10 - nTiO_2, Δ_{50}=50 nm; 11 - nTiO_2, Δ_{50}=90 nm; 12 - nTiO_2, Δ_{50}=350 nm; 13 - TiO_2, Δ_{50}=10 mkm; 14 - nSi_3N_4, Δ_{50}=80 nm; 15 - nSiC, Δ_{50}=95 nm; 16 - BT-50; 17 - MCD-material

It is possible to solve this problem using more sensitive test-functions based on the behavioural and physiological responses of entomostracans. One of such functions is trophic activity (TA) *Daphnia magna Straus*.

The change in the trophic function in comparison with the validation batch is illustrated in the group of NPs of metals and metal oxides for which no biological effects on the daphnids' mortality indicator were registered (figure 4).

We considered studying the trophic activity while adding nAg and nZnO into GM composition to be unreasonable because of the high mortality rate of entomostracans under the influence of DS of the indicated NPs. Moreover, in accordance with our data, the mortality rate of entomostracans grows because of additional impact of NPs after their bio-accumulation and adsorption by the cells of chlorella added as feed-stuff. The possibility of NPs accumulation in unicellular algae and ciliates is indicated by the data obtained by the authors in the experiments with nAg, nZnO, nTiO_2, nAu and quantum dots [14, 16, 17].

Our research showed that the trophic activity is inhibited in 5 out of 10 tested NMs (figure 4). The trophic activity of *Daphnia magna Straus* decreases with exposing entomostracans into DS nTiO_2 (Δ_{50}=5 nm) by 20.6 %, nSi_3N_4 (Δ_{50}=80 nm) – by 25.1% and nAl_2O_3-electrocorundum (Δ_{50}=70 nm) - by 30.9 %. The addition of NPs α-nAl_2O_3 (Δ_{50}=7 nm) or nTiO_2 (Δ_{50}=50 nm) into the test-system stimulates the trophic activity by 17.6 % and 10.0 % correspondingly.

Consequently, the entomostracans *Daphnia magna Straus* react on the addition of most SFM into GM by changing the physiological function – the trophic activity. Daphnids are known to be classical test-organisms while studying water medium contamination. It is related to the fact that daphnids being active filterers react quickly on the change in the medium quality by means of changes in their physiological functions such as locomotory, breathing and trophic activity, heart-beating rhythm, fertility, etc., which predict more serious anthropogenic loading [15, 18].

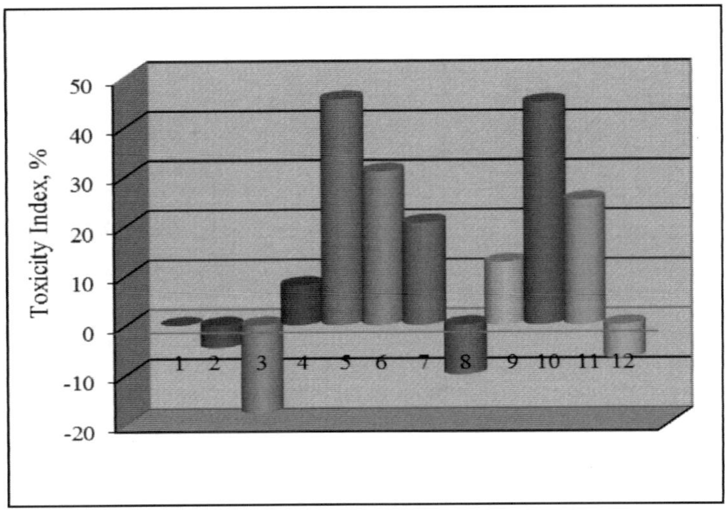

Figure 4. The toxicity index of NMs with various chemical composition and dispersion rate determined in accordance with the indicator of trophic activity of entomostracans *Daphnia magna Straus* (C=1.0 mg/l): 1 - nPt, Δ_{50}=3-5 nm; 2 - nNi, Δ_{50}=3-5 nm; 3 - α-Al_2O_3, Δ_{50}=7 nm; 4 - α-nAl_2O_3, Δ_{50}=70 nm; 5 - α-Al_2O_3, Δ_{50}=4 mkm; 6 - nAl_2O_3-corundum, Δ_{50}=70 nm; 7 - nTiO_2, Δ_{50}=5 nm; 8 - nTiO_2, Δ_{50}=50 nm; 9 - nTiO_2, Δ_{50}=90 nm; 10 - nTiO_2, Δ_{50}=350 nm; 11 - nSi_3N_4, Δ_{50}=80 nm; 12 - nSiC, Δ_{50}=95 nm

However, the researches into medium quality assessment using the physiological functions of entomostracans are represented in literature by few works. The works on studies of the influence of heavy metals on the heart beating of *Daphnia Cucullata* G.O.SARS, 1862 [18] and the trophic activity of *Daphnia magna Straus* [15] are of great interest. We have not found any works studying the bio effects of SFM on the physiological parameters in the literature available.

We observed the decrease in the trophic activity of *Daphnia magna Straus* while exposing entomostracans into GM with the substances in macro form such as NPs α-nAL_2O_3 (4000 nm) - by 45,5 %, nTiO_2 (350 nm) - by 44,8 %. Entomostracans are known to have a filter feeding type. They can filter great amounts of water and swallow the particles whose size is 0.4-40.0 mkm, consisting of algae cells, large bacteria and other organic and inorganic particles. Entomostracans can seize small particles especially the ones that are less than 50 micron in diameter and suspended aggregates NPs (TiO_2 and C_{60}), which have micro- rather than nano-size [19]. Large particles are too difficult to digest, and entomostracans can prevent them from penetrating to a filtration chamber. It is possible that while filtrating GM with added substances in macro form the filtration mechanism of entomostracans becomes cluttered, which results in the decrease in their trophic activity and often in the death of entomostracans.

We determined ecotoxicological parameters of DS SFM: NOEC, (L(E)C_{20} and L(E)C_{50} for each test-organism. Table 1 contains biologically active tested SFM and their concentrations L(E)C_{20}, which result in toxicity effect appearing. We showed that L(E)C_{20} depends on the test-organism type and the physical and chemical properties of SFM.

We revealed 6 types of the tested SFM that have a toxic effect on chlorella growth in smaller concentrations than on the chemotaxis of ciliates and mortality of entomostracans: nAg Δ_{50}=2-5 nm, nAu Δ_{50}=2-5 nm, nZnO Δ_{50}=12 nm, nAl_2O_3 (α-form) Δ_{50}=7 nm, nTiO_2 Δ_{50}=90 nm, nSi_3N_4 Δ_{50}=80 nm and BT-50. We found 3 types of the tested SFM that have toxic impact on the chemotaxis of ciliates in fewer amounts than on chlorella growth, mortality and trophic activity: nNi Δ_{50}=3-5 nm; α-nAl_2O_3 Δ_{50}=70 nm and nTiO_2 Δ_{50}=5 nm.

Table 1. The value L(E)C_{20} for the biologically active, tested SFM

NPs	*Chlorella vulgaris Beijer*	*Paramecium caudatum*	*Daphnia magna Straus*	
			Dead	Trophic activity
Ag, Δ_{50}=2-5 nm	0.00009	0.00010	0.00003	-
Au Δ_{50}=2-5 nm	0.043	>100	-	-
Ni Δ_{50}=3-5 nm	0.067	0.042	15.21	32.77
ZnO Δ_{50}=12 nm	0.0087	0.044	0.029	-
Al_2O_3 (α-form)Δ_{50}=7 nm	0.0087	>100	>100	0.062
Al_2O_3 (α-form) Δ_{50}=70 nm	3.121	0.056	>100	0.208
Al_2O_3 (corund) Δ_{50}=7 nm	>100	>100	>100	0.041
TiO_2 Δ_{50}=5 nm	0.053	0.000009	>100	0.274
nTiO_2 Δ_{50}=50 nm	>100	>100	>100	0.465
nTiO_2 Δ_{50}=90 nm	0.279	>100	>100	2.965
nSi_3N_4 Δ_{50}=80 nm	>100	>100	>100	0.122
BT-50	1.796	>100	>100	-

We found 3 SFM types that disrupt the trophic activity of entomostracans in lower concentrations, than the growth of algae and the chemotactic response of ciliates: nAl_2O_3 (corund) Δ_{50}=7 nm, nTiO_2 Δ_{50}=50 nm and nSi_3N_4 Δ_{50}=80 nm.

4. Conclusion

Our research showed the vulnerable elements of the ecosystem resistance to water contamination with SFM. We stated that phytoplankton (unicellular algae) and zooplankton (unicellular animals and entomostracans) representatives are susceptible to harmful influence of SFM. For some SFM kinds algae are the most vulnerable element in the chain «SFM-algae-water animals». Water animals (ciliates and entomostracans) are more sensible to other SFM kinds. The toxic effect of SFM is reflected in species composition of communities and the correlation of species populations, which may bring about a change in the structure of trophic connections and the distortion of the food pyramid and the ecosystem as a whole.

Acknowledgments
This study (research grant No 8.2.57.2015) was supported by Tomsk State University Academic D.I. Mendeleev Fund Program in 2015.
Work was conducted with the application of the Tomsk regional common use centr technical equipment acquired thanks to a grant of the Russian Ministry of the Agreement No.14.594.21.0001 (RFMEFI59414X0001).

References

[1] Griffitt R, Luo J, Gao J, Bonzongo J and Barber D 2008 *Environmental Toxicology and Chemistry* **27** (9) 1972-8
[2] Gong N, Shao K, Feng W, Lin Z, Liang C and Sun Y 2011 *Chemosphere* **83** (4) 510-6
[3] Kovrižnych J, Sotnikova R, Zeljenkova D, Rollerova E and Szabova E 2014 *Interdiscip Toxicol.* **7** (1) 23-6

[4] Li T et. al. 2010 *Anal Bioanal Chem.* **398** 689-700
[5] Renault S, Baudrimont M, Mesmer-Dudons N, Gonzalez PM, Brisson A 2008 *Gold Bulletin* **41** 116-26
[6] [6] Pendashte H, Shariati F, Keshavarz A and Ramzanpour Z 2013 *World Journal of Fish and Marine Sciences* **5** (5) 563-70
[7] Lopes S, Ribeiro F, Wojnarowicz J, Lojkowski W, Jurkschat K, Crossley A, Soares A and Loureiro S 2014 *Environmental Toxicology And Chemistry* **33** 190-8
[8] Liu J, Fan D, Wang L, Shi L, Ding J, Chen Y and Shen S 2014 *Environment Protection Engineering* **40** (1) 139-49.
[9] Svetlichnyi V, Lapin I 2013 *Russian Physics Journal* **56** (5) 581-7
[10] Morgalev Yu, Khoch N, Morgaleva T, Dunaevsky G and Morgalev S 2010 *Methodological Guide* Tomsk
[11] Morgalev Yu, Morgaleva T and Grigoriev Yu 2010 *FR.1.39. 2010.09103* (Moskow:Federal Reestr) p 45
[12] Morgalev Yu, Morgaleva T and Grigoriev Yu 2010 *FR.1.39.2010.09102* (Moskow:Federal Register) p 45
[13] Morgalev S, Morgaleva T, Morgalev Yu and Gosteva I 2015 *Advanced Materials Research* **1085** 424-30
[14] Morgalev Yu., Khoch N, Morgaleva T, Gulik E, Borilo G, Bulatova U,. Morgalev S and Ponyavina E 2010 *Nanotechnologies in Russia* **5** 11-2
[15] Shashkova T 2010 *Int. Conf. Problems of ecology: read the memory of Professor M.M. Kozhova* (Irkutsk) 483
[16] Fabrega J, Luoma S, Tyler Ch, Galloway T and Lead J 2011 *Environment International* **37** 517-31
[17] Renault S, Baudrimont M, Mesmer-Dudons N, Gonzalez Pm. and Brisson A 2008 *Gold Bulletin* **41** 116-26
[18] Leontiev V 2013 *http://www.sworld.com.ua/konfer30/233.pdf*
[19] Hund-Rinke K and Simon M 2006 *Environ Sci. Pollut. Res.* **13** (4) 225–32

Effect of AL_2O_3 and TiO_2 nanoparticles on aquatic organisms

I Gosteva, Yu Morgalev, T Morgaleva, S Morgalev
Centre "Biotest-Nano", Tomsk State University, 36, Lenina ave., Tomsk, 634050, Russia

E-mail: ia.gosteva1@gmail.com

Abstract. Environmental toxicity of aqueous disperse systems of nanoparticles of binary compounds of titanium dioxides (with particle size Δ_{50}=5 nm, Δ_{50}=50 nm, Δ_{50}=90 nm), aluminum oxide alpha-forms (Δ_{50}=7 nm and Δ_{50}=70 nm) and macro forms (TiO_2 Δ_{50}=350 nm, Al_2O_3 Δ_{50}=4000 nm) were studied using biological testing methods. The bioassay was performed using a set of test organisms representing the major trophic levels. We found the dependence of the toxic effect concentration degree of $nTiO_2$ and nAl_2O_3 on the fluorescence of the bacterial biosensor "Ekolyum", the chemotactic response of ciliates *Paramecium caudatum*, the growth of unicellular algae *Chlorella vulgaris Beijer* and mortality of entomostracans *Daphnia magna Straus*. We revealed the selective dependence of $nTiO_2$ and nAl_2O_3 toxicity on the size, concentration and chemical nature of nanoparticles. The minimal concentration causing an organism's response on $nTiO_2$ and nAl_2O_3 effect depends on the type of the test-organism and the test reaction under study. We specified $L(E)C_{50}$ and acute toxicity categories for all the studied nanoparticles. We determined that $nTiO_2$ (Δ_{50}=5 nm) belong to the category «Acute toxicity 1», $nTiO_2$ (Δ_{50}=90 nm) and nAl_2O_3 (Δ_{50}=70 nm) – to the category «Acute toxicity 2», nAl_2O_3 (Δ_{50}=7 nm) - to the category «Acute toxicity 3». No acute toxicity was registered for $nTiO_2$ (Δ_{50}=50 nm) and macro form TiO_2.

1. Introduction

Development and implementation of nanotechnologies in many industries increases the probability of harmful effects of nanoparticles (NPs) on the environment. Wide usage of NPs resulted in a great attention paid lately to NPs of metal oxides; titanium dioxide ($nTiO_2$) and aluminum oxide (nAl_2O_3) that are very popular among them.

$nTiO_2$ application is determined by its antibacterial and photocatalytic properties in the first place [1, 2]. nAl_2O_3 has a number of properties – thermal stability, high hardness, resistance and chemical inertness, which makes it possible to use it in various industries. The most wide-spread modification is α-Al_2O_3 used in optics and laser technologies [3].

It is common knowledge that NP size substances increase in their response ability, which is potentially hazardous for the environment. As there are various ways of polluting water bodies with nanoindustry products, it is necessary to assess NPs toxicity for water ecosystems. The requirement documents in Russia and abroad require this property of NPs to be determined.

The published data concerning the biological effects of $nTiO_2$ and nAl_2O_3 on aqueous environment are deficient, segmentary and discrepant. This is caused not only by the lack of necessary information about physical and chemical properties of the tested NPs, the method of their synthesis and experiment conditions, but also by the absence of a unified analytical system providing valid ecotoxicological assessment of the hazard the NPs cause to natural objects.

Content from this work may be used under the terms of the Creative Commons Attribution 3.0 licence. Any further distribution of this work must maintain attribution to the author(s) and the title of the work, journal citation and DOI.
Published under licence by IOP Publishing Ltd

As a result of the programs and research grants implemented earlier (2008-2013) we developed a methodological approach to assessing ecological toxicity that helps to study test-reactions of a wide range of organisms with one algorithm.

2. Materials and methods

We tested $nTiO_2$ with mean particle size Δ_{50}=50 nm, Δ_{50}=90 nm, nAl_2O_3 (α-form) with particle size Δ_{50}=70 nm and macro forms TiO_2 Δ_{50}=350 nm and Al_2O_3 Δ_{50}=4000 nm obtained with the help of the pneumo circuital method («Mipor» Ltd, RF). The disperse systems $nTiO_2$ Δ_{50}=5 nm and nAl_2O_3 Δ_{50}=7 nm were obtained with the help of the method of laser ablation in distilled water, the initial concentration 50 mg/l [4].

Preparation of DS $nTiO_2$ and nAl_2O_3 on growth medium (GM) with the concentrations 0.000001, 0.00001, 0.0001, 0.001, 0.01, 0.1 and 1.0 mg/l was done in accordance with the technique [5] with further ultrasonic material dispersion (30 W/l, 5 minutes) just before biological testing.

The toxicity of $nTiO_2$ and nAl_2O_3 was studied by the change in luminescence of bacterial biosensor «Ekolyum» [6], the chemotactic response of ciliates *Paramecium caudatum* [6], the growth of unicellular algae *Chlorella vulgaris Beijer* [7] and mortality of entomostracans *Daphnia magna Straus* [8]. The assessment of toxicity of the analysed DS $nTiO_2$ and nAl_2O_3 with various particle sizes was done in accordance with the criteria given in certified methods' outlines [7, 8]. The research design is similar to the described in [9, 10].

The properties of $nTiO_2$ and nAl_2O_3 were verified with the help of transmission electronic microscop «Philips CM-30» (France), using the method of dynamic light scattering (DLS) with the help of the disperse particles sizes analyser «Zetasizer Nano ZS» (USA), the method of BET («TriStar 3000», USA) to measure specific surface and the method of X-ray structure analysis of the phase composition and structural characteristics (X-ray diffractometer Shimadzu XRD-6000, Japan). The research was conducted with the application of metrologically gauged instruments: Autolumat LB953 (Germany), spectrofluorometer «Shimadzu» RF-5301 PC (Japan), photoelectric colorimeter DMS-03 (Russia), concentration meter BIOTESTER-2 (Russia). The statistical processing of the obtained data was done with probit analysis.

3. Results

The results of our investigation showed that the dynamics of the fluorescence of bacterial biosensor «Ekolyum» in GM with $nTiO_2$ concentration did not depend on the size and concentration of NPs. The toxicity of particles of $nTiO_2$ (Δ_{50}=5 nm, Δ_{50}=50 nm and Δ_{50}=350 nm) practically for the whole tested concentration range (C) changed insignificantly (figure 1). The exception was the tested DS $nTiO_2$ Δ_{50}=5 nm with C=0.1 mg/l, the toxicity of which was above "permissible level" and, in accordance with the criteria of a standard method, was defined as a sample with "moderate" degree of toxicity (I=27.5±0.5 %). In DS TiO_2 (Δ_{50}=350 nm) we noted "suppression" of bio fluorescence within the interval of the tested concentrations. The fluorescence of the biosensor decreased with the increase in DS TiO_2 concentration, but the toxicity index value was above the "permissible level" by 1.0-2.0 % only.

Similar effects were registered during introducing in GM NPs Al_2O_3 Δ_{50}=7 nm and Δ_{50}=70 nm: no sufficient changes in the fluorescence of bacterial biosensor «Ekolyum» (figure 2) were found. In contrast to $nTiO_2$ Δ_{50}=5 nm the «moderate» degree of toxicity (I=23.5±8.5 %) was registered for DS with lower concentration of nAl_2O_3 Δ_{50}=70 nm (C=0.001 mg/l).

The macro form Al_2O_3 Δ_{50}=4000 nm unlike the nanoform Al_2O_3 of all the analysed concentrations of DS quenched the fluorescence of the biosensor. Within the interval C from 0.0001 mg/l to 1.0 mg/l the toxicity index exceeded the «permissible» level and changed within the range from I=20.5±0.5 % to I=27.9±0.4 %, which is defined as the sample with the «moderate» toxicity.

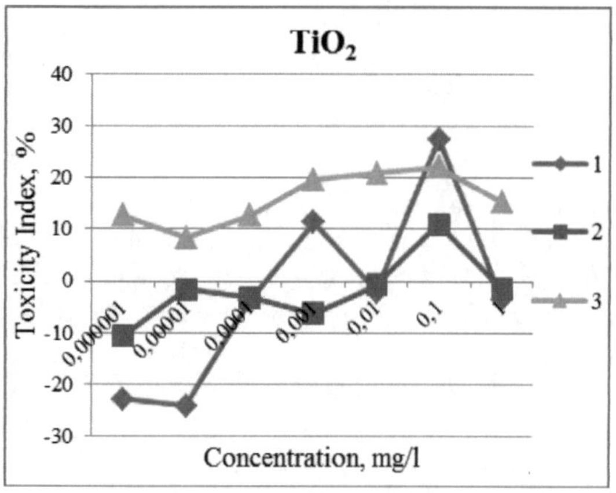

Figure 1. Toxicity index of DS TiO_2 for bacterial biosensor «Ekolyum»: 1 - Δ_{50} = 5 nm, 2 - Δ_{50} = 50 nm, 3 - Δ_{50} = 350 nm

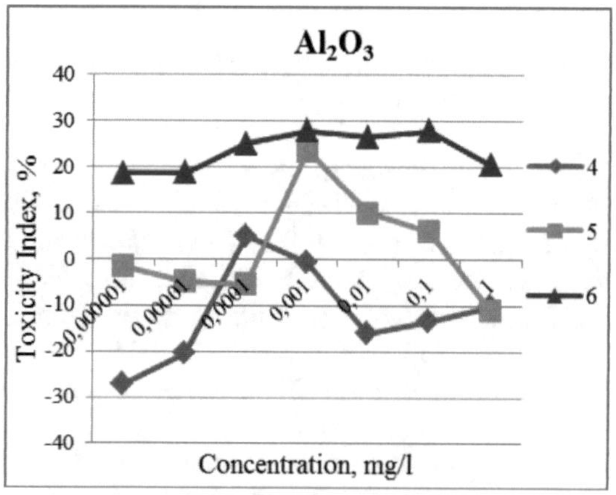

Figure 2. Toxicity index of DS Al_2O_3 for bacterial biosensor «Ekolyum»: 4 - Δ_{50} =7 nm, 5 - Δ_{50} = 70 nm, 6 - Δ_{50} = 4000 nm

On exposing *Paramecium caudatum* to DS $nTiO_2$ Δ_{50}=5 nm with C=0.000001 mg/l and C=0.00001 mg/l the toxicity index value corresponded to control values (figure 3). With the increase in the concentration of $nTiO_2$ Δ_{50}=5 nm in GM the percentage of ciliates outflow in DS was not monotonic, but decreased definitely, which demonstrated the NPs toxic effect. In DS with the concentrations 0.01, 0.1 and 1.0 mg/l the toxicity index changed within the range from 38.8±6.2 % to 69.9±16.8 % and according to the standard method [6] corresponded to the «moderate» degree of toxicity (0.40<I<0.70).

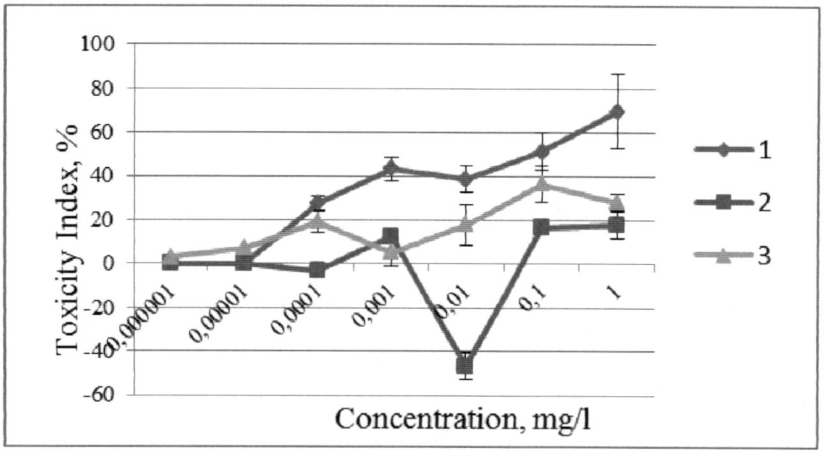

Figure 3. Toxicity index of DS $nTiO_2$ and nAl_2O_3 for *Paramecium caudatum:* 1 - $nTiO_2$ Δ_{50} = 5 nm, 2 - $nTiO_2$ Δ_{50} = 50 nm, 3 - nAl_2O_3 Δ_{50} = 70 nm

We registered the dependence of the chemotactic response of *Paramecium caudatum* on the particle size of TiO_2 in GM. The introduction of particles TiO_2 Δ_{50}=50 nm did not caused toxic impact on *Paramecium caudatum.* We found DS $nTiO_2$ Δ_{50}=50 nm with C=0.0001 mg/l and C=0.01 mg/l, in which the ciliates outflow exceeded the control values (figure 3). The highest stimulating effect was registered with C=0.01 mg/l (I= -46.6±9.4 %). We stated that the introduction of nAl_2O_3 Δ_{50}=70 nm to GM within the concentration range from 0.000001 to 1.0 mg/l resulted in the change in the chemotactic response of *Paramecium caudatum* in the range of «permissible» values. The maximal toxicity index was registered with C=0.1 mg/l I=36.4 ±1.1 % (figure 3).

Our investigations resulted in revealing the dependence of $nTiO_2$ effect on the particle size. The particles $nTiO_2$ Δ_{50}=5 nm and Δ_{50}=90 nm inhibited the growth of *Chlorella vulgaris Beijer*, with $nTiO_2$ Δ_{50}=5 nm in contrast to $nTiO_2$ Δ_{50}=90 nm producing a pronounced toxic effect (figure 4).

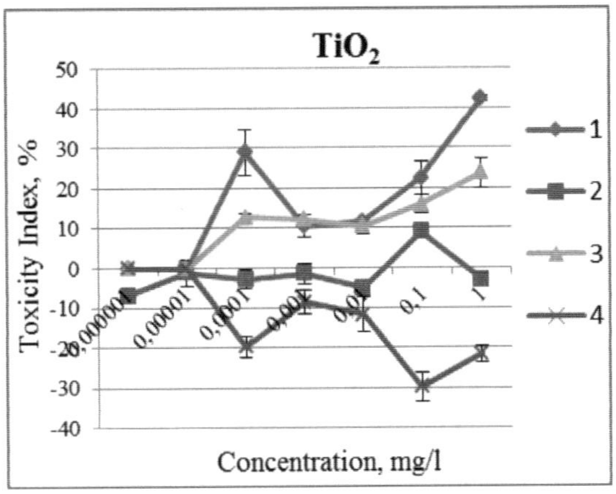

Figure 4. Toxicity index of DS TiO_2 for *Chlorella vulgaris Beijer:* 1 - Δ_{50}=5 nm, 2 - Δ_{50} = 50 nm, 3 - Δ_{50} = 90 nm, 4 - Δ_{50} =350 nm

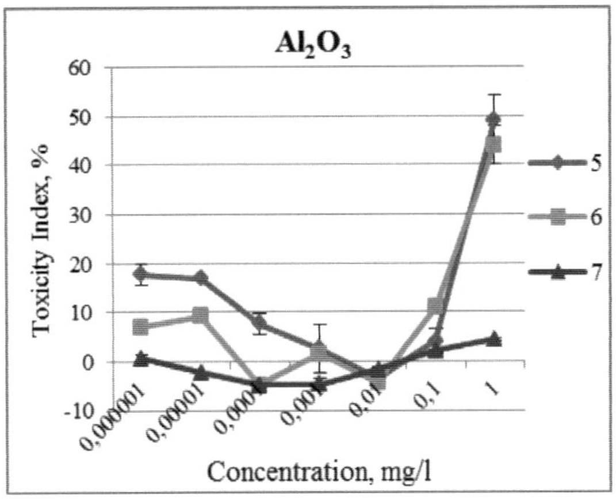

Figure 5. Toxicity index of DS Al_2O_3 for *Chlorella vulgaris Beijer:* 5 - Δ_{50}=7 nm, 6 - Δ_{50}=70 nm, 7 - Δ_{50}=4000 nm

The introduction into GM of the algae DS $nTiO_2$ Δ_{50}=50 nm caused insignificant stimulation of *Chlorella vulgaris Beijer* cells' growth. We registered a non-monotonic decrease in the growth of alga cells while cultivating *Chlorella vulgaris Beijer* in the analysed DS $nTiO_2$ Δ_{50}=5 nm. The toxic effect of $nTiO_2$ Δ_{50}=5 nm showed itself with C=0.0001 mg/l (I=29.0±5.9 %) already. With further increase in the concentration the toxicity of DS decreased (I=10.4±2.8 % with C=0.001 mg/l and I=11.7±0.5 % with C=0.01 mg/l), and then increased again to gain the maximal effect with C=1.0 mg/l. The growth of *Chlorella vulgaris Beijer* decreased by 33.8 % compared to the control values. The introduction into GM of the particles TiO_2 Δ_{50}=350 nm in macro form had a stimulating effect on *Chlorella vulgaris Beijer* cells' growth in concentration dependence. The maximal response was registered during introducing particles into C=0.1 mg/l. The toxicity of DS TiO_2 Δ_{50}=350 nm was close to significant manifestation.

In contrast to $nTiO_2$ the introduction of nAl_2O_3 into GM did not cause the expected changes in the growth of *Chlorella vulgaris Beijer* depending on the particle size. The exposure of *Chlorella vulgaris Beijer* to DS nAl_2O_3 Δ_{50}=7 nm and Δ_{50}=70 nm within the range C=0.000001 mg/l – 0.1 mg/l caused the inhibition of alga cells' growth вызывала ингибирование роста клеток водоросли within tolerance (figure 5). A pronounced toxic effect was observed only with C=1.0 mg/l: *Chlorella vulgaris Beijer* mass growth decreased compared to the control values by 49.1±5.9 % for DS nAl_2O_3 Δ_{50}=7 nm and 44.0±3.9 % for DS nAl_2O_3 Δ_{50}=70 nm. The macro form Al_2O_3 Δ_{50}=4000 nm did not have a toxic effect on *Chlorella vulgaris Beijer,* the toxicity index changing in range of control values.

The researches in the test-organism of *Daphnia magna Straus* in contrast to biosensor «Ekolyum», *Chlorella vulgaris Beijer* and *Paramecium caudatum* did not reveal toxic effects of $nTiO_2$ and nAl_2O_3. The exposure of *Daphnia magna Straus* in DS with particles TiO_2 (Δ_{50}=5 nm, Δ_{50}=50 nm and Δ_{50}=350 nm) and into DS with particles Al_2O_3 (Δ_{50}=7 nm, Δ_{50}=70 nm and Δ_{50}=4000 nm) did not cause the mortality of more than 10 % of the entomostracans for 48 hours, which indicated the absence of the toxic effect for all the tested concentrations DS TiO_2 and Al_2O_3 (figure 1, 2, 3).

On the basis of the data of the eco toxicological analysis obtained with the help of test-systems with the basic set of test-organisms we did the integral assessment of the toxicity of TiO_2 and Al_2O_3 with various dimension characteristics (table 1).

According to GHS, while making the conclusion on the ecotoxicity of the tested samples of NPs we took into account the principle of assessment by the lowest index available $L(E)C_{50}$ [11].

Table 1. Values of N-dimensional $L(E)C_{50}$ TiO_2 and Al_2O_3

$L(E)C_{50}$, (mg/l) / Test-organism	TiO_2 NPs			TiO_2 macro form	nAl_2O_3 NPs		nAl_2O_3 macro form
	5 nm	50 nm	90 nm	350 nm	7 nm	70 nm	4000 nm
«Ekolyum»	>100	>100	-	47.6	>100	>100	>100
Paramecium caudatum	0.0256	>100	>100	>100	-	1.22	-
Chlorella vulgaris Beijer	1.153	>100	7.7	>100	30.8	15.15	-
Daphnia magna Straus	>100	>100	>100	>100	>100	>100	>100
Conclusion	Acute toxic class 1	No acute toxicity	Acute toxic class 2	Acute toxic class 3	Acute toxic class 3	Acute toxic class 2	No acute toxicity

Consequently, the research results showed that the toxicity of DS $nTiO_2$ with dimension Δ_{50}=5 nm were sufficiently higher than the toxicity of DS $nTiO_2$ with particle size Δ_{50}=50 nm, Δ_{50}=90 nm. A different dependence is observed for nAl_2O_3. Per totality of the test-parameters under study nAl_2O_3 Δ_{50}=70 nm are most toxic. nAl_2O_3 with dimension Δ_{50}=7 nm are less toxic. The comparison of the toxicity of $nTiO_2$ and nAl_2O_3 with close dimensional characteristics indicated that the toxicity of $nTiO_2$ (Δ_{50}=5 nm) is much higher than the toxicity of nAl_2O_3 Δ_{50}=7 nm. At the same time, nAl_2O_3 (Δ_{50}=70 nm) have a more pronounced toxic effect compared to $nTiO_2$ (Δ_{50}=50 nm) correspond to $nTiO_2$ (Δ_{50}=90 nm) by toxicity category. The macro form TiO_2 per totality of the test-responses under study belongs to the category «Acute toxicity 3», and the macro form Al_2O_3 is «non-hazardous substance».

4. Discussion

The absence of standardized toxicological methods of assessing the hazard NPs can cause to living objects of nature together with little information in published papers about the size, crystal structure, morphology of NPs, methods of their synthesis, experiment conditions, methods of preparation of DS NPs makes it difficult to compare the obtained results with the literature data.

In the literature available the data on the influence of $nTiO_2$ and nAl_2O_3 on the test-parameters of aquatic organisms are few. Most researches in the recent years have been devoted to the genotoxic and cytotoxic effect of $nTiO_2$ and nAl_2O_3 as well as the cancerogenic effect of $nTiO_2$ [1, 2, 3, 12, 13, 14].

The data we obtained are in accord with the data of Liu et. al. [15], who studied the effect of $nTiO_2$ (Δ_{50}<20 nm and Δ_{50}=30 nm) on the survival rate of *Daphnia magna Straus* and stated that $nTiO_2$ with particle size Δ_{50}<20 nm induced a higher toxicity compared to $nTiO_2$ Δ_{50}=30 nm, which contradicted the data of L. Adams et. al. [16] and J. Verran [17], who showed in their studies that the particle size of $nTiO_2$ does not influence its toxicity. The inconsistency of the data on the dependence of $nTiO_2$ toxicity on the particle size may be caused by the authors' employing different methods of assessing the toxicity of NPs. For example, Liu et. al. are known to have studied the survival rate of *Daphnia magna Straus* in a long-term experiment. The research in a long-term toxicity of $nTiO_2$ for 21 days displayed the disorder of fertility and the increase in the mortality of entomostracans up to 70 % [18, 19]. The analysis of toxicity of nAl_2O_3 with dimensional characteristics Δ_{50}=40-100 nm on immobilisation and further death of *Ceriodaphnia dubia* displayed a clear temporal dose-dependent trend to reveal a toxic effect taking into account aggregation and bioavailability reduction of NPs [20]. One of the basic mechanisms of the toxic effect of $nTiO_2$ and nAl_2O_3, in the authors' view, is an oxidative stress that arises in the condition of NPs impact on aquatic organisms.

According to the literature, the absence of toxic effect of $nTiO_2$ with particle size Δ_{50}<50 nm was found for luminescent bacteria *Vibrio fischeri* [21], which is in accord with the data we obtained.

V. Aruoja et.al. [22] studied the influence of $nTiO_2$ on the freshwater alga *Pseudokirchneriella subcapitata.* It was stated that the macro form TiO_2 (L(E)C_{50}=35.9 mg/l) was less toxic than $nTiO_2$ (L(E)C_{50}=5.83 mg/l). I.M. Sadiq et.al. [23] revealed the inhibitory effect of nAl_2O_3 with particle size Δ_{50}<50 nm on the growth of microalgae *Scenedesmus sp.* and *Chlorella sp.* The value of L(E)C_{50} for 72 hours of incubation was 45.4 mg/l for *Chlorella sp.* and 39.4 mg/l for *Scenedesmus sp.* The macro form Al_2O_3 (Δ_{50}<5 mkm) also displayed toxicity, though to less extent (L(E)C_{50}= 110.2 mg/l for *Chlorella sp.* and L(E)C_{50}=100.4 mg/l for *Scenedesmus sp.*). The researchers revealed selective sensitivity of various algae species to nAl_2O_3. The authors suppose that $nTiO_2$ and nAl_2O_3 form specific aggregates settling on the surface of algae cells, which causes toxic effect.

We were the first to show the influence of $nTiO_2$ and nAl_2O_3 with various size characteristics on the chemotactic response of ciliates *Paramecium caudatum.*

5. Conclusion

Eco toxicological analysis showed that $nTiO_2$ Δ_{50}=5 nm belong to the category « Acute toxicity 1», $nTiO_2$ Δ_{50}=90 nm and nAl_2O_3 Δ_{50}=70 nm - to the category «Acute toxicity 2», macro form TiO_2 Δ_{50}=350 nm and nAl_2O_3 Δ_{50}=7 nm - to the category «Acute toxicity 3». The absence of acute toxicity was found for $nTiO_2$ Δ_{50}=50 nm and the macro form Al_2O_3 Δ_{50}=4000 nm.

In most test-systems with both $nTiO_2$, and nAl_2O_3 having various size characteristics there is no monotonic dependence between the concentration of DS and the level of the test-response change, the general trend being the increase in the test-response with the concentration increased. The degree of hazard of DS containing $nTiO_2$ and nAl_2O_3 is determined by the size of NPs, their chemical nature and concentration in DS.

The toxicity of $nTiO_2$ and nAl_2O_3 is different for various test-organisms. Most sensible to the presence of $nTiO_2$ Δ_{50}=5 nm and nAl_2O_3 Δ_{50}=70 nm in the environment are the ciliates *Paramecium caudatum*, of $nTiO_2$ Δ_{50}=90 nm and nAl_2O_3 Δ_{50}=7 nm – the unicellular algae *Chlorella vulgaris Beijer*, of macro form TiO_2 Δ_{50}=350 nm – bacterial sensor «Ekolyum».

6. Acknowledgments

This study (research grant No 8.2.57.2015) was supported by Tomsk State University Academic D.I. Mendeleev Fund Program in 2015.
Work was conducted with the application of the Tomsk regional common use centr technical equipment acquired thanks to a grant of the Russian Ministry of the Agreement No.14.594.21.0001 (RFMEFI59414X0001).

References

[1] Literature review on the safety of titanium dioxide and zinc oxide nanoparticles in sunscreens 2013 V1.0 *TGA* (Australian) p 19
[2] Sharma V 2009 *J. of Environmental Science and Health* **44** 1485–95
[3] Pakrashia S, Dalaia S, Prathna T, Trivedia S, Mynenia R, Raichurb A, Chandrasekarana N and Mukherjeea A 2013 *Aquatic Toxicology* **132** 34–45
[4] Svetlichnyi V, Lapin I 2013 *Russian Physics Journal* **56** (5) 581-7

[5] Morgalev Yu, Khotch N, Morgaleva T, Gulik E, Borilo G, Bulatova U, Morgalev S and Ponyavina E 2010 *Nanotechnologies in Russia* **5** (11-12) 851–6

[6] Morgalev Yu, Khoch N, Morgaleva T, Dunaevsky G and Morgalev S 2010 *Methodological Guide* Tomsk

[7] Morgalev Yu, Morgaleva T and Grigoriev Yu 2010 *FR.1.39. 2010.09103* (Moskow:Federal Reestr) p 45

[8] Morgalev Yu, Morgaleva T and Grigoriev Yu 2010 *FR.1.39.2010.09102* (Moskow:Federal Register) p 45

[9] Morgalev Yu, Morgaleva T, Gosteva I, Morgalev S, Kulizhskiy S, Astafurova T 2015: submitted to *J. of Advanced Materials Research*

[10] Morgaleva T, Morgalev Yu, Gosteva I and Morgalev S 2015: submitted to *J. of Advanced Materials Research*

[11] Globally Harmonized System of Classification and Labelling of Chemicals (GHS) The third revised edition 2009 UN

[12] Shi H, Magaye R, Castranova V and Zhao 2013 *J Particle and Fibre Toxicology* **10** 33

[13] Zhang X, Yin L, Tang M and Pu Y 2011 *Biomed Environ Sci.* **24** (6) 661-9

[14] Hueiwang A and Swanson J 2006 *Journal of Environmental Science and Health* **41** (12) 2699-711

[15] Verran J, Sandoval G, Allen N, Edge M, Stratton J 2007 *Dyes and Pigments* **73** 298–304

[16] Liu J, Fan D, Wang L, Shi L, Ding J, Chen Y and Shen Sh 2014 *Environment Protection Engineering* **40** (1) 139-49

[17] Adams L, Lyon D and Alvarez P 2006 *Water Research* **40** 3527

[18] Zhu X, Chang Y and Chen Y 2010 *Chemosphere* **78** 209-15

[19] Kim K, Klaine S, Cho J, Kim S and Kim S 2010 *Science of the Total Environment* **408** 2268-72

[20] Pakrashi S, Dalai S, Humayun A, Chakravarty S, Chandrasekaran N, Mukherjee A 2013 *Aquatic Toxicology* **132-133** 34-45

[21] Heinlaana M, Ivaska A, Blinova I, Dubourguier H and Kahru A 2008 *Chemosphere* **71** (7) 1308-16

[22] Aruoja V, Dubourguier H, Kasemets K and A. Kahru 2009 *Science of The Total Environment* **407** (4) 1461–68

[23] Sadiq I, Pakrashi S, Chandrasekaran N and Mukherjee A 2011 *J. Nanopart Res.* **13** 3287-99

Multi-walled carbon nanotubes: biodegradation by gastric agents *in vitro* and effect on murine intestinal system

A Masyutin[1], M Erokhina[1], K Sychevskaya[1], A Gusev[2], I Vasyukova[2], E Smirnova[1], G Onishchenko[1]

[1]Department of Cell Biology and Histology, Biology Faculty, Lomonosov Moscow State University, Leninskie Gory 1/12, Moscow, 119234 Russian Federation, tel./fax: +7(495)9391794

[2]NanoCenter of Nanotechnologies and Nanomaterials, Tambov Derzhavin State University, Zashchitnii pereulok 7, Tambov, Russian Federation, tel./fax: +7 (4752) 53-26-80

E-mail: squiggoth@yandex.ru

Abstract. One of the main questions limiting application of fibrous carbon nanomaterials (CNM) in medicine and food industry concerns presumptive degradation of CNM in living organisms. In this study, we have investigated biodegradation of multi-walled carbon nanotubes (MWCNTs) by gastric agents *in vitro* and influence of ingested MWCNTs on murine intestine. Using scanning, conventional transmission and analytical electron microscopy, we demonstrated that industrial MWCNTs treated *in vitro* by 0.1 M hydrochloric acid (pH=1) and gastric juice (pH=2-3) isolated from murine stomach, are subjected to incomplete degradation. After 30 days of oral administration to experimental mice, we did find MWCNTs in the cells of small intestine, and it may indicate that agglomerates of MWCNTs do not penetrate into colon epithelia and do not accumulate in enterocytes. However, we observed local areas of necrotic damages of intestinal villi. It seems likely, therefore, that MWCNTs end up leaving gastrointestinal tract by excretion with the feces. Our results suggest that MWCNTs do not undergo complete degradation in gastrointestinal tract of mice, and passing through non-degraded particles may negatively affect intestinal system.

1. Introduction

Nowadays multi-walled carbon nanotubes (MWCNTs) are widely used as components of filters for drinking water, various adsorbents, carriers of medical drugs for anti-cancer therapy and addressed delivery to the target organs [1-4] through intravenous or oral administration [5, 6]. In spite of growing perspectives of the MWCNTs' applications, there is not enough evidence on the consequences of their prolonged contact with living organisms.

Contemporary research data are controversial and reported effects of carbon nanotubes depend on model organism, types of nanotubes, exposure conditions, dispersion state and concentration. It was shown that MWCNTs are able to penetrate into plant cells and accumulate in tissues [7, 8]. At the same time, rodents excrete at least 75% of ingested MWCNTs with the feces [9]. In vitro studies demonstrated that MWCNTs affect genes expression [10] and may induce different types of cell death [11, 12]. It has been suggested that fibrous carbon nanotubes may show effects similar to asbestos and

Content from this work may be used under the terms of the Creative Commons Attribution 3.0 licence. Any further distribution of this work must maintain attribution to the author(s) and the title of the work, journal citation and DOI.
Published under licence by IOP Publishing Ltd

cause inflammatory processes in lungs and blood cells [13]. Notably, uptake and accumulation of MWCNTs in the cells often results in production of reactive oxygen species (ROS) *in vitro* and *in vivo* [14]. At the same time, other studies demonstrated, that carbon nanotubes degrade by ROS-related pathways and via other biological oxidizers, which are produced by the cells of immune system [15, 16]. Thus, the aim of this study was to investigate biodegradation of MWCNTs *in vitro* and evaluate possible changes to the intestinal system caused by prolonged administration of MWCNTs with drinking water to mice.

2. Object of research

The object of the study was an industrial carbon nanomaterial Taunit (NanoTechCenter Ltd., Tambov, Russia). This material is a loose black powder, composed of grainy agglomerates with a size of several mkm. Taunit is produced by chemical vapor deposition with the 98.5% purity. 1.5% of additions includes 0.3-0.5% of amorphous carbon and 1-1.2% of remaining metallic catalysts (Ni/Mg).

3. Gastric juice extraction

Three males of C57Bl/6×DBA/2 mice were fasted overnight with free access to water. The gastric juice (pH=2-3) was collected using pyloric ligation procedure: the abdomen was incised under anesthesia, and the pylorus was ligated for 5 hours. Then mice were sacrificed using cervical dislocation, and the gastric juice was collected.

4. Experimental biodegradation of MWCNTs *in vitro*

For investigation of MWCNTs biodegradation, we have used 0,1M hydrochloric acid (pH=1) and isolated from murine stomach gastric juice. Nanotubes were suspended (400 mkg/ml) in gastric juice or hydrochloric acid and incubated at 38°C for 24 hours. Then samples were consecutively centrifuged at 5000 rpm and washed in distilled water. Suspended in water MWCNTs were placed on copper grid coated with Formwar film and dried at room temperature for 24 hours. Dried specimens were further used for TEM imaging.

5. Experiments *in vivo*

Detection of MWCNTs in cells and tissues was carried on mice C57Bl/6×DBA/2 (9 males in each group, approximately 10 weeks old). The average weight at the beginning of experiment was 29±1.7g and 31.8±1.8 g at the end of experiment. Control group of animals was getting pure drinking water. Experimental group was exposed for 30 days to suspension of industrially produced nanomaterial «TAUNIT» composed of MWCNTs, added to drinking water (30 mg/kg). Suspension of MWCNTs in water was prepared by sonication.

6. Electron microscopy analysis

For transmission (TEM) and analytical electron microscopy, segments of small intestine were cut out and washed in fixative (2.5% glutaraldehyde and 2% neutral formalin mixture in 0.1 mol·L–1 Na-K-phosphate buffer, pH 7.2), cut into fragments of 1×1 mm and again plunged in 2.5% glutaraldehyde with 2% neutral formalin in 0.1 mol·L–1 Na-K-phosphate buffer (pH 7.2) for 2 h at room temperature. Afterwards the samples were washed in PBS for 30 min and post-fixed in 1% osmium tetroxide for 2 h at 4°C. Then samples were dehydrated and embedded in epon raisin according to standard procedure. Ultrathin sections (60-80 nm) were made with a diamond knife and mounted on copper grids, coated with Formvar film. Ultrathin sections were additionally stained with uranyl acetate and lead citrate (Reynolds stain) and used for TEM imaging («JEM 1011», Jeol, Japan)). Detection of MWCNTs was made without additional staining. Electron diffraction patterns of MWCNTs were observed by analytical electron microscope «JEM 2100» (Jeol, Japan). For scanning electron microscopy, one drop of aqueous suspension of MWCNTs was placed on glass cover slip and dried at room temperature for 24 hours. Dried specimens were sputter-coated with a thin film of gold and examined under scanning electron microscope «JSM-6380 LA» (Jeol, Japan).

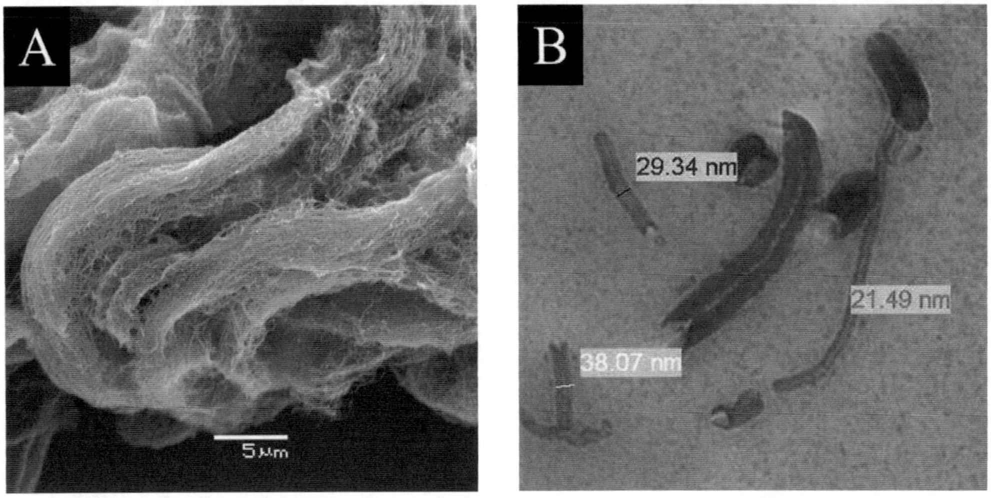

Figure 1. Industrial carbon nanomaterial «Taunit». a) Scanning electron microscopy. Aggregates of the MWCNTs; b) Transmission electron microscopy. Variety of shapes and forms of MWCNTs.

7. Results and discussion

Intact nanomaterial is a black hydrophobic powder with low ability to suspend in water. MWCNTs from nanomaterial «Taunit» usually form aggregates up to 40 mkm (figure 1a). Single MWCNTs are seen as hollow cylinders with length from 100 nm up to few micrometers, with inner diameter 5-10 nm and outer diameter 15-100 nm (figure 1b). The diameter and length of the MWCNTs were determined from n=100 random observations.

Analytical study of intact nanotubes with TEM is demonstrated in figure 2. Electronograms of single nanotubes show diffraction pattern typical for MWCNTs (obtained from n=100 independent observations).

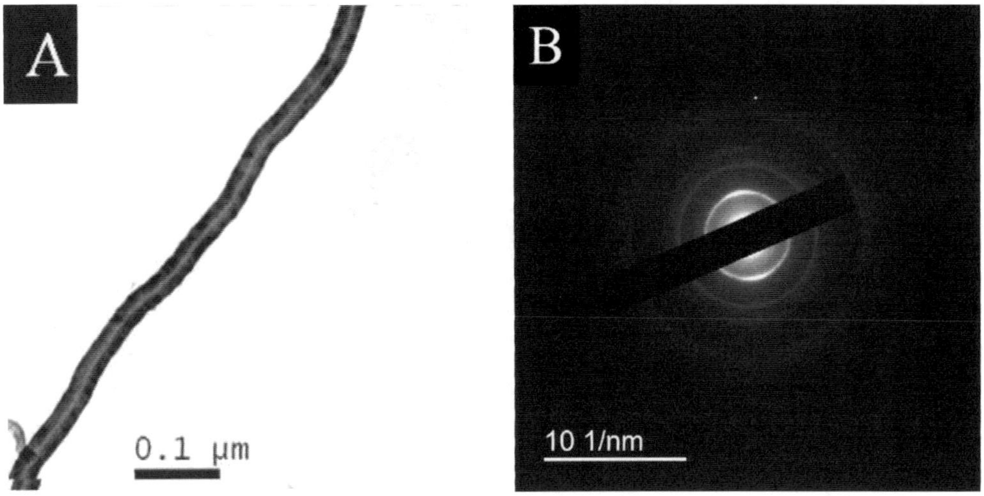

Figure 2. a) Transmission electron microscopy. Multiwalled carbon nanotube. Inner and outer diameters are clearly seen. c) Electron diffraction pattern.

For assessment of possible degradation of MWCNTs in gastrointestinal tract, we have studied the effect of murine gastric juice on the structural characteristics of this nanomaterial. The data on structural changes of MWCNTs exposed to gastric juice are shown on figure 3. Thus after gastric juice treatment, we detected few nanotubes without significant structural changes (figure 3a), yet numerous nanotubes had distorted structure. Some nanotubes had disruptions in the inner channel and intact outer walls (figure 3b). However, most MWCNTs had significantly altered structure, were ruptured or malformed, and often lost their tubular appearance (figure 3c). Such distorted nanotubes had rough surface and heterogeneous electron density throughout the length, probably due to the coating with organic material.

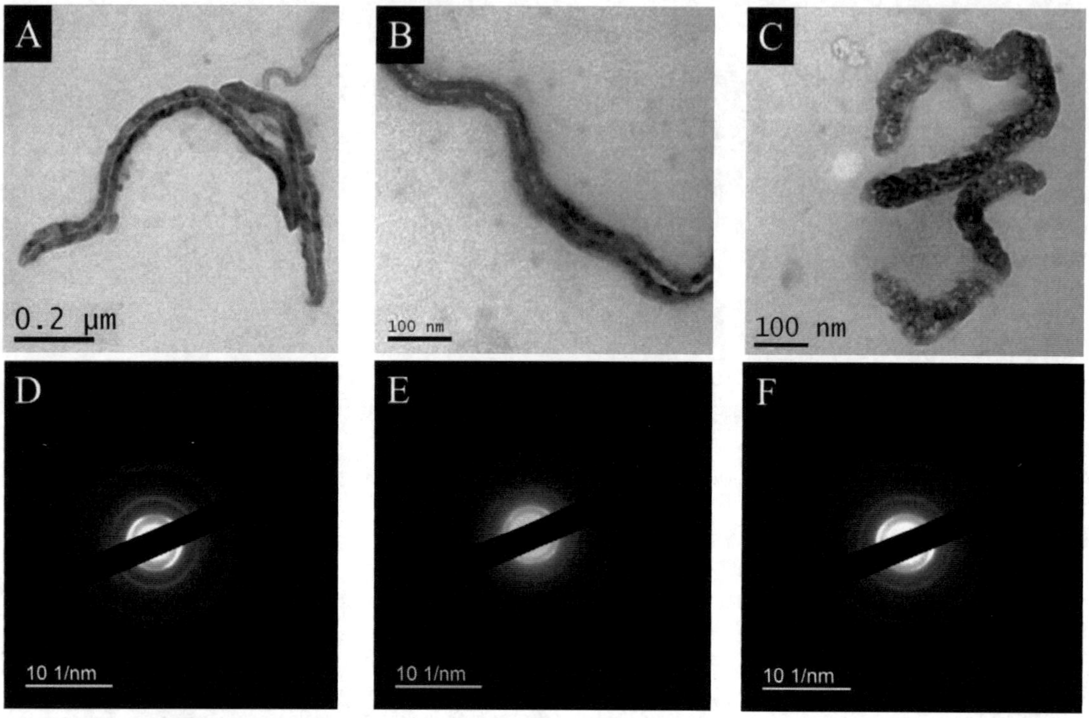

Figure 3. Gastric juice-induced distortion of MWCNTs (24 hours) a) Intact MWCNT. Walls and inner channel are well seen; b) MWCNT with ruptured inner channel; c) MTCNT with severely distorted structural appearance. No signs of inner channel can be observed.

Since the content of gastric juice is highly heterogeneous, we tested the main active component, which may affect the structural integrity of MWCNTs. For this purpose, nanomaterial was treated with 0.1M hydrochloric acid. TEM analysis showed that hydrochloric acid caused structural changes of MWCNTs, inducing rupture of the surface and/or complete distortion of the appearance of individual nanotubes within agglomerates (figure 4a), although some MWCNTs retained almost intact structure (figure 4b). In contrast to gastric juice treatment, MTCNTs treated with hydrochloric acid did not have heterogeneity in electron density throughout their length, probably, due to the absence of mucus and proteins (figure 4c). It is worthwhile to note, that distorted nanotubes appeared to be more curved than intact ones. Thus, the inconsistent distortion of MWCNTs exposed to gastric juice or hydrochloric acid may be explained by the fact that during industrial manufacturing, MWCNTs have different structural imperfections, which determine their ability to undergo degradation during the treatment.

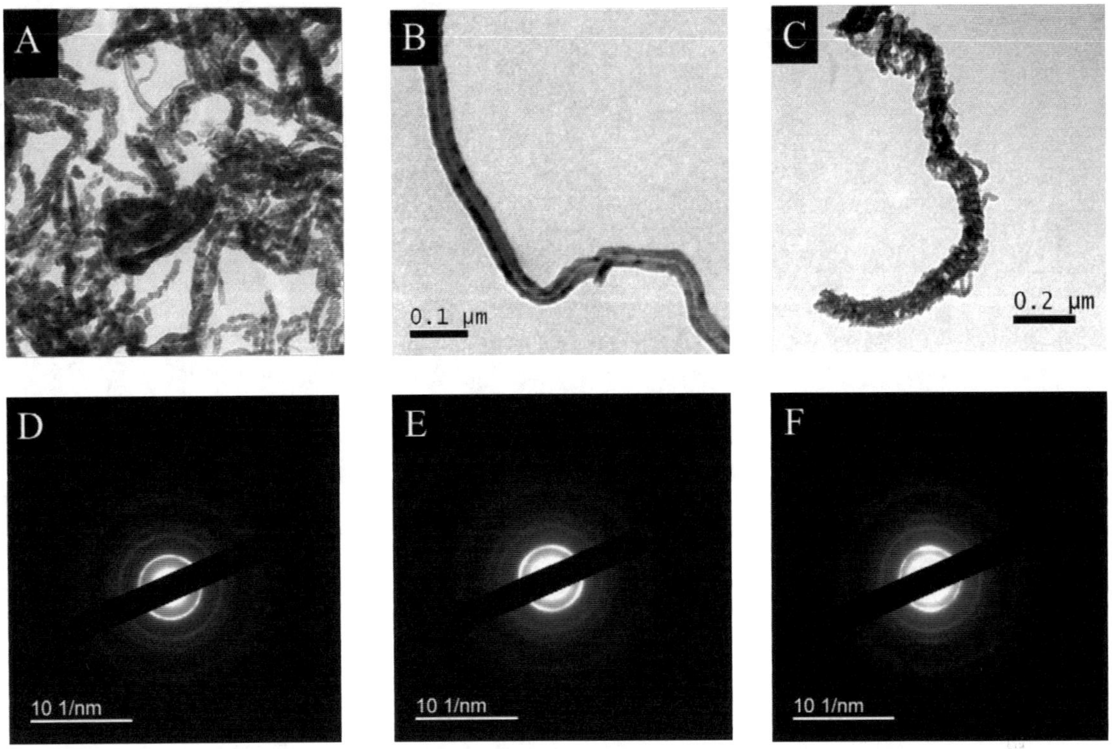

Figure 4. Hydrochloric acid-induced distortion of MWCNTs (24 hours) a) Aggregated MWCNTs with damaged surface. b) MTCNT with normal structure. Inner/outer diameters and walls are intact. c) MWCNT with distorted structure. No signs of inner channel or walls; d, e, f – electron diffraction patterns obtained for a, b, c, respectively.

We believe that distortion of MTCNT structure in the presence of gastric agents reflects the steps in progressive degradation of MWCNTs. Nanotubes with smooth surface and straight form (probably the most stable and chemically inert) seem to be less degradable, then rough and twisted ones. Analysis of electron diffraction patterns of intact and treated MWCNTs showed that crystal structure of degrading nanotubes remains unchanged. It seems that during biodegradation, MWCNTs decompose into smaller particles without significant changes in chemical bonds. Therefore degrading MWCNTs can be detected in biological material by conventional and analytical electron microscopy. According to some authors, smaller and shorter impurity-free carbon nanotubes show more tendencies to penetrate through the cell membranes and accumulate within the cells [17].

Since only partial degradation of MTCNTs treated with gastric ingredients *in vitro* was confirmed, we investigated whether MWCNTs could be absorpted passing through the small intestine of mice and accumulate in the enterocytes. TEM studies showed that small intestine epithelium cells maintain their normal structure, intact organelles (figure 5a) and microvilli (figure 5b). However, we found local areas of necrosis in small intestine of experimental mice (figure 5c). These areas were visualized as disturbances of epithelial layer of intestinal villi. Ultrastructural analysis has shown that groups of enterocytes in epithelium layer were undergoing different stages of cell death by necrosis.

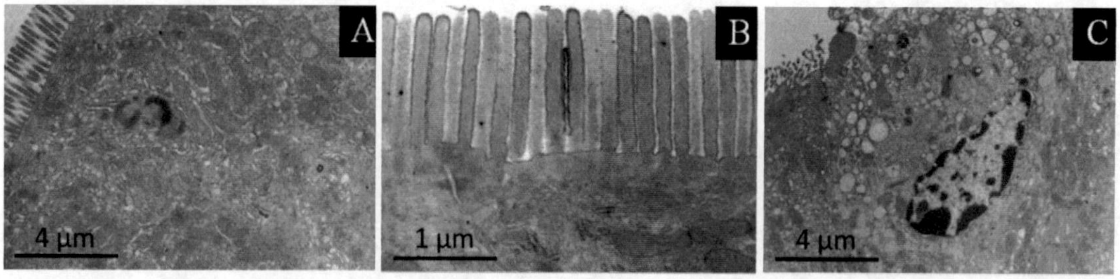

Figure 5. Transmission electron microscopy. Ultrastructure of enterocyte layer in small intestine of mice exposed to MWCNTs. a) Intact enterocytes with no pathological changes. b) Intact zone of microvilli c) Zone of necrosis. Enterocyte is vacuolated, apical membrane is destroyed.

Epithelium cells of small intestine were also analyzed for the presence of single or aggregated nanotubes. We did not find detectable clusters of MWCNTs in intact enterocytes, while tubular nano-sized structures, morphologically similar to MWCNTs (40-60 nm in diameter and 200-500 nm in length) were found at the areas of necrosis. These nanoparticles had inner cavity, distinct walls, outer and inner diameters (figure 6a, b) similar to MWCNTs, and noticeable layered structure (figure 6c). As MWCNTs without structural imperfections may be less prone to degrade within murine stomach, small fragments of stable nanotubes may be able to penetrate into epithelium cells, causing necrosis.

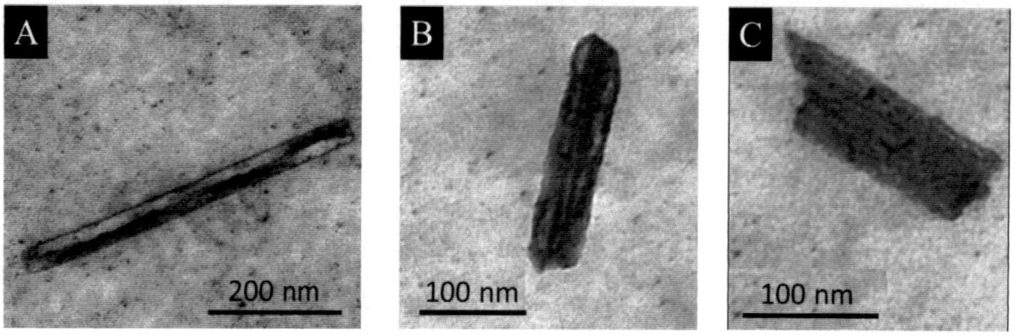

Figure. 6. Nanoparticles detected in the cytoplasm of enterocytes. Transmission electron microscope. a,b) Tubular particles. Walls and lumen are seen. c) Nano-sized particle with layered structure.

8. Conclusions

We demonstrated that treatment of MWCNTs with gastric digestion reagents (gastric juice and hydrochloric acid) induced distortion of MWCNT structure, which may reflect the stages of their degradation. However, in our experiments, MWCNTs do not undergo complete destruction during the gastric stage of the murine digestive cycle. It is known that complete digestive cycle of mice lasts up to 3-24 hours [18], and gastric phase lasts from 1 till 6 hours [19]. Interestingly, MWCNTs can facilitate gastric functioning, including excessive production of gastric juice components, including hydrochloric acid. This may increase the reactivity of gastric juice and enhance the degradation of MWCNTs [20]. We believe that upon oral administration, MWCNTs only partially degrade in the murine stomach, and the remnants of distorted nanomaterial transit further to small intestine and then excrete with the feces. Moreover, the direct contact of epithelial cells with partially distorted carbon nanotubes may induce severe damage of cellular layers and tissues, resulting in necrotic changes of the intestinal villi. We also detected individual tubular nanoparticles, similar to carbon nanotubes, located in the cells at the areas of necrosis. However, the identity of these particles needs to be confirmed. We assume that MWCNTs in the presence of digestive agents undergo distortion and consecutive disruption, followed by degradation, which is not fast enough to be completed during digestive cycle. Thus, nanomaterial composed of MWCNTs may cause negative effect on digestive system and should not be considered as harmless to living organisms.

Acknowledgments
This work was supported by Russian Science Foundation (# 14-50-00029).

References

[1] Chen S, Hu S, Smith EF, Ruenraroengsak P, Thorley AJ, Menzel R, Goode AE, Ryan MP, Tetley TD, Porter AE and Shaffer MS. 2014 *Biomaterials* **35(17)** 4729
[2] Zhang L, Wang S, Zhang M and Sun J. 2013 *J Drug Target* **21(6)** 515
[3] Zhang Q, Huang JQ, Qian WZ, Zhang YY and Wei F. 2013 *Small* **9(8)** 1237
[4] Du J, Wang S, You H and Zhao X. 2013 *Environ Toxicol Pharmacol* **36(2)** 451
[5] Lee S, Khang D and Kim SH. *Int J Nanomedicine* 2015 **(10)** 2697
[6] Ito Y, Venkatesan N, Hirako N, Sugioka N and Takada K. 2007 *Int J Pharm* **337(1-2)** 357
[7] Smirnova EA, Gusev AA, Zaitseva ON, Lazareva EM, Onishchenko GE, Kuznetsova EV, Tkachev AG, Feofanov AV, Kirpichnikov MP. 2011 *Acta Naturae* **3(1)** 99
[8] Khodakovskaya MV, de Silva K, Nedosekin DA, Dervishi E, Biris AS, Shashkov EV, Galanzha EI, Zharov VP. 2011 *Proc Natl Acad Sci USA* **108(3)** 1028
[9] Deng X, Jia G, Wang H, Sun H, Wang X, Yang S, Wang T and Liu Y. 2007 *Carbon* **45(7)** 1419
[10] Ju L, Zhang G, Zhang X, Jia Z, Gao X, Jiang Y, Yan C, Duerksen-Hughes PJ, Chen FF, Li H, Zhu X and Yang J. 2014 *PLoS One* **9(1)** e84974
[11] Hussain S, Sangtian S, Anderson SM, Snyder RJ, Marshburn JD, Rice AB, Bonner JC and Garantziotis S. 2014 *Part Fibre Toxicol* **11**:28
[12] Ye S, Jiang Y, Zhang H, Wang Y, Wu Y, Hou Z and Zhang Q. 2012 *J Nanosci Nanotechnol* **12(3)** 2101
[13] Donaldson K, Poland CA, Murphy FA, MacFarlane M, Chernova T and Schinwald A. 2013 *Adv Drug Deliv Rev* **65(15)** 2078
[14] Patlolla A, Berry A and Tchounwou PB. 2011 *Mol Cell Biochem* **358(1-2)** 189
[15] Vlasova II, Vakhrusheva TV, Sokolov AV, Kostevich VA, Gusev AA, Gusev SA, Melnikova VI and Lobach AS. 2012 *Toxicol Appl Pharmacol* **264(1)** 131
[16] Kagan VE, Kapralov AA, St Croix CM, Watkins SC, Kisin ER, Kotchey GP, Balasubramanian K, Vlasova II, Yu J, Kim K, Seo W, Mallampalli RK, Star A and Shvedova AA. 2014 *ACS Nano* **8(6)** 5610
[17] Fujita K, Fukuda M, Endoh S, Kato H, Maru J, Nakamura A, Uchino K, Shinohara N, Obara S, Nagano R, Horie M, Kinugasa S, Hashimoto H and Kishimoto A. 2013 *Toxicol Mech Methods* **23(8)** 598
[18] Padmanabhan P, Grosse J, Asad AB, Radda GK and Golay X. 2013 *EJNMMI Res* **3(1)** 60
[19] Schwarz R, Kaspar A, Seelig J and Künnecke B. 2002 *Magn Reson Med* **48(2)** 255
[20] Li Z, Qi W, Geng YX, Pan DQ, Lu Y, Xu JZ and Wu WS. 2011 *Nanoscale Res Lett* **6** 63

IOP Conf. Series: Materials Science and Engineering **98** (2015) 012009 doi:10.1088/1757-899X/98/1/012009

Nanomaterials in consumer's goods: the problems of risk assessment

I V Gmoshinski[1], **S A Khotimchenko**[2]

[1]Senior researcher, *Federal State Budgetary Scientific Institution «Institute of Nutrition»* 2/14 Ust'insky proezd Moscow, 109240, Russia
[2]Head of laboratory, *Federal State Budgetary Scientific Institution «Institute of Nutrition»* 2/14 Ust'insky proezd Moscow, 109240, Russia

E-mail: gmosh@ion.ru

Abstract. Nanotechnology and engineered nanomaterials are currently used in wide variety of cosmetic products, while their use in food industry, packaging materials, household chemicals etc. still includes a limited number of items and does not show a significant upward trend. However, the problem of priority nanomaterials associated risks is relevant due to their high production volumes and an constantly growing burden on the environment and population. In accordance with the frequency of use in mass-produced consumer goods, leading priority nanomaterials are silver nanoparticles (NPs) and (by a wide margin) NPs of gold, platinum, and titanium dioxide. Frequency of nanosized silica introduction into food products as a food additive, at the moment, seems to be underestimated, since the use of this nanomaterial is not declared by manufacturers of products and objective control of its content is difficult. Analysis of literature data on toxicological properties of nanomaterials shows that currently accumulated amount of information is sufficient to establish the safe doses of nanosized silver, gold and titanium dioxide. Data have been provided in a series of studies concerning the effect of oral intake of nanosized silica on the condition of laboratory animals, including on the performance of the immune system. The article examines the existing approaches to the assessment of population exposure to priority nanomaterials, characteristics of existing problems and risk management.

1. Introduction

The unique properties of nanoparticles (NPs) and nanomaterials (NMs) open up broad prospects for their targeted use in such areas of science and technology as materials science, microelectronics, optics, energy, chemical technology, environmental protection, medicine, consumer goods, including household chemicals, perfumes, cosmetics and food products. In accordance with the forecasts made in the early 2000s, the use of artificial NPs and NMs in food technology promises breakthrough achievements in the field of packaging materials with improved properties, new forms of nutrients with high bioavailability, effective and low-toxic food additives [1, 2].

Physicochemical properties and biological activity of the compounds in the nano-dispersed or nano-structured form may be significantly different from their "traditional forms", presented solutions, bulk phases and macroscopic dispersion [3-5]. Since many of the important new properties of NMs currently are still insufficiently studied, the problem rises their risks assessment to human health and the environment. Suggesting the observed progress of nanotechnology industries it can be assumed

Content from this work may be used under the terms of the Creative Commons Attribution 3.0 licence. Any further distribution of this work must maintain attribution to the author(s) and the title of the work, journal citation and DOI.
Published under licence by IOP Publishing Ltd

that in the short-term exposure of human to NMs grows rapidly, both in terms of quantity and in terms of the number of affecting nanoscale factors [6,7].

Research on the risks of nanotechnology (NT) and NMs is conducted since the early 2000s in the United States (under the supervision of the FDA and EPA), in the EU (IEC, EFSA, ECETOC et al.), as well as in a number of international organizations (OECD, WHO, FAO, ILSI) [8-10]. In Russia, the problem is studied of NT and NM related risk since the end of 2006. In 2007 the "Concept of toxicological research, risk assessment methodology, identification and quantification of nanomaterials" has been adopted [11], which defined the methodological principles and direction of research in this area in the next future. A total of 50 regulatory and guidance documents were developed and subsequently approved as a part of NTs and NMs safety assessment system in progress of the Federal Target Program "Development of infrastructure nanotechnology industry in the Russian Federation for 2008-2011".

Consensus is achieved in modern scientific literature that the traditional four-tier system of risk assessment is applicable to the risk assessment of NMs and NTs which includes stages of 1) dangerous factor identification (which is a source of risk?), 2) hazard estimation (how and in what quantities of the factor being studied produce the harmful effects for the human body?), 3) exposure estimation (how large is the load of harmful factors on the human body in the real conditions, and how the load is distributed in the population), and 4) the risk estimation (its quantitative expression followed by scaling as negligible small, moderate, high and very high) [12-14]. A block diagram is shown in figure 1 of the system of risk assessment in relation to the NPs and NMs peculiarities

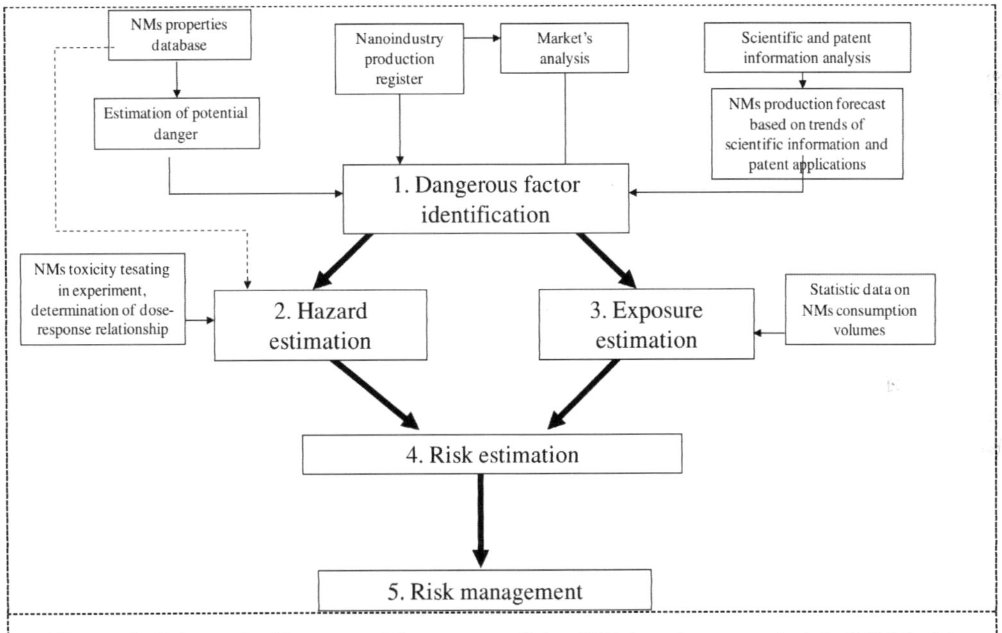

Figure 1. Schematic diagram of the nanoparticles (NPs) and nanomaterials (NMs) risk assessment

It is believed that there is no fundamental difference between the applicability of this scheme to the NPs and NMs on the one hand, and to the toxic chemicals of traditional dispersion stage - on the other [15], but the implementation of these steps should be specific in all cases to the properties of NPs and NMs defined in the first place by a high degree of dispersion of their structures.

The objectives of this article include consideration of problems of risk assessment of NPs and NMs used in consumer products.

2. Hazard identification

Variety of NPs and NMs produced with modern nanotechnology industry is extremely high. Thus, according to the resource NanoWerk (http://www.nanowerk.com/phpscripts/n_dbsearch.php), registering only NPs and NMs as individual substances, as of April 2015, production in the world includes 616 varieties of carbon nanotubes, 122 - fullerenes, 107 - derivatives of graphene, 567 – NPs of elementary substances (metals and non-metals), 609 – NPs of binary compounds (excluding quantum dots), 161 – NPs of ternary and more complex compounds, 276 – quantum dots, 42 – nanofibers, 5 – non-carbon nanotubes (2504 positions in total). Since the same NM can be used in practice in several types of products, the potential number of species of nanotechnology products can now exceed 10^4 items. However, this estimate does not take into account large differences in the degree of commercialization of the some kinds of nanotechnology products in the manufacturing of consumer goods. It is interesting from this point of view to consider the actual state of the nanotechnology products on market, identifying the types of NPs and NMs, which are priorities from the perspective of the consumer product exposure.

A computer database (registry) is maintained in ION to solve this problem since 2010, that accumulates information on NT products presented in turnover and use in the Russian Federation. The database is updated from public sources, including registries certificate of state registration of products in the Russian Federation and the Customs Union (http://fp.crc.ru); Websites of "RUSNANO" Fund of Infrastructure and Educational Programs (part of the Group RUSNANO) and the Russian nanotechnology network (http://www.rusnanonet.ru/products/list/; http://www.rusnanonet.ru/ goods /; http://www.portalnano.ru/), sites of companies that implement in the Russian Federation different kind of consumer nanotechnology products (food supplements, perfumes, cosmetics and others.). Database on NMs is in free access on the website of institute, www.ion.ru.

Analysis of the data provided by the products applicants declared as nanotechnology products, does not allow to reliably identify the NPs and NMs in the composition of the product in a significant number of cases. This may be due, on the one hand, to the desire of producers to maintain technological secrets and know-how, and on the other hand, to the common practice of using the notation "nanotechnology" or "nanotechnology products" in advertising and marketing purposes without a sufficient factual basis.

The database contains information on 1,035 kinds of products manufactured using NTs and NMs as of April 2015. 958 items included (92 % of the total number of titles of nanotechnology products) presented variety of consumer products, and 77 (12%) – non-consumer (pure materials, industrial raw materials, components, manufacturing equipment). The predominant place in the structure of nanotechnological consumer products take perfumes and cosmetics (792 items, 83%) as can be seen from the data obtained. They are followed by food products (47 items, 5%); it is total 55 titles (6%) combining this class of products with packaging materials for food products. Then follow household chemicals (28 items, 3%), fuel components, additives and converters for motor vehicles (24, 2.5%), car cosmetics (20 titles, 2%) and construction materials presented mostly varnishes and paints for exterior use - (17 items, 2%) in descending order. The share is insignificant of other classes of products which production is represented only by single names.

The most common commodity groups are a means for skin care (creams, masks and balms), toilet water, coloring agents, decorative cosmetics and nail polish among perfumery and cosmetic products manufactured using NTs and NMs. In many cases applicants of these products declare the production to be manufactured "with use of NMs and NTs" for purposes of state registration but the composition of these NMs can't be disclosed on the basis of available information. The products for which the composition of the NM can be set include 90 kinds of perfumes and cosmetic products with silver NPs, 35 – gold NPs, 17 – platinum NPs. In 33 kinds of perfume and cosmetic products (skin care) the applicant declares the presence of “nanosomes”, which are presumably nanometer-sized capsules jacketed with molecules of hyaluronic acid and/or β-cyclodextrin containing inside various biologically active agents. Biologically active supplements dominate (36 items) in the structure of food nanoindustry production. Half of their number (18) presented products, which are the sources of colloidal (nanoscale) silver. The remaining commodity groups of products (food additives, complex food additives, processing aids) are represented by single items.

Analysis is of great interest of the frequency of various NPs and NMs use in products in turnover, since it allows to identify priority objects for toxicology and hygienic studies aimed by hygienic regulation of NMs in products and environmental samples. Unfortunately, as already noted above, in many cases (for 612 kinds of perfumes and cosmetic products, represented in the database) strict identification of NMs use is not possible. The kind of NMs could be established for 346 types of consumer products, presented in the database. Total 24 types of individual NMs were identified in them. The top of the most "popular" NMs in the consumer goods is as follows (table 1) according to these data.

Nanosized titanium dioxide presented 29 kinds of production. This is, firstly, sunscreen and, secondly - catalytic air purifier. Use of nanosomes and nanoliposomes is sufficiently widespread as components of creams and other skin care products; silicone nanofilms are applied as a water- and dirt-repellent coating on various surfaces - parts of vehicles, building construction, furniture, housewares, textiles and more. Scope of diamond NPs includes decorative cosmetics, metal alloys NPs - remetallizing fuel additive for motor vehicles.

Table 1. Ranking of engineered nanomaterials in accordance with the frequency of their use in consumer products.

Nanomaterial:	Number of products	Ranking position
Silver nanoparticles	133	1
Gold nanoparticles	35	2
Nanosomes	33	3
Nanoparticles of titanium dioxide	29	4
Platinum nanoparticles	20	5
Silicone nanofilms	19	6
Surfactant micelles	10	7
Silica nanoparticles	8	8
Diamond nanoparticles	7	9-10
Alloy nanoparticles	7	9-10

Nanoscale amorphous silica (SiO_2) contains in 8 product items included in the database. However, reason is to assume that the frequency is strongly underestimated of this NM use and hence the degree of the consumers' exposure to it. As is known, amorphous silica is widely used as a food additive E551, and also as a part of a large number of medicinal tablets together with many types of cosmetic products. There is no information about the size of its particles in the present specification of JECFA on this food additive [16], which allows the use of amorphous superfine silica, obtained by gas-phase hydrolysis of pure tetrachlorosilane. This material known as "Aerosil®" and under a number of other trade names, is characterized by specific surface margined 200-350 m^2/g and a primary particle size of 6-30 nm, forming loose aggregates of submicron size, easily destroyable by sonication. Use of this NPs is often not declared by the manufacturers of food production.

Thus, the list of priority NM identified as sources of risk in consumer products includes **silver NPs** and (by a wide margin in the smaller side in order of importance), **NPs of gold, platinum and titanium dioxide**. This list should be supplemented with **amorphous silica NPs** in view of the considerations above. The resulting schedule of NMs coincides in the most number of positions with the list of OECD [17] with the exception that it does not include carbon NMs, such as single- and multi-walled carbon nanotubes and fullerenes. Despite the presence of a plenty potentially useful properties in these NMs, their use as part of commercialized consumer products is missing or includes single names. In the list of priority NMs nanosomes aren't presented which are widely used in the composition of cosmetic products. This is due to the fact that this type of NPs is biocompatible according to modern concepts, and within the body they quickly degrade in processes of bioassimilation.

3. Hazard characterization

Toxicological evaluation of NPs and NMs is based currently on a huge variety of biological test systems, including models *in vitro*, *in vivo*, as well as *in silico* (computer simulation). The problem is a difficult task of o these approaches choice for priority NPs and NMs needing in hazard characterization. Complex hierarchical approaches are developed for this based on the achievements of genomics, transcriptomics and proteomics, for example, such as program NeoGen in the USA [18].

More traditional methodology is based on the selection of the optimal strategy of toxicological and hygienic studies based on preliminary data on the potential hazards of the NPs and NM. Criterion of the potential hazards of nanomaterials should recognize in the first approximation their insolubility and the ability therefore to the long-term persistence in the biological environment [19]. So, in the drafting of Annex 18 to the REACH Regulation of the European Union there was proposed [20] a preliminary assessment criteria of potential danger of NPs and NMs basing only on two indicators: the volume of annual production and solubility in water and in various biological fluids. The publication of the European Commission [7] proposed the criteria of NM's potential danger based on the physico-chemical characteristics of the particles. Use of the five major external NM properties was recommended - (1) agglomeration (weakly bound particles) and aggregation (strongly bound particles), (2) reactivity, (3) the presence of active functional groups on the surface, (4) the particle size, and (5) the presence of contaminants bound to the surface or in the bulk of the NPs. Factors include bioavailability, bioaccumulation, potential translocation and potential toxicity that may affect the risks of artificial NM. A system was developed of pre-classification of NMs according to their hazard level including very high, high, medium, low and very low risk on this basis.

A more sophisticated system of hazard assessment approach must include information about structure -activity relationships of artificial NMs [21]. Similar structured approach is applicable in the Russian Federation to the determination of the required volumes of toxic hygienic and special studies of NMs, wherein the integrated assessment of the potential hazard is based on forecasting and analysis procedure that takes into account the entire volume of accessible, reliable and relevant to this NM scientific and technical information, including data of physico-chemical characteristics of the NPs, information on effects in cell-free systems and cell cultures, the results of toxicological and environmental studies, the estimated volume of production [22]. Selected information about the properties of NM is evaluated, grouped into six functional blocks, ranked in order of importance, as determined in accordance with peer review. Each estimate is assigned according to its rank with weighting factor after which the resulting value (index) is calculated of the potential hazard. The algorithm used includes built-in procedure for evaluating of the uncertainty of the estimation.

As a result, designation of a nanostructured object to the low level of potential hazard require no any special toxicological-hygienic and medical-biological assessment, and the evaluation should use the same criteria and approaches applicable to "traditional" analogues of a material manufactured without use of NT. NPs and NM with an average potential danger are a subject of standard toxicological and hygienic assessment of their impact on the most important functions of the body. NPs and NMs characterized by a high degree of potential danger are recommended to be a subject of more sophisticated studies involving not only the general toxicological research, but also special types of testing (organotoxicity, immunotoxicity, neurotoxicity, allergenicity, embryotoxicity, teratogenicity, mutagenicity, carcinogenicity), as well as, possible experiments lifelong laboratory animals (1.5-2 years), or in a few generations.

One should always take into account in the analysis of dose-effect obtained on biological models that the difficulties and uncertainties are associated with the specific properties of the NPs and NMs. So far, there is no consensus about by what units the acting dose of NM should be expressed [15,23].

Value of the results should be considered as a very relative which were obtained in experiments with *in vitro* models, because there is no information in most cases about whether the dose (concentration) of NPs used in these studies is relevant to situation *in vivo* related to exposing the organism by the natural ways of intake (inhalation, through the skin, and especially with food and water consumption). A research can play a major role on the penetration (absorption) of NPs through biological barriers, their bioaccumulation, biotransformation and excretion (ADME - research) in the removal of these uncertainties, but the volume is very limited of such information in relation to the practically important NPs and NMs currently [22,24] . Information is "scarce" especially about the possibility of absorption of the main types of NPs in the gastrointestinal tract, as well as through uninjured skin [25,26].

A strong dependence exists of NMs properties including toxic one on the state of the interface (so-called "surface chemistry") [27]. It should be borne in mind also that partial solubility of toxic chemical compounds (which, by itself, is not a constant but is a function of particle size) plays a large role in the toxic properties of NPs and NMs, and the ability is important of NPs to release toxic species adsorbed or chemically bonded to them, such as bacterial endotoxins, heavy metal ions, etc (all of which is defined as the so-called effect of "Trojan horse") [15].

The values of safe doses (concentrations) are calculated after the establishment of non-observed effect maximum exposing dose (NOAEL) for critical organs and systems, as well as whole body by the introduction of appropriate safety coefficients. As a rule, two consecutive 10-fold safety coefficients are introduced when extrapolating to human data obtained in small animals (rats or mice).

The world's scientific literature has accumulated to date a large amount of experimental data on the toxicology of the most important NPs and NMs. The amount of information experienced exponential growth over the last decade on the relevant biological and physico-chemical properties of the NPs and NMs meeting the criteria of scientific accuracy and completeness as evidenced by the results of scientometric tests. The total number of publications registered in specialized databases (Web of Science, PubMed and others) exceeded 55,000 positions already in 2012 [15]. However, the distribution of publications on the NMs species studied does not correlate with their practical importance as constituents of consumer products. In addition there is usually a significant disparity between the use of biological models and test systems and their relation to the most likely scenario of exposure within each sufficiently studied NM. In particular among 63 articles concerning toxicological characterization of the anatase form of titanium dioxide, 27 of papers (43%) were performed on *in vitro* models of cell cultures and 36 (57%) - *in vivo*, including 18 papers (29%) on inhalation model, 13 (21%) - when administered parenterally, 4 (6%) - if administered to the gastrointestinal tract, and only one (1%) - when applied to skin according to a meta-analysis [25].

However, sufficient amount of information is now accumulated that enable one to establish safe doses of silver, gold, titanium dioxide NPs, which was reflected in the publications [28-33]. Properties of platinum NPs were studied in much smaller extent; at that almost no data were obtained on biological effects at transdermal way of exposure. The same is true to the NPs of amorphous silica, for which the volume of exposure through food is apparently significantly underestimated.

The majority of nanosized amorphous silica studies were carried out in systems *in vitro* (cell cultures of various types). Thus, the possibility was demonstrated of catalytic generation of reactive oxygen species in a cell-free system [34] and in keratinocytes [35] and in human alveolar epithelial cells in culture [36]. The ability was set of this NPs type to accumulate in cells of different lines in culture showing a variety of cytotoxic effect. Authors of paper [37] described the actions of amorphous nano-silica on human bronchial epithelial cell line Beas-2B. Reduction in cell viability was close to 20% accompanied with development of peroxidation, changes in the proteomic profile of cells (increase in the expression of several enzymes of intracellular kinase cascade). The summary is presented in table 2 of some other results obtained in cell cultures.

Table 2. Some results of in vitro toxicological evaluation of nanosized amorphous silica

#	Experimental model	Threshold NPs concentration in media, μg/ml	References
1	Normal human bronchial epithelial cells, Beas 2 B	1.0	[38]
2	Normal human lung epithelial and endothelial cells	300	[39]
3	Normal human endothelial cells	10,0	[40]
4	Platelets	10,0	[41]
5	Stem cells	1,0	[42]
6	Alveolar epithelial cells of human lung	≥150 (evaluation)	[43]
7	Hepatocytes, Kupfer cells	100	[44]
8	Hepatocytes, Kupfer cells	10	[45]
9	Human hepatocytes L-02	50	[46]
10	Endothelial cells	25	[47]
11	Normal and malignant cells of the alveolar epithelium of human lung, fibrosarcoma cells	10	[48]

The mechanism of the cytotoxic effects of silica NPs, as might be expected, is associated with a non-specific deleterious effects on the cell structure of free radicals and reactive oxygen species, catalytically generated on SiO_2/water interface [49]. These effects are much more pronounced in NPs when compared to silica of "traditional" degree of dispersion (such as fumed silica, silica gel and so on) in view of silica NPs highly developed interface and a small radius of curvature (which entails an increase in the chemical potential). However, it should be borne in mind that the majority of effects were detected at large concentrations of NPs in cell culture (exceeding 1.0 μg / ml, see table 2).

Quantitative data is now extremely scarce on the extent of absorption, biodistribution and bioaccumulation of nanosized silica due to the high technical difficulties related to the identification and quantification of these NPs in the biological matrix. Accumulation of label was detected after intravenous administration of [^{125}I]-labeled NPs to mice mainly in cells of the reticulo-endothelial system of liver and spleen [50]. An attempt was undertaken in single study [51] to quantify the absorption and bioavailability of the silica NPs in the organs and tissues of rats after prolonged oral administration using ICP-MS. However, the results obtained were ambiguous due to a high background level of silicon in the bodies of animals of the control group. Significant increase of the silicon content was shown on a qualitative level in the liver and spleen of animals treated with said NPs at doses of 1000 mg / kg body weight or more. Thus, at present data is insufficient which would allow extrapolating the results of silica NPs toxicity studies in cell cultures to the situation *in vivo*, using the ADME properties of this NM.

Several studies have characterized toxic properties of amorphous silica *in vivo*. NPs were injected to mice intraperitoneally at very high doses (up to 2 g/kg body weight) in the study [52]. Marked changes were noticed in the function of peritoneal macrophages, increasing production IL-1β, TNF-α, NO, gene expression of IL-1,6, TNF-α, nitric oxide synthase, cyclooxygenase-2. Signs of cell death were observed with the introduction of NM at only the highest doses. 70 nm nanosized silica was hepatotoxic to rats at a dose 30 mg / kg when administered intravenously [53]. Nanoscale SiO_2 with particle diameter of 50-100 nm was captured by the liver and kidneys, excreted in the urine and bile when injected intravenously according to [54]. Aim of the study [55] was characterization of silica NPs aerosol toxicity in neonatal rats. Single or 3-day exposure to the aerosol showed no significant pulmonary inflammatory, genotoxic or histopathological effects in rats exposed to particles.

A number of works was devoted to the identification of possible allergenic properties of nanosized silica. For example, these NPs were able to enhance nasal allergic sensitization of mice by ovalbumin allergen according to [56]. Similar data was obtained in a study [57] using a model of asthma in rats sensitized by ovalbumin. A recent study [58] evaluated immunotoxic properties of amorphous SiO_2 NPs intraperitoneally administered to mice at a dose of 2-50 mg/kg body weight. Changing proportions were shown of CD3+, CD45+, CD4+ and CD8+ cells in the spleen, shifts in the levels of total IgG and IgM among the effects studied.

Silica NPs were not carcinogenic in rats unlike quartz and carbon NPs according to [59].

Information is scarce in the literature about the possible effects of the silica NPs administered by oral route. Van der Zande M et al. [51] studied subacute (in 84-day experiment) toxicity of two kinds of nanostructured silica administered orally to rats at high doses (100-2500 mg/kg body weight daily). Biochemical and immunological parameters haven't been studied in blood of animals but fibrosis increase was noticed in a dose-dependent manner and expression of genes responsible for this process also took place in the liver. Authors estimated the threshold subacute oral dose (LOAEL) of silica NPs close to of 2500 mg/kg body weight and NOAEL exceeding 100 mg/kg body weight according to these indicators.

Tananova ON et al [60] examined proteome of microsomal fraction of liver cells from rats, orally receiving nanostructured silica in three doses (from 1 to 100 mg / kg body weight) during 28 days. NPs caused both the disappearance and appearance of new components of the proteome (protein spots on the *2D*- electrophoregram) compared to the control group of animals in all doses studied. Mass spectrometric identification revealed the dominant protein peak which disappeared under the influence of NPs at dose 10-100 mg / kg body weight, which was identified as the glucose-regulated protein precursor (GRP 78).

Zaitseva NV and co-workers [61,62] studied the acute oral toxicity in mice of silica NP obtained by method of liquid crystal templating in the presence of cetyltrimethylammonium bromide. LD_{50} was 4638 mg/kg body weight by oral route of administration. Nano disperse silica has expressed high cumulatively (the cumulative index Ik = 0,45). Said NM demonstrated also marked toxic effect on animals, judging by the complex biochemical, hematological and morphological parameters at a dose of 0,1-0,3 LD_{50}

Thus, the data available are inconsistent on the toxicity of silica NP *in vivo*. It should be noted that most of the samples of nanosized silica studied in the literature do not correspond to nanosized silica such as "AEROSIL ®», which is most widely used as a food additive. In view of this, joint research has been conducted by FSBI "Institute of Nutrition", FSBI "FNTS health-care technology risk management to public health" of Rospotrebnadzor and A.N.Bach Institute of biochemistry in 2013-2014 on establishment of maximum no observed effect and threshold doses of this particular form of NM [63-66].

The object of investigation was the amorphous silica "Orisil 300". The specific surface area of the sample was equal to 300 m^2/g when determined by method of BET according to the manufacturer data, which corresponds to calculated diameter of the primary nonporous spherical particles of 7.5 nm. This material presented a white, light, x-ray amorphous powder giving after ultrasonic development opalescent colorless solution in water, which was stable for at least 2 days. Examination of this silica slurry by TEM showed at concentration of 1 mg/ml (figure 2 a-c) that particles of the disperse phase presented mainly large aggregates consisting of distinguishable primary particles sized from 5 to 100 nm, and a very small amount of free particles sized 5-20 nm. AFM showed on the scanned image of the sample slurry dried on a substrate the presence of NPs predominantly in the aggregated state. NPs aggregates varied in size on all the scans from area 20x20 mm^2, reaching a maximum of up to 2 μm in one dimension. The structural element was a particle sized in diameter less than 100 nm for all the aggregates. NP's population had uniform morphology consisting of spherical particles with sizes from 20 to 60 nm (figure 2 d,e). Spektroacoustic study of sonicated aqueous silica slurry showed a bimodal particle size distribution with a predominance of the NP's fraction with an average diameter of 20-40 nm at concentration of the 5% by weight. Particle size analysis by means of laser dynamic light

IOP Conf. Series: Materials Science and Engineering **98** (2015) 012009 doi:10.1088/1757-899X/98/1/012009

scattering showed that the same sample presented predominant fraction of particles in preparation with number average hydrodynamic diameter of 56.6 ± 32.1 nm with 90th percentile - 91.7 nm (figure 3 a,b). The content of particles with a diameter greater than 100 nm did not exceed 10% of the total number of particles after sonication.

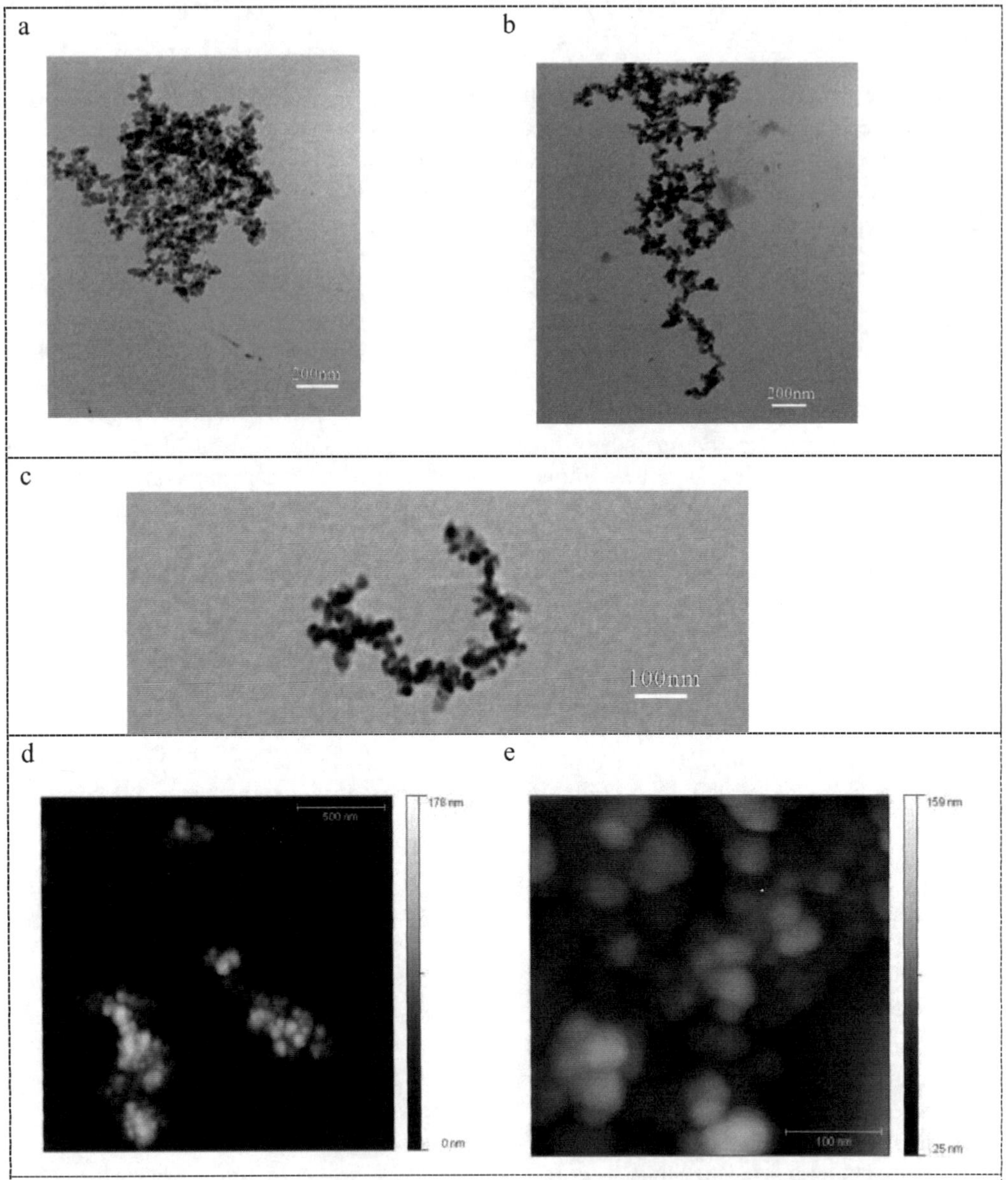

Figure 2. Characteristics of nanosized silica particles by transmission electron microscopy, device «JEM-100CX» («JEOL», Japan); (a-c), and atomic force microscopy, device «SmartSPM» («AIST-NT», Russia), (d, e) (data taken from [63,65])

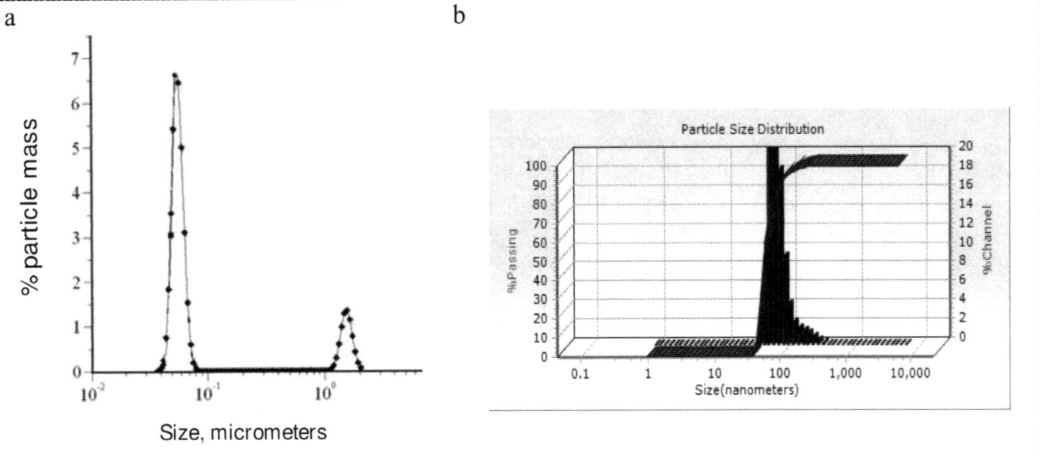

Figure 3. The size distribution of silica NPs in the sonicated aqueous dispersion according to spektroacoustic studies on device «DT-1202» («Dispersion technology Inc.», USA) (a); and dynamic light scattering on device Nanotrack Wave (Microtrack Inc., USA) (b) (taken from [63,65])

Biological experiments were conducted in accordance with Russian and international rules for work in laboratory animals. Acute toxicity of nanosized silica was studied by a single point method in 20 adult male BALB/C mice weighing 25.0 ± 2.0 g aged 2 months (data taken from [65]). Analysis of the results obtained led to conclusion that an aqueous suspension of nanosized silica did not cause death and of intoxication at a dose of 10,000 mg / kg. Changes in the integrated performance of experimental animals weren't observed. No any morphological changes were identified in histological preparations of colon and jejunum. Thus, the value of LD_{50} of the studied nanosized silica exceeded 10,000 mg / kg body weight in oral route.

Subacute oral toxicity of nanosized silica was studied in the growing male Wistar rats in 3-month (92 days) experiment. Animals were divided into five groups (№№ 1-5), treated with this NM at doses, respectively, 0 (control); 0.1; 1.0, 10 and 100 mg / kg of body weight daily as aqueous sonicated slurry. Within 1 month NM was administered intragastrically by gavage, and then added to the experimental semisynthetic balanced diet.

Death occurred of one rat in group 4 and three – in group 5 during the 1st month of intragastric administration. The autopsy of animals showed that the death was caused by bilateral pneumonia, which developed as might be expected due to accidental inhalation of traces of suspension containing a high concentration of NPs when administered by gavage. Furthermore, one rat died in group 3 on the 3rd month of experiment. Other animals of all experimental groups had well appearance of skin, hair and mucous membranes, normal motor activity, behavior and stool and did not differ from the control group. Determination of average monthly gains of body weight showed that animals of all groups grew by almost the same rate (p> 0.1, ANOVA) during the 1st and 2nd month, however a slight (less than 15%) but significant lag in weight gain was noticed in all four experimental groups as compared with control group at the end of the third month. This effect was not dose-dependent and was apparently associated with a reduction of fat mass gain in these groups of aged animals.

Testing showed no negative impact on cognitive function studied by the method of the conditioned reflex of passive avoidance (CRPA) in terms of CRPA resistance at all doses of nanosized silica studied, but at the largest dose of 100 mg/kg the elevation could be suggested of animals discomfort (anxiety), which resulted in significant reductions in the number of initial visits to dark compartment of the experimental setup.

Integral indices (permeability of the intestinal barrier, the relative weight of internal organs), biochemical, hematological, immunological parameters, the composition of colon micro biota, oxidative DNA damage by excretion of 8-oxo-2-deoxy guanosine (8-oxoG) in urine, liver hepatocyte apoptosis by flow cytometry were evaluated after removal of the animals from the experiment. Effects were studied in separate series of experiments of silica NPs on allergic sensitivity in rats immunized with chicken ovalbumin using model of systemic anaphylaxis (duration of treatment in this case made 28 days).

The results obtained were used to determine NOAEL, taking into consideration for each effect it's statistical significance, absolute value, the direction (i.e., the interpretation as an adverse effect like), as well as the presence of a monotonic dependence on the dose of NM. Table 3 presents summary of basic experimental data. The results showed that the most sensitive target for silica NPs is the immune system of animals (figure 4). The total number of leukocytes was significantly lowered in animals of group 5 treated with silica NPs at 100 mg/kg body weight. At the same time, the qualitative composition of Ly was significantly modified despite only a marginal and insignificant decrease of the relative number of lymphocytes (Ly) (7%, $p > 0.05$) in animals of group 5. Proportion significantly decreased of T-helper cells (13%, $p < 0.05$), increased of cytotoxic Ly (19%, $p < 0.05$), all of this leading to the decrease (27%, $p < 0.05$) in the immunoregulatory index (dimensionless ratio CD4/CD8) as shown by flow cytometry. TNF-α level in serum of animals of group 5 increased by an average of 590%, $p < 0.05$, and the concentration of IL-10 showed pronounced (36%) trend to decrease.

Table 3. Summary table of the effects of subacute oral administration of nanosized silica to rats in the 92- and 28-day trials (taken from [63,64,66])

№№	Indices studied	Presense (+) or absense (-) of effect	Effect can (+) or can't (-) be considered as harmful	NOAEL evaluation	Note
1	Body weight gain	+	-	>100 mg/kg	
2	Relative organ's mass	+	+	?	No dose-dependent
3	Intestinal wall macromolecular permeability	-	-	>100 mg/kg	
4	Liver microsome system of xenobiotic detoxication				
	Total CYP450	-	-	>100 mg/kg	
	Total Cytochrome b5	-	-	>100 mg/kg	
	CYP 1A1 activity	+	-	>100 mg/kg	
	CYP 1A2 activity	-	-	>100 mg/kg	
	CYP 2B1 activity	+	+	?	No dose-dependent
	Glutathione-S-transferase activity	+	-	>100 mg/kg	
	UDP-glucuronosyl transferase activity	-	-	>100 mg/kg	
5	Activity of lysosomal hydrolases				
	total	-	-	>100 mg/kg	
	non-sedimentable	-	-	>100 mg/kg	
6	Serum dyene conjugates of PUFA	+	-	>100 mg/kg	
	Antioxidant enzymes activity	-	-	>100 mg/kg	

IOP Conf. Series: Materials Science and Engineering **98** (2015) 012009 doi:10.1088/1757-899X/98/1/012009

7	Liver pool of non-protein thioles	-	-	>100 mg/kg	
8	Blood biochemistry				
	AlAT activity	+	-	>100 mg/kg	
	AsAT activity	-	-	>100 mg/kg	
8	Serum albumin	+	+	?	No dose-dependent
	Total serum protein	+	+	?	same
	Serum glucose	+	+	?	same
	Serum creatinine	-	-	>100 mg/kg	
	Serum uric acid	-	-	>100 mg/kg	
	Alcaline phosphatase activity	-	-	>100 mg/kg	
9	Total hemoglobin	-	-	>100 mg/kg	
10	Hematological indices (red blood cells)				
	Hemoglobin concentration in cell	-	-	>100 mg/kg	
	Red blood cells count	+	-	>100 mg/kg	
	Hematocrit	-	-	>100 mg/kg	
	Mean red blood cell volume	+	-	>100 mg/kg	
	Hemoglobin amount in cell	+	-	>100 mg/kg	
11	Hematological indices (white blood cells)				
	Total white cells count	+	+	10-100 mg/kg	
	Lymphocytes ratio	-	-	>100 mg/kg	
	Same	+	+	<100 mg/kg	
	Other (monocytes, neutrophyles, basophiles etc)	-	-	>100 mg/kg	at sensitization
12	Platelets	-	-	>100 mg/kg	
13	DNA oxidative damage (urinary 8-oxoG)	+	-	>100 mg/kg	
14	Hepatic cells apoptosis (flow cytometry)	-	-	>100 mg/kg	
15	Allergic sensitivity (active anaphylactic shock)	-	-	>100 mg/kg	
16	IgG to ovalbumine level, ELISA, in sensitized animals	-	-	>100 mg/kg	
17	Cytokines				
	TNF-α	+	+	10-100 mg/kg	
18	Neutrophilic leucocytes phagocytosis activity	-	-	>100 mg/kg	
19	Cellular immunity				
	CD45RA+ (B-Ly)	-	-	>100 mg/kg	
	CD161a+ (Natural killers)	+	-	>100 mg/kg	
	CD3+ (T-Ly)	-	-	>100 mg/kg	
	CD3+CD4+ (T-helpers)	+	+	10-100 mg/kg	
	CD3+CD8+ (T-cytotoxic)	+	+	10-100 mg/kg	
	CD4+/CD8+ ratio	+	+	10-100 mg/kg	
20	Cognitive function (CRPA)	-	-	>100 mg/kg	
	Anxiety level (CRPA)	+	+	10-100 mg/kg	

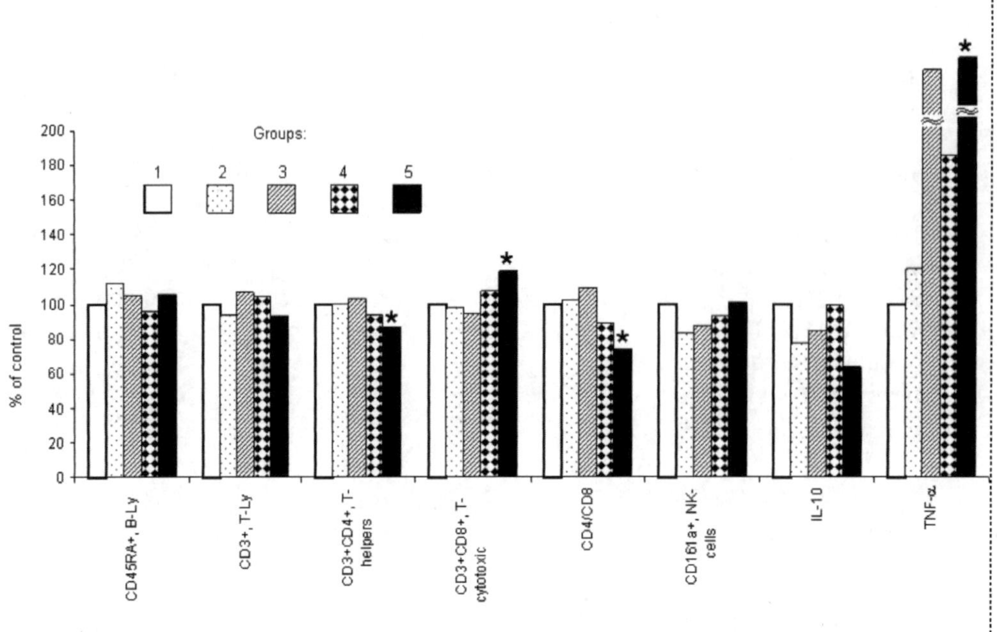

Figure 4. Effect of 92-day receiving various doses of nanoscale silica on the performance of the immune system of rats (data taken from [66])

Data obtained in animals sensitized with ovalbumin demonstrated similar yet less pronounced shifts in immunological parameters induced by silica and no significant increase was noticed in the severity of the allergic reaction after administration of antigen challenge dose ($p>0.05$). The same could be noted for the level of antibodies to ovalbumin.

Thus, these data allow estimating LOAEL of nanosized amorphous silica corresponding to used as a food additive E551, by the value of 100 mg / kg body weight / day, and NOAEL value of 10 mg / kg body weight / day. Size of safe dose of amorphous nanosized silica must be reduced for human ingestion in accordance with the relevant safety coefficients to 100 times and thus is equal to 0.1 mg / kg of body weight per day, or 7-10 mg per day for an adult.

4. Exposure assessment

Assessing exposure to NPs and NMs needs measuring their number (quantity) entering the body by different ways (inhalation, oral, dermal) as a result of contact with various objects of the environment (air, water, soil) and products (food, perfume and cosmetic products, goods household chemicals, etc). The most important steps in the evaluation of exposure are: identification of scenarios and routes of exposure; identification of the medium that carries the NPs/NMs; determination of their concentrations in the control points; timing, frequency, and duration of exposure; identification of affected population. Exposure scenario can often be simplified in the real conditions of risk assessment and reduced to the flow of NPs/NMs by only one way (eg, ambient air, drinking water, food, cosmetic products, etc.). In some cases an exposure scenario may be limited to NPs/NMs income from certain source of emissions (eg waste of nanoindustry enterprises stationed in the area, the packaging material containing NPs and so on).

With regard to NMs used in the composition of consumer's goods direct impact of production on population should be considered both in the case of food products - by oral exposure, perfume and cosmetics - primarily transdermal. Indirect impact should also be considered through objects of the environment. The significance increases of the latter group of scenarios with the volume of priority NM production. Thus annual production of silver NP totaled in the world more than 500 tons in terms of Ag in 2011, and could exceed 1,000 tons in 2015 according to [34], which corresponds to about 140 mg per year for every person on Earth. Silica NPs annual production is estimated at 25,000 tons in the world [67], which is about 40% of global issue volume of nanopowders. Daily load on human with silica NPs from food can reach 1.8 mg / kg body weight according to recent evaluation [51]. A more accurate estimate of the load is not possible at the moment, because the specification of silica used as a food additive E551 isn't disclosed in many cases by the manufacturers of products, and its nanoscale nature isn't declared. Selective control of a number of food and food supplements samples containing E551 and its products held in ION by means of dynamic light scattering showed that this dietary additive is known to present NPs having particle size in the range below 100 nm in many cases.

It's important to bear in mind when evaluating the exposure that the distribution of artificial mineral NP in the production and in the biosphere can be extremely uneven. It is assumed that the general population is exposed to NPs mainly by oral and transdermal route when consuming foodstuffs, food supplements and cosmetics containing these NPs, as well NPs migrating into food from packaging materials [68, 69], whereas inhaled route of exposure apparently predominates in employees of enterprises producing products with NMs [70]. The assumption does not receive experimental confirmation that a large number of mineral NPs can enter the atmosphere in the form of aerosols during the disposal of waste of textiles, medical supplies and other products of NT on waste-incineration plants. In fact, a major amount of NPs is accumulated in ash together with the sludge from wet treatment system of exhaust gas clarification [71], which can be supplied subsequently to fields as a surrogate fertilizer, washed away into water, accumulate in different soil and water organisms [72], and eventually come to man through the food chain as a part of agricultural production. However, data is currently insufficient to quantify the exposure that proceeds in such a scenario.

5. Risk characterization

Several models are available to characterize quantitatively the risks of exposure to NPs and NMs depending on various scenarios of exposure and the rendered character of harmful effects [73-77]. Special attention should be given when choosing a model that is adequate to the current situation to information about the presence of cumulative effects produced by NPs/NMs (accumulation in organs and tissues, accumulation of toxic action) and certain types of remote effects on the body (mutagenic, carcinogenic effects, etc.), synergistic effects of various NPs and NMs in conjunction with toxic substances of traditional dispersity.

The risk profiling of NPs/NMs is qualitative in nature when based on a model that takes into account deterministic effects, since all the risks are considered insignificant that are associated with the current dose (concentration) of the NPs/NMs at a level below the reference values (norms). If one exceed the reference value there is a significant risk, but it's not currently possible to establish clearly the nature of the quantitative relationship between excess dose over reference level and value of risk (probability of harmful effect). In view of this, the risk assessment is based on qualitative model using so-called "Hazard quotinent" (*HQ*) and "hazard index" (*HI*) [78]. This model is basically the same that is applicable to non-carcinogenic risk of chemicals for which one can set the threshold for safe exposure. In this case, the hazard ratio is calculated by the formula

$$HQ = Ed \,/\, RfD$$

when *Ed* – exposure, a *RfD* – safe referent level of exposure (dose or concentration). It is clear that both values used in the calculation must match in dimension, so that the resulting hazard ratio remains dimensionless.

The aggregative risk is characterized of a number of nanoscale factors in the case of their combined action by the hazard index (*HI*) which is calculated by the formula

$$HI = \sum_{i=1}^{N} HQ_i$$

when i- is a serial number of nanoscale factors taken into account.

The total hazard index (*THI*) is calculated when the complex flow take place of nanoscale chemicals into the human body from the environment by a number of ways, as well as in the multimedia and multipath impact as follows:

$$THI = \sum_{j=1}^{M} HI_j$$

when HI_j - hazard index for way or route of action number j from M possible.

If risk calculated according to the specified parameters (coefficient index or the total hazard index) is 0.1 or less, the significance is negligible of harmful effects in humans, hence this impact is characterized as acceptable and does not require any measures for its decline. If the calculated hazard index of NPs/NMs is between 0.1 and 1.0 it's possible that the risk of harmful effects in man throughout life must be regarded as significant taking into consideration the uncertainty in evaluating the exposure setting and the reference levels, and decision-making must provide a set of measures to reduce the impact of NPs/NMs on population/person . If the calculated ratio exceeds 1.0, the risk of harmful effects in humans is high (unacceptable), but it's apparently not possible to pinpoint the precise value of this due to the qualitative nature of this model.

Authors of paper [79] have shown with respect to the silver NPs migrating into the food from packaging materials, using the data of mass spectrometry confirmed by TEM, AFM and DLS, that exposure amount of consumer through packaged products didn't not exceed the upper allowable levels of Ag consumption as a chemical element in the conditions when the packaging material was used by proper destination (e.g. for package of products in accordance with the manufacturer's recommendations). It is essential to note in this regard, that the silver nanoparticles could be successfully detected in the food product by DLS in modern method modification [80].

The results of the risk assessment of the NPs and NM should be taken into account in the design of decision-making to reduce risk levels for the population by controlling the production, trafficking and use of nanotechnology products, which is the generally termed as "risk management".

6. Risk management

Management is possible of risks posed by the NMs and NPs presented in consumer products on the basis of the assessment of the risks and in the presence of the regulatory framework to control NT and NM on the state, municipal (local) and the corporate level.

Regulation of NT, which is now developing in most industrialized countries in the world, can be divided into the so-called "Horizontal" and "Vertical" regulation [81]. The subject of "Horizontal" regulation is formulated extremely broad and covers all kinds of products of certain functional purpose, for example, all food products, all cosmetics, medicines, etc., in the total amount of which NM and NT products occupy usually only a small section. In the case of "Horizontal" legislation specific properties of NPs and NMs are commonly not specified, and the rules are not stated applicable specifically to them. NM and nanotechnology products are covered by this legislation mainly because of their functional purpose to the extent that they perform the same tasks and have in some sense similar consumer properties as products not containing NM and produced by "traditional" technologies [82-84]. "Vertical" legislation is specifically designed for the purpose of NT and its products regulation and is applicable only in those industries. In view of this specificity is provided of the "Vertical" regulations, as opposed to the "horizontal", as a rule, by well-developed terminology and definitions, which allows one's to apply nanotechnology products with much greater uniqueness and fewer discrepancies than "Horizontal" legislation.

The analysis of legal acts and special literature has shown that currently "vertical" legislation is developed abroad in a very small extent in relation to NPs and NT [81,85,86]. Perhaps the only example, to fully meet these criteria, is Regulation EC No 1223/2009 on cosmetic products, in which extensive sections are, providing specific procedures governing placing on the market and labeling of this type of NT products [87].

In the EAEC explicitly formulated articles are defined in a number of Technical regulations of the Customs Union adopted during the 2011-2012 that apply to NM and NT products specified to their procedures-to-market (associated usually with the state registration of products). Thus, the elements of "Vertical" "nano" - legislation are already presented in these documents. As for lower level regulations, according to which the state registration and necessary expertise of products is held a whole system of recommendations and guidance documents was approved by Rospotrebnadzor that consider all aspects of control procedures implemented to nanotechnology products, including classification, examination procedures, state registration and supervision, quantitative analysis and valuation, toxicological and special studies, risk assessment. The system allows risk management of NMs and nanotechnology products at level of the most possible certainty at actual scientific knowledge in accordance with uniform, unified approach.

7. Conclusion

Genuine “flowering” of NT and NM application has been expected in the manufacturing of a great many consumer products, including food according to the forecasts for the period 2012-2014 which had been made at the beginning of the 2000s, mainly on the basis of trend analysis of scientific and patent information. However, studies have shown of commercialized range of consumer products, that there is the massive introduction of NT and NM predominantly in the manufacture of cosmetics, while their use in food products, packaging materials, household chemicals, etc. still includes a limited number of items and shows no significant upward trend. However, the problem is relevant of exposing population to priority NM and the associated risks due to high production volumes of some NMs and their entrance to the environment. Silver NPs and (by a wide margin) NPs of gold, platinum, and titanium dioxide have leading priority in this respect in accordance with the frequency of use in mass-produced consumer goods. Nanosized silica introduction is now apparently underestimated into food products as a food additive since the use of this NM is not declared by manufacturers of products and objective control of its content is difficult.

Analysis of literature data on toxicological properties of NMs shows that sufficient amount of information is currently accumulated enabling establishment of safe exposing doses of silver, titanium dioxide and gold NPs. In series of recent studies evidence have been provided on the effect of silica NPs oral intake on the state of the organism of laboratory animals, including the performance of the immune system. These results allow establishing a NOAEL for this practically important NM and move on to their hygienic regulation.

References

[1] Bouwmeester H, Brandhoff P, Marvin HJP, Weigel S, Peters RJB. 2014 *Trends in Food Science & Technology* **40**(2) 200,
[2] Popov KI, Filippov AN and Khurshudyan SA 2009 *Russian Chem. J.* **53** 86
[3] Borm PJ, Robbins D, Haubold S , Kuhlbusch T, Fissan H, Donaldson K, Schins R, Stone V, Kreyling W, Lademann J, Krutmann J, Warheit D and Oberdorster E 2006 *Part. Fibre Toxicol.* **3**, 11
[4] Oberdörster G, Maynard A, Donaldson K, Castranova V, Fitzpatrick J, Ausman K, Carter J, Karn B, Kreyling W, Lai D, Olin S, Monteiro-Riviere N, Warheit D and Yang H 2005 *Part. Fibre Toxicol.* **2** 8
[5] Oberdörster G, Oberdorster E. and Oberdorster J 2005 *Environ. Health Perspect.* **113** 823
[6] Gmoshinski IV, Smirnova VV and Khotimchenko SA 2010 *Nanotechnologies in Russia* **5** 6

[7] Henkler F, Tralau T, Tentschert J, Kneuer C, Haase A, Platzek T, Luch A and Gotz ME 2012 *Arch.Toxicol.* **86** 1641

[8] Risk Assessment of Products of Nanotechnologies Scientific Committee on Emerging and Newly Identified Health Risks (SCENIHR) 2009 *European Commission Health and Consumers DG.* Brusseles 71 p.

[9] Brief P 2009 Appropriate Risk Governance Strategies for Nanotechnology Applications in Food and Cosmetics. *Geneva: International Risk Governance Council* 44 P.

[10] Grobe A, Renn O and Jaeger A 2008 Risk Governance Strategies for Nanotechnology Applictions in Food and Cosmetics . *Geneva: International Risk Governance Council* 52 P

[11] Onishchenko GG and Tutrelyan VA 2007 *Voprosy pitaniya (Problems of nutrition)* **76**(6) 4

[12] Zaitseva NV, Trusov PV, Shur PZ, Kyryanov DA, Chigvintsev VM and Cinker MJu 2013 *Analyz riska zdorov'yu (Health risk analysis)* **1** 15.

[13] Yokell RA and MacPhail RC 2011 *J.Occupational Med. Toxicol.* **6** 74

[14] Stander L and Theodore L 2011 *Int. J. Environ. Res. Public Health.* **8** 470

[15] Maynard RL 2012 *Emerg. Health Threats J* **5** 10

[16] Silicon dioxide, amorphous 1973-1992 *Rome: JECFA.* 2 P. http://www.fao.org/food/food-safety-quality/scientific-advice/jecfa/jecfa-additives/en/

[17] OECD 2010 List of manufactured nanomaterials and list of endpoints for phase one of the sponsorship programme for the testing of manufactured nanomaterials: revision.-*Paris: OECD Environment Directorate* 13

[18] Cote I, Anastas PT, Birnbaum LS, Clark RM, Dix DJ, Edwards SW and Preuss PW 2012 *Environ. Health Perspect* **120** 1499

[19] Utembe W, Potgieter K, Byron SA and Mary G 2015 *Part. Fibre Toxicol.* **12** 11

[20] Oksel C, Ma CY and Wang XZ 2015 *Procedia Engineering* **102**, 1500

[21] Nanomaterials and REACH 2012 *Background Paper on the Position of German Competent Authorities* 52 P. http://www.bfr.bund.de/cm/349/ nanomaterials-and-reach.pdf

[22] Onishchenko GG, Tutrelyan VA, Gmoshinski IV and Khotimchenko SA 2013 *Gigiena & sanitariya (Hygiene and Sanitation)* **1** 4

[23] Draft scientific opinion. Guidance on risk assessment concerning potential risks arising from applications of nanoscience and nanotechnologies to food and feed 2011 *EFSA Scientific Committee.- Parma, Italy.: European Food Safety Authority (EFSA)* 32 P. http://www.efsa.europa.eu/en/consultationsclosed /call/scaf110114.pdf

[24] Oostingh GJ, Casals E, Italiani P, Colognato R, Stritzinger R, Ponti J, Pfaller T, Kohl Y, Ooms D, Favilli F, Leppens H, Lucchesi D, Rossi F, Nelissen I, Thielecke H, Puntes VF, Duschl A and Boraschi D 2011 *Part. Fibre Toxicol* **8** 18

[25] Chang X, Zhang Y, Tang M and Wang B 2013 *Nanoscale Res Lett.* **8**(1) 51

[26] Fröhlich E and Roblegg E 2012 *Toxicology* **291** 10

[27] Levard C, Hotze EM, Lowry GV, Gordon E and Brown GE 2012 *Environ Sci Technol.* **46** 6900

[28] Stensberg MC, Wei Q, McLamore ES, Porterfield DM, Wei A and Sepúlveda MS 2011 *Nanomedicine (London)* **6** 879

[29] Schluesener JK and Schluesener HJ 2013 *Arch Toxicol* **87** 569

[30] Gerber A, Bundschuh M, Klingelhofer D and Groneberg DA 2013 *J.Occupational Med. Toxicol.* **8** 32

[31] Khlebtsov N and Dykman L 2011 *Chem. Soc. Rev.* **40** 1647

[32] Shi H, Magaye R, Castranova V and Zhao J. 2013 *Part. Fibre Toxicol.* **10** 15

[33] Gaillet S, Rouanet J-M. 2015 *Food and Chemical Toxicology* **77** 58

[34] Thomassen LC, Aerts A, Rabolli V, Lison D, Gonzalez L, Kirsch-Volders M, Napierska D, Hoet PH, Kirschhock CE and Martens JA 2010 *Langmuir* **26** 328

[35] Nabeshi N., Yoshikawa T., Matsuyama K. Nakazato Y, Tochigi S, Kondoh S, Hirai T, Akase T, Nagano K, Abe Y, Yoshioka Y, Kamada H, Itoh N, Tsunoda S and Tsutsumi Y 2011 *Part. Fibre Toxicol.* **8** 1

[36] Eom HJ and Choi J 2009 *Toxicol. In Vitro* **23** 1326
[37] Eom HJ and Choi J 2011 *Environ. Health Toxicol* **26** e2011013
[38] Kasper J, Hermanns MI, Bantz C, Maskos M, Stauber R, Pohl C, Unger RE and Kirkpatrick JC 2011 *Part. Fibre Toxicol.* **8** 6
[39] Corbalan JJ, Medina C, Jacoby A, Malinski T and Radomski MW 2011 *Int. J. Nanomedicine* **6** 2821
[40] Shi J, Karlsson HL, Johansson K Gogvadze V, Xiao L, Li J, Burks T, Garcia-Bennett A, Uheida A, Muhammed M, Mathur S, Morgenstern R, Kagan VE and Fadeel B. 2012 *ACS Nano* **6** 631
[41] Napierska D, Thomassen LC, Rabolli V, Lison D, Gonzalez L, Kirsch-Volders M, Martens JA and Hoet PH 2009 *Small* **5** 846
[42] Ye Y, Liu J, Xu J, Sun L, Chen M and Lan M 2010 *Toxicol. In Vitro* **24** 751
[43] Sohaebuddin SK, Thevenot PT, Baker D, Eaton JW and Tang L 2010 *Part. Fibre Toxicol.* **7** 22
[44] Yang X, Liu J and He H 2010 *Part. Fibre Toxicol.* **7** 1
[45] Xue Y, Chen Q, Ding T and Sun J 2014 *Int. J. Nanomedicine* **9** 2891
[46] Wang W, Li Y, Liu X, Jin M, Du H, Liu Y, Huang P, Zhou X, Yuan L and Sun Z 2013 *Int. J. Nanomedicine* **8** 3533
[47] Duan J, Yu Yongbo, Yu Yang, Li Y, Huang P, Zhou X, Peng S and Sun Z 2014 *Part. Fibre Toxicol* **11** 50
[48] Fede C, Selvestrel F, Compagnin C, Mognato M, Mancin F, Reddi E and Celotti L 2012 *Anal. Bioanal. Chem.* **404** 1789
[49] Napierska D, Thomassen LCJ, Lison D, Martens JA and Hoet PH 2010 *Part Fibre Toxicol.* **7** 39
[50] Yu T, Hubbard D, Ray A and Ghandehari H 2012 *J. Control Release* **163** 46
[51] Van der Zande M , Vandebriel RJ , Groot MJ , Kramer E , Rivera ZEH, Rasmussen K, Ossenkoppele JS, Tromp P, Gremmer ER, Peters RJB, Hendriksen PJ, Marvin HJP, Hoogenboom RLAP, Peijnenburg AAM and Bouwmeester H 2014 *Part. Fibre Toxicol.* **11** 8
[52] Park EJ and Park K 2009 *Toxicol. Lett.* **184** 18
[53] Nishimori H, Kondoh M, Isoda K, Tsunoda S, Tsutsumi Y and Yagi K 2009 *Eur. J. Pharm. Biopharm* **72** 496
[54] Cho M, Cho WS, Choi M, Kim SJ, Han BS, Kim SH, Kim HO, Sheen YY and Jeong J 2009 *Toxicol. Lett.* **189** 177
[55] Sayes CM, Reed KL, Glover KP , Swain KA, Ostraat ML, Donner EM and Warheit DB 2010 *Inhal Toxicol.* **22** 348
[56] Yoshida T, Yoshioka Y, Fujimura M Yamashita K, Higashisaka K, Morishita Y, Kayamuro H, Nabeshi H, Nagano K, Abe Y, Kamada H, Tsunoda S, Itoh N, Yoshikawa T and Tsutsumi Y. 2011 *Nanoscale Res. Lett.* **6** 192
[57] Han B, Guo J, Abrahaley T Qin L, Wang L, Zheng Y, Li B, Liu D, Yao H, Yang J, Li C, Xi Z and Yang X 2011 *PLoS One* **6** e17236
[58] Lee S, Kim M-S, Lee D, Kwon TK, Khang D, Yun H-S and Kim S-H 2013 *Int. J. Nanomedicine* **8** 147
[59] Kolling A, Ernst H, Rittinghausen S, Heinrich U 2011 *Inhal. Toxicol.* **23** 544
[60] Tananova ON, Arianova EA, Gmoshinski IV, Toropygin IYu, Kryapova EV, Trusov NV, Khotimchenko SA and Tutelyan VA 2014 *Biochemistry (Moscow) Suppl.Ser.B: Biomedical Chem.* **8** 125
[61] Zaitseva NV, Zemlyanova MA, Zvezdin VN and Saenko EV 2013 *Analyz riska zdorov'yu (Health risk analysis)* **1** 65
[62] Zaitseva NV, Zemlyanova MA, Lebedinskaya OV, Zvezdin VN, Melekhin SV and Akaf'yeva TI 2013 *Morfologya (Morphology)* **144**(5) 78
[63] Shumakova AA, Arianova EA, Shipelin VA, Sidorova JuS, Selifanov AV, Trushina EN, Mustafina OK, Safenkova IV, Gmoshinski IV and Tutelyan VA 2014 *Voprosy pitaniya (Problems of nutrition)* **83**(3) 52

[64] Shumakova AA, Avren'eva LI, Guseva GV, Trusov NV, Maltsev GYu, Soto SH, Vorozhko IV, Solomatina VI, Gmoshinski IV, Khotimchenko SA and Tuteluan VA 2014 *Voprosy pitaniya (Problems of nutrition)* **83**(4) 58
[65] Zaitseva NV, Zemlyanova MA, Zvezdin VN, Dovbish AA, Gmoshinski IV, Khotimchenko SA, Safenkova IV and Akafyeva TI 2014 *Voprosy pitaniya (Problems of nutrition)* **83**(2) 42
[66] Gmoshinsky IV, Khotimchenko SA, Trushina EN and Khanferyan RA 2014 Toxicological assessment of silica nanoparticles in subacute experiment in rats *7 International Nanotoxicology Congress (Nanotox 2014) Antalya-Turkey, April 23-26, 2014. Program & abstracts* 160
[67] http://www.abercade.ru/research/analysis/67.html
[68] Hansen SF, Michelson ES, Kamper A, Borling P, Stuer-Lauridsen F and Baun A 2008 *Ecotoxicology* **17** 438
[69] Keller AA, Vosti W, Wang H, Lazareva A. 2014 *J. Nanoparticle Res.* **16**, 1
[70] Christensen FM, Johnston HL, Stone V, Aitken RJ, Hankin S, Peters S and Aschberger K 2010 *Nanotoxicology* **4** 284
[71] Vejerano EP, Leon EC, Holder AL and Marr LC 2014 *Environ. Sci.: Nano* **1** 133
[72] Fabrega J, Luoma SN, Tyler CR, Galloway TS and Lead JR 2011 *Environ. Int.* **37** 517
[73] Shatkin JA 2008 Nanotechnology: health and environmental risks. *Boca Raton, London, NY: CRC Press* 167 P.
[74] Linkov I, Steevens J, Adlakha-Hutcheon G., Bennett E, Chappell M, Colvin VJ, Davis M, Davis T, Elder A, Hansen SF, Hakkinen PB, Hussain SM, Karkan D, Korenstein R, Lynch I, Metcalfe C, Ramadan AB and Satterstrom K 2009 J.Nanoparticle Res. 11 513
[75] Tolaymat T, El Badawy A, Sequeira R, and Genaidy A. 2015 Journal of Hazardous Materials 298, 270
[76] Kazak AA, Stepanov EG, Gmoshinski IV and Khotimchenko SA 2012 Voprosy pitaniya (Problems of nutrition) 81(4) 11
[77] Sotiriou GA, Singh D, Zhang F, Wohlleben W, Chalbot M-CG, Kavouras IG, Demokritou P. 2015 Environmental Science: Nano 2(3) 262
[78] Risk assessment of the impact of nanomaterials and nanoparticles on the human body 2011 *Guidelines MR 1.2.0038-11* Moscow: Rospotrebnadzor 98 P.
[79] Smirnova VV, Kracnoyarova OV, Pridvorova SM, Zherdev AV, Gmoshinski IV, Kazydub GV, Popov KI and Khotimchenko SA 2012 *Voprosy pitaniya (Problems of nutrition)* **81**(2) 40
[80] Rykhik OV., Panferov VG, Kotova NN and Popov KI 2013 *Khranenie i pererabotka selkhozsyrya* **6** 43
[81] Cushen M, Kerry J, Morris M, Cruz-Romero M and Cummins E 2012 *Trends Food Sci. Technol.* **24** 30
[82] Directive 2011/65/EU of the European Parliament and of the Council of 8 June 2011 on the restriction of the use of certain hazardous substances in electrical and electronic equipment (recast). http://eur-lex.europa.eu/LexUriServ/LexUriServ.do?uri=OJ:L:2011:174:0088:0110:en:PDF.
[83] Falkner R and Jaspers N 2012 *Global Environmental Politics* **12** 30
[84] Malloy TF Soft Law and Nanotechnology. A Functional Perspective. Essay. http://www.nanolawreport.com/tags/soft-law-and-nanotechnology-a/#axzz 24N3I7Zjo.
[85] Classification, labeling and packaging of nanomaterials in REACH and CLP. Annex II: Final version. Doc. CA/90/2009 Rev2. 2009. *European Commission, Environment Directorate-General Water, Chemical and Biotechnology, Brussels, 3 December* 6 P http://ec.europa.eu/environment/ chemicals/reach/pdf/classif_nano.pdf.
[86] Preparing for our future: Developing a common strategy for key enabling technologies in the EU. 2009 *Communication from the Commission to the European Parliament, the Council, the European Economic and Social Committee and the Committee of the regions.- Brussels:*

IOP Conf. Series: Materials Science and Engineering **98** (2015) 012009 doi:10.1088/1757-899X/98/1/012009

Commission of the European Communities 12 P. http://ec.europa.eu/enterprise/ sectors/ict/files communication_key_enabling_technologies_en.pdf.
[87] Cosmetics companies to face new rules in Europe. 2009 *International Herald Tribune*, 24 March, http://iht.com/bin/printfriendly.php?id=21022812

Abbreviations list:

ADME - absorption, distribution, metabolism and excretion (research); AFM - atomic force microscopy; BET - Brunauer–Emmett–Teller (method); CRPA- conditioned reflex of passive avoidance; DLS - dynamic light scattering; HI - hazard index; HQ - hazard quotinent; ICP-MS - mass-spectrometry with inductively coupled plasma; ION- Institute of nutrition (Moscow); LOAEL - the lowest observable adverse effect dose (threshold dose); NM(s) - nanomaterial(s); NOAEL - not observable adverse effect dose (safe dose); NP(s) -nanoparticle(s); NT - nanotechnology; TEM - transmission electron microscopy; THI - total hazard index;

Risk assessment of silver nanoparticles

V A Shipelin[1], I V Gmoshinski[1], S A Khotimchenko[1]
[1]Federal State Budgetary Research Institution «Institute of Nutrition», 2/14, Ust'inskiy proezd, Moscow, 109240 Russia

E-mail: v.shipelin@ya.ru

Abstract. Nanoparticles of metallic silver (Ag) are among the most widely used products of nanotechnology. Nanosized colloidal silver (NCS) is presented in many kinds of production as solutions of particles with diameter less than 100 nm. NCS is used in a variety of fields, including food supplements, medicines, cosmetics, packaging materials, disinfectants, water filters, and many others. Problems of toxicity and related safety of NCS for humans and environmental systems are recently overestimated basing on data of numerous toxicological studies *in vitro* and *in vivo*. The article discusses the results of current studies in recent years and the data of author's own experiments on studying the safety of NCS, that allows to move on to risk assessment of this nanomaterial presented in consumer products and environmental samples.

1. Introduction

Among the most widely used nanotechnology objects there are metallic silver nanoparticles (Ag NPs) with dimension less than 100 nm that are manufactured in the form of colloidal solutions of nano-sized colloidal silver (NCS). Despite the fact that the history of colloidal Ag production dates back to antiquity, it was in the past 2 decades, that the use of this nanomaterial as part of consumer products became massive, which was in most cases justified by the presence of potent biocidal properties in Ag NPs against pathogenic bacteria, fungi and even some viruses [1].

Biocidal effect of Ag NPs by themselves is likely to be negligible [2] according to carefully conducted recently control experiments, however, the ability of this NP to release high concentrations of Ag + ions in oxidative condition gives them in practice pronounced antimicrobial properties including that in respect of microorganisms having resistance to the vast majority of modern antibiotics. [3] The notion has been overestimated that the NCS is a hazard to humans and higher animals based on the recent works on numerous toxicological data obtained in studies *in vitro* and *in vivo* [4]. In addition, the growing concern is caused by the possible consequences of entering Ag NPs into the environment as a part of various waste products. In some foreign countries NCS is currently considered as a pesticide that is a subject of the regulation applicable to the highly dangerous chemicals [5]. In Russia, despite a significant increase in production volumes and the use of the NCS, its regulation is missing, which makes actual the task of the risk assessment of this priority nanomaterial.

2. Hazard identification

NCS is commercially available in several forms, differing in size and shape of NPs, and the composition of their surface coatings (functionalization). The most common are, firstly, the so-called "Classical" NCS stabilized by adsorbed citrate ions, second, preparation obtained by biochemical

Content from this work may be used under the terms of the Creative Commons Attribution 3.0 licence. Any further distribution of this work must maintain attribution to the author(s) and the title of the work, journal citation and DOI.
Published under licence by IOP Publishing Ltd

synthesis in the reverse micelle system comprising sodium dioctyl sulfosuccinate (food additive E480), thirdly, so-called "Cluster" NCS produced by photocatalytic method in the presence of polyvinylpyrrolidone (PVP, food additive E1201) and, fourthly, containing no stabilizers "ion-free" colloidal Ag, formed under the action of high-energy dispergation in distilled water (under the influence of the laser pulse on metal silver target or electrical explosion of silver wire) [6]. When assessing the risks of all these forms of NCS, it is important to keep in mind that their actions in relation to toxic doses of biological objects can vary significantly. Thus it is the "cluster" NCS stabilized with PVP that is apparently one of the most popular forms of this product, because of the relative ease of preparation technology, stability and biocompatibility of the stabilizer used.

Applications of NCS include food supplements, medicines, bandage, packaging materials, cosmetic products, disinfectants, paints and varnishes, textiles, water filters [7-9]. There are registered in the Russian Federation more than 120 types of consumer products containing the Ag NPs as of 2015/ Cosmetic products predominate among them (70%). In addition to the direct impact in the composition of products, the Ag NPs may expose the population through various environmental objects, such as water, air, soil, agricultural raw materials.

3. Hazard characterization

NCS has toxic effects on the eukaryotic cells in culture [10-13], water and soil organisms [4,13], in laboratory animals at inhalation [14,15] epicutaneous [16] and oral [17-20] administration. Mechanism of Ag nano-toxicity, according to modern concepts, lies in Ag NPs ability to penetrate through biological barriers [21], accumulate in cells and release silver ions (Ag^+) in very high local concentrations under the influence of oxidizing agents [22] including endogenous. This mode of action has been called in the literature as a "Trojan horse effect" [23]. Ag^+ is a potent cellular poison due to its ability to inhibit enzymes and membrane transporters containing thiol groups in irreversible manner [13].

Our analysis of numerous literature data on the cytotoxic effect of the Ag NPs on models of cell cultures *in vitro* showed that the threshold concentration of NPs, corresponding to the minimum observable effect on cells is in any case not less than 3 μg/g of the incubation medium. The question remains open whether such a concentration of NPs can arise in the tissue during the natural ways of exposure (inhalation, oral ingestion or skin resorption). In several studies it has been shown by means of mass spectrometry [21] radioisotope methods [24], transmission electron microscopy [25,26] that the Ag NPs were bioavailable when entering through the gastrointestinal tract, capable to penetrate through the intestinal wall, reach blood circulation and accumulate in internals, primarily in the liver and spleen. Application of methods of radioactive tracers and neutron activation analysis allowed to quantify the transfer of orally administered Ag NPs from pregnant and nursing rats to their offspring through the fetoplacental barrier and via breast milk [27], as well through the blood-brain barrier into the brain [28].

Mathematical modeling of Ag NPs administered to the gastrointestinal tract bioaccumulation and biodistribution was performed in [29] using kinetic equations of 1-st order for describing of interorgan exchange of this NPs. In this way it was shown that potentially dangerous levels of Ag NPs can be achieved in critical organs (liver, spleen) after acute or subacute oral administration of these NPs in a daily dose of at least 5-10 mg / kg body weight.

Direct estimations of toxic doses of Ag NP in experiments on mammals are inconsistent, which may be due to differences in particle size and functionalization, insufficient duration of the experiment and a limited set of biomarkers of toxicity. In a series of papers by single or relatively short repeating (28 days) oral administration of various kinds of surface functionalized Ag NP adverse effects were detected only in extremely high doses ranging from 125 mg/kg body weight; critical organ, apparently, was the liver of animals. In particular, the analysis of the morphological changes of the liver tissue and the biliary system of mice after exposure to Ag NP showed vacuolization, focal necrosis of liver, bile duct hyperplasia, increased cell infiltration, inflammation, central veins extension [19].

Ag NP caused marked increase in expression of IL-1, IL-6, IL-4, IL-10, IL-12 and TGF-β, together with signs of histopathological effects on kidney when administered to mice intragastrically in a dose of 1 mg/kg body weight for 14 days[20]. In research [17] it was shown that the 28-day administration of Ag NP to growing male rats, aged about one month at the beginning of the experiment, resulted in decreased fasting glucose, increased gut macromolecular permeability and suppression of gut symbiotic micro flora. In NCS doses of 250 mg / kg body weight and more no significant signs have been detected of reproductive toxicity [18]. Unlike this in other studies, unfunctionalized Ag NPs, orally administered to rats for 6 months, expressed marked cumulation and had a threshold dose for systemic effect much smaller than 0.1 mg / kg body weight per day [30].

So named "cluster" NCS presenting the most common and practically important form of such nanomaterial in Russia has been studied in joint research of FSBI "Institute of Nutrition", FSBI "FNTS health-care technology risk management to public health" of Rospotrebnadzor and A.N.Bach Institute of biochemistry. NCS preparation stabilized with PVP was introduced into the gastrointestinal tract of laboratory animals (rats and mice) in 92-day lasting experiment using a wide variety of integral, functional, morphological, biochemical, cytological (ex vivo) and proteomic indicators. The aim of this current research is to evaluate the safe doses of Ag NPs, which will further create a framework for their hygienic rationing in production.

4. Exposure assessment, risk characterization

Accurate assessment of human exposure to Ag NPs is currently hindered by the lack of reliable statistical information on the volume of production and use of this nanomaterial, as well as reliable exposure scenarios.

According to [31] the annual quantity of NCS production in the world totaled more than 500 tones in terms of Ag in 2011 whereas in 2015 it could exceed 1,000 tones, which corresponded to about 140 mg of silver for each inhabitant of the Earth for one year. It is important to bear in mind that the distribution of Ag NPs in the consumer's production as well as in biosphere can be extremely uneven. It is assumed that the general population is exposed to Ag NPs mainly by the oral and epicutaneous route when consuming food supplements, cosmetics and also by NPs migrating into food from packaging materials [32], while for the employees of enterprises producing products with NCS apparently predominates inhaled route of exposure [33].

The assumption did not receive experimental confirmation that large amounts of Ag NPs could enter atmosphere in the form of aerosols during disposal of wastes containing NCS textiles, medical supplies and other products at waste incineration plants. In fact, a major amount of these NPs accumulated in the ash and in some extent in sludge formed in wet cleaning systems of flue gases [34]. Subsequently this NCS can be transferred to the field in the composition of surrogate fertilizers, rinsed in water and accumulate in different soil and water organisms [31], and, ultimately, come to human through the food chain as part of agricultural production. However, data is currently insufficient to quantify the exposure proceeds in such scenarios.

Ag NPs migrating into the food from packaging materials was studied in [35] using data of mass spectrometry confirmed by electron and atomic force microscopy and it was shown that the exposure value of consumer via packaged products does not exceed current upper allowable level of Ag consumption as a chemical element if using the packaging material for intended purposes (i.e. for packing products in accordance with the manufacturer's recommendations).

Apart this question is raised of risk assessment of NCS for the environment, related to the impact of said nanomaterial on aquatic and soil micro- and macro-organisms - normal components of ecosystems. To solve all these problems, further accumulation and systematization of data is necessary on the toxic effects of Ag NPs at the level of organism and population that enables refinement of their safe doses.

5. Conclusion

Thus, the analysis of nanotechnology products assortment presented in the turnover in the Russian Federation indicates that various forms of nano-sized colloidal silver are the most widely used in consumer products far superior in this regard to all other types of mineral engineered nanomaterials. This makes of paramount importance the assessment of risks, produced by Ag NPs for health of consumers and workers employed in nanotechnology facilities, as well as for the environment.

Analysis of the literature indicates that currently nanoscale silver is one of the most widely studied nanomaterials from the point of view of its toxic effect on biological objects including laboratory animals *in vivo*. However, the available results of toxicological studies are largely inconsistent, which may be due either to differences in the samples of used NPs (with different particle size and shape, functionalization of the surface), and lack of a unified methodology in the planning the biological experiment. Upon completion of the ongoing 3-month study on the effect of Ag NP stabilized with polyvinylpyrrolidone on laboratory animals using a combination of laboratory tests included in official guidelines approved for the safety assessment of nanomaterials in Russia, it will be possible to establish safe levels of exposure with this widely manufactured and used form of nano-sized silver in Russia. In the future, it will go to its hygienic rationing in products and environmental objects.

References

[1] EPA document 2012 *Nanomaterial Case Study: Nanoscale Silver in Disinfectant Spray. EPA/600/R–10/081F* 423
[2] Xiu ZM, Zhang QB, Puppala HL, Colvin VL, Alvarez JJ 2012 *Nano Lett.* **12**(8) 4271
[3] Sheehy K, Casey A, Murphy A, Chambers G 2015 *J Colloid Interface Sci* **443**(1) 56
[4] Stensberg MC, Wei Q, McLamore ES, Porterfield DM, Wei A, Sepúlveda MS 2011 *Nanomedicine* **6** 879.
[5] Faunce T, Watal A 2010 *Nanomedicine* **5**(4) 617
[6] Krutyakov Yu A, Kudrinskiy A A, Olenin A Yu, Lisichkin G V 2008 *Russ Chem Rev* **77**(3) 233
[7] Blaser S A, Scheringer M, MacLeod M, Hungerbühler K 2008 *Sci Total Environ* **390** 396
[8] Marambio-Jones C, Hoek E M 2010 *J. Nanopart. Res.* **12**(5) 1531–51
[9] Savage N, Diallo M S 2005 *J Nanopart Res* **7** 331
[10] Gliga AR, Skoglund S, Wallinder IO, Fadeel B, Karlsson HL 2014 *Particle and Fibre Toxicol* **11** 11
[11] Hussain SM, Hess KL, Gearhart JM, Geiss KT, Schlager JJ 2005 *Toxicol. in vitro* **19** 975
[12] Powers CM, Badireddy AR, Ryde IT, Seidler FJ, Slotkinet TA 2011 *Env. Health Perspectives* **119** 37
[13] Lapresta-Fernández A, Fernández A, Blasco J 2012 *Trends Anal. Chem.* **32** 40
[14] Sung JH, Ji JH, Park JD, Yoon JU, Kim DS, Jeon KS, Song MY, Jeong J, Han BS, Han JH, Chung YH, Chang HK, Lee JH, Cho MH, Kelman BJ, Yu IJ 2009 *Toxicol Sci* **108** 452
[15] Sung JH, Ji JH, Yoon JU, Kim DS, Song MY, Jeong J, Han BS, Han JH, Chung YH, Kim J, Kim TS, Chang HK, Lee EJ, Lee JH, YuIJ 2008 *Inhal. Toxicol* **20** 567
[16] Korani M, Rezayat S M, Gilani K, Arbabi Bidgoli S, Adeli S 2011 *Int. J. Nanomed.* **6** 855
[17] Shumakova AA, Smirnova VV, Tananova ON, Trushina EN, Kravchenko LV, Aksenov IV, Selifanov AV, Soto SKh, Kuznetsova GG, Bulakhov AV, Safenkova IV, Gmoshinsky IV, Khotimchenko SA 2011 *Voprosy pitaniya (Problems of nutrition)* **80**(6) 9
[18] Hong JS, Kim S, Lee SH, Jo E, Lee B, Yoon J, Eom IC, Kim HM, Kim P, Choi K, Lee MY, Seo YR, Kim Y, Lee Y, Choi J, Park K 2014 *Nanotoxicology* **8** 349
[19] Kim YS, Kim JS, Cho HS, Rha DS, Kim JM, Park JD, Choi BS, Lim R, Chang HK, Chung YH, Kwon IH, Jeong J, Han BS, Yu IJ 2008 *Inhal. Toxicol.* **20**(6) 575–83
[20] Park EJ, Bae E, Yi J, Kim Y, Choi K, Lee SH, Yoon J, Lee BC, Park K 2010 *Environ Toxicol Pharmacol* **30** 162
[21] Van der Zande M, Vandebriel RJ, Doren EV, Kramer E, Rivera ZH, Serrano-Rojero CS, Gremmer ER, Mast J, Peters RJB, Hollman PCH, Hendriksen PJM, Marvin HJP, Peijnen-

IOP Conf. Series: Materials Science and Engineering **98** (2015) 012010 doi:10.1088/1757-899X/98/1/012010

burg AACM, Bouwmeester Y 2012 *ACS Nano* **6** 7427
[22] Lubick N 2008 *Environ. Sci. Technol* **42** 8617
[23] Choi O, Deng KK, Kim NJ, Ross L, Surampalli RY, Hu ZQ 2008 *Water Res.* **42** 3066
[24] Buzulukov JuP, Gmoshinski IV, Raspopov RV, Demin VF., Soloviev VJu, Kuzmin PG, П.Г., Shafeev GA, Khotimchenko SA 2012 *Meditsinskaya radiologiya I radiatsionnaya bezopasnost' (Medical radiology and radiation safety)* **57**(3) 5
[25] Platonova TA, Pridvorova SM, Zhewrdev AV, Vasilevskaya LS, Arianova EA, Gmoshinski IV, Khotimchenko SA, Dzantiev BB, Popov VO, Tutelyan 2013 *Bulletin of experimental biology and medicine* **155** 204
[26] Gmoshinski IV, Khotimchenko SA , Popov VO , Dzantiev BB, Zherdev AV, Demin VF, Buzulukov YuP 2013 *Russian Chemical Reviews* **82**(1) 48
[27] Melnik EA, Buzulukov YuP, Demin VF,Demin VA, Gmoshinski IV,Tyshko NV,Tutelyan VA 2013 *Acta Naturae* **5**(3) 25
[28] Buzulukov YuP, Arianova EA, Demin VF, Safenkova IV, Gmoshinski IV, Tutelyan VA. 2014 *Biology Bulletin.***41** 255.
[29] Demin VA, Gmoshinsky IV, Demin VF, Anciferova AA, Buzulukov YuP, Khotimchenko SA, Tutelyan VA 2015 *Nanotechnologies in Russia* **10**(3–4) 288
[30] Khodykina NV., Gorshenin AV., Klauchek VV., Pocheptsov AYa., Sroslov MS, Tochilkina LP., Shalagina TA., Filatov BN 2014 *Nanotoxicology Achievements, Problems and Prospects Volgograd, September 2014* 65.
[31] Fabrega J, Luoma S N, Tyler C R et al. 2011 *Environ. Int.* **37**(2) 517–31.
[32] Hansen S F, Michelson E S, Kamper A et al. 2008 *Ecotoxicology* **17**(5) 438–47
[33] Christensen F M, Johnston H L, Stone V et al. 2010 *Nanotoxicology* **4**(3) 284–95
[34] Vejerano E P, Leon E C, Holder A L, Marr L C 2014 *Environ. Sci.: Nano.* **1**(2) 133–143
[35] Smirnova VV, Kracnoyarova OV, Pridvorova SM, Zherdev AV, Gmoshinski IV, Kazydub GV, Popov KI and Khotimchenko SA 2012 *Voprosy pitaniya (Problems of nutrition)* **81**(2) 40

IOP Conf. Series: Materials Science and Engineering **98** (2015) 012011 doi:10.1088/1757-899X/98/1/012011

Application of Traditional and Nanostructure Materials for Medical Electron Beams Collimation: Numerical Simulation

I A Miloichikova[1], S G Stuchebrov, G K Zhaksybayeva and A R Wagner
Tomsk Polytechnic University, Tomsk, Russia

E-mail: miloichikova@gmail.com

Abstract. Nowadays, the commercial application of the electron accelerators grows in the industry, in the research investigations, in the medical diagnosis and treatment. In this regard, the electron beam profile modification in accordance with specific purposes is an actual task. In this paper the model of the TPU microtron extracted electron beam developed in the program "Computer Laboratory (PCLab)" is described. The internal beam divergence influence for the electron beam profile and depth dose distribution in the air is considered. The possibility of using the nanostructure materials for the electron beam formation was analyzed. The simulation data of the electron beam shape collimated by different materials (lead, corund-zirconia nanoceramic, gypsum) are shown. The collimator material influence for the electron beam profile and shape are analyzed.

1. Introduction

At the present time particle accelerators and X-ray sources have a wide range of application [1–4]. One of the most promising radiations is the electron beams with different energies, which are widely used in such clinical areas as medical diagnosis, external-beam radiotherapy and intraoperative radiotherapy [5]. The clinical electron beam application needs to have an exact representation of the beam profile and shape and be able to manage the parameters in accordance with specific purposes. As it is known the numerical simulation allows estimating the electron beam parameters, making it faster and easier to obtain beam characteristics than the practical measurements. In this regard, the electron beam model development is an topical issue.

As it is known the nanostructure materials can be used in cancer treatment. Nanotechnology research in this aria ranges from the diagnostics and therapeutics with the use of nanoparticles to the production of the dendrimers for boron neutron capture therapy [6–7]. One of the perspectives is to use the nanostructured materials for the electron beam formation.

In this research the suitability of the industrial nanoceramics was analyzed for these purposes. Up to date, submicrocrystalline nanoceramics based on zirconia and alumina compositions are intensively studied because of their service properties [8–9]. In this research the corund-zirconia nanoceramic was analyzed as a collimation material.

Nowadays, there are a lot of foundations of the new collimation materials for the electron beam. One of the perspectives is to use the 3D-printer materials for the accelerator beam collimator production. The 3D printed structures are useful in the different medical and industrial areas, because of the possibility to change the material property for the task-specific [10–12]. In this paper the gypsum composite was chosen from a wide range of 3D printing materials, since it is widely distributed and is relatively cheap raw material.

Content from this work may be used under the terms of the Creative Commons Attribution 3.0 licence. Any further distribution of this work must maintain attribution to the author(s) and the title of the work, journal citation and DOI.
Published under licence by IOP Publishing Ltd

Within the framework of this investigation the theoretical analysis of the Tomsk Polytechnic University (TPU) microtron extracted electron beam was carried out. The models of the accelerator electron beam shape modulation weredeveloped in the program "Computer Laboratory (PCLab)". The following factors, affecting of the electron beam profile are analyzed: the internal beam divergence, beam collimation by different materials (lead, corund-zirconia nanoceramic, gypsum).

2. Materials and methods

2.1. Collimation materials
We analyzed the corund-zirconia nanoceramic for the electron beam formation. The lead was used as the classical material for the electron beam collimator production. As the cheaper alternative the gypsum was selected.

2.2. Emitting source
The following parameters of the TPU microtron extracted electron beam were used as the emitting source: electron energy – 6.1 MeV; beam size at output≈ 2.0 mm2; beam divergence – 0.1 rad.

2.3. Simulation program
The program "Computer laboratory (PCLab)" version 9.5 was used for the TPU microtron extracted beam model creation. Simulation is carried out by applying the Monte Carlo method. The software package allows calculating the propagation process of electrons, positrons, protons and photons in matter with specified characteristics [13].

2.4. Experiment geometry.
The normal plane disc (diameter – 2.0 mm) monoenergetic electron source with energy of 6.1 MeV, corresponding to the actual TPU microtron beam, was used in the simulation. The source was located in front of the beryllium output window (thickness – 50 μm; diameter – 40 mm). The beam shape analysis was carried out in the air.

In the simulation with a collimated electron beam in the first instance the output window was overlapped by the plates from different materials (collimator channel length – 5 mm and 10 mm) with a taper hole (taper diameter increased from 0.5 mm to 1.5 mm). The collimator materials are lead, corund-zirconia nanoceramic and gypsum.

The figure 1 (a, b) illustrates the calculated path particles in the geometries with noncollimated and collimated electron beams, in view of photon production, correspondingly.

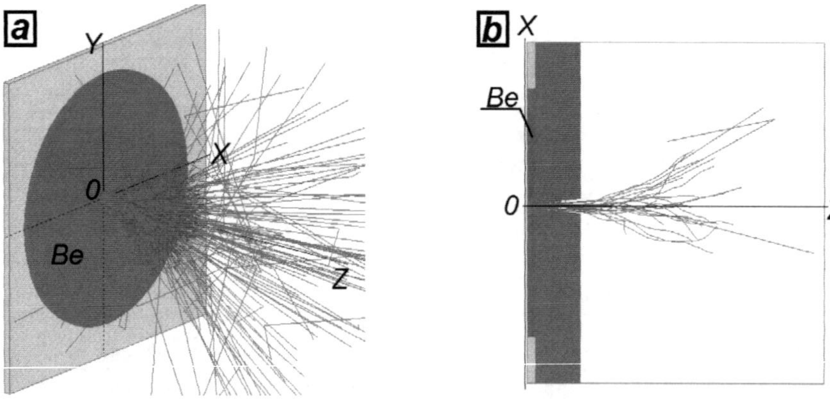

Figure 1. The calculated path particles: a – noncollimated electron beam; b – collimated electron beam; blue lines – electron path; red lines – photon path.

3. Results and discussions

In the figure 2 the simulation data of the TPU microtron extracted electron beam profile and shape at the 2 cm distance from the output window ignoring the internal beam divergence and taking into account the internal beam divergence are shown. The dose results were averaged and normalized to the maximum simulation dose.

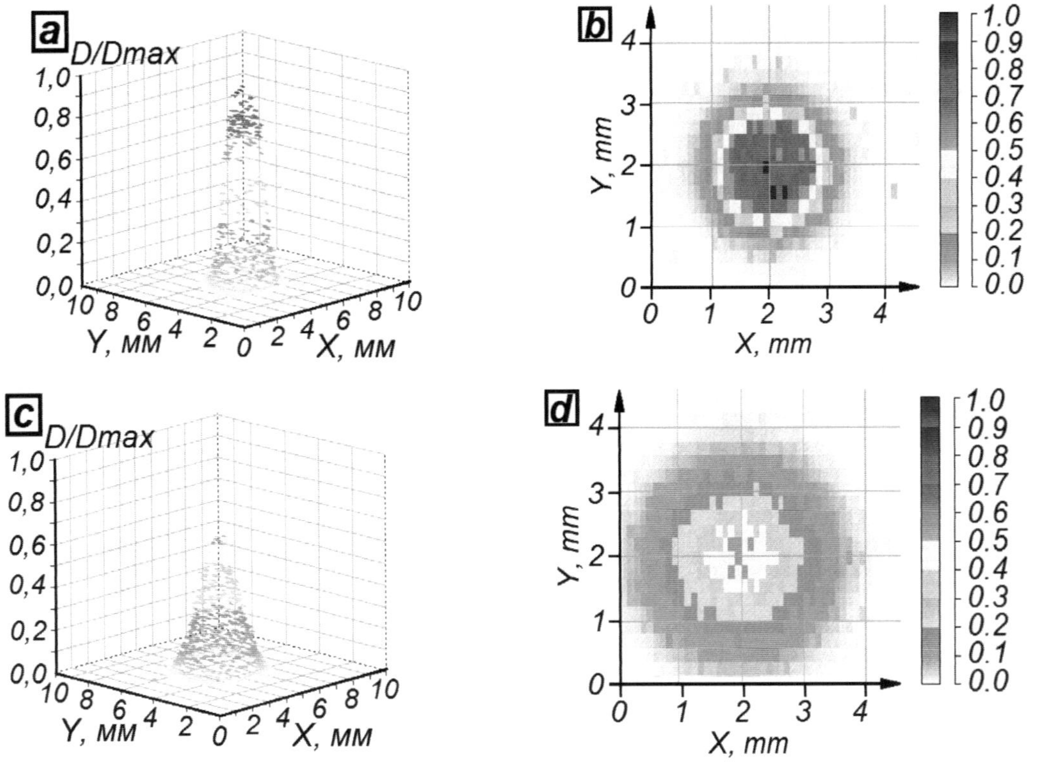

Figure 2. The TPU microtron extracted electron beam profile and shape at the 2 cm distance from the output window: a, b – ignoring the internal beam divergence; c, d –taking into account the internal beam divergence.

The figure 2 illustrates that increasing the internal beam divergence of the electron in the accelerator, the radiation dissipates faster then monodirectional beam. As a result it can be observed the dramatic drop of the dose and the beam broadening.

The figure 3 presents the simulation data of the TPU microtron extracted electron beam depth dose distribution in the air taking into account and ignoring the internal beam divergence.

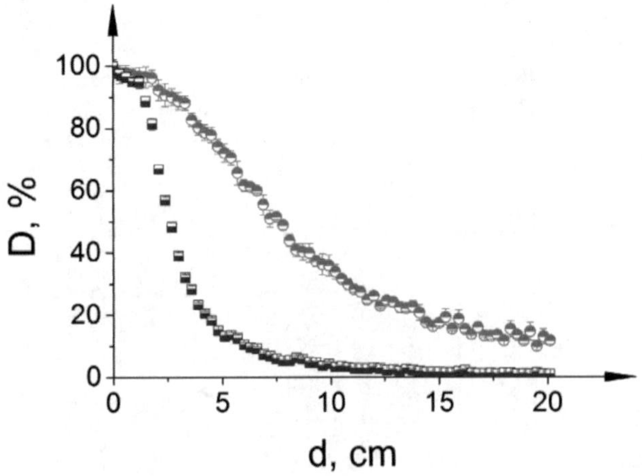

Figure 3. The TPU microtron extracted electron beam depth dose distribution in the air: ◓ – ignoring the internal beam divergence;
⬓ –taking into account the internal beam divergence.

The figures 2, 3 show that accounting the internal beam divergence of the electron in the accelerator, which is typical for real machines, significantly affects for the calculations results.

In the figure 4 (a, b) the simulation data of the TPU microtron collimated extracted electron beam shape with the lead collimator channel length equal to 5 mm and 10 mm at the collimator output with taking into account the internal beam divergence are shown, correspondingly.

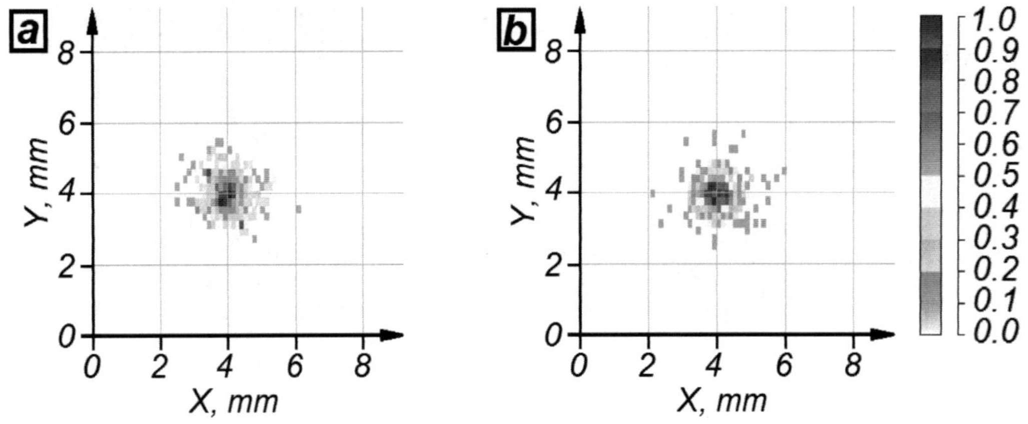

Figure 4. The TPU microtron collimated extracted electron beam shape at the lead collimator output: a – collimator channel length equal to 5 mm;
b – collimator channel length equal to 10 mm.

The figure 4 illustrates that within the collimation window the dose distribution doesn't change significantly, but with the increasing lead collimation channel length the scattered radiation contribution is greatly reduced because of the large absorption.

In the figure 5 (a, b) the simulation data of the TPU microtron collimated extracted electron beam shape with the corund-zirconia nanoceramic collimator channel length equal to 5 mm and 10 mm at the collimator output taking into account the internal beam divergence are shown, correspondingly.

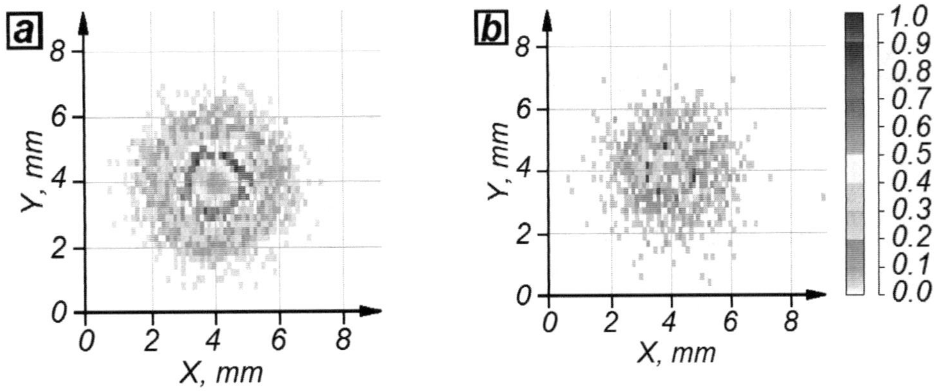

Figure 5. The TPU microtron collimated extracted electron beam shape at the corund-zirconia nanoceramic collimator output: a – collimator channel length equal to 5 mm; b – collimator channel length equal to 10 mm.

The simulation model (figure 5) shows that the dose maximum is observed in the region of the scattered and direct radiation contribution. The dose reduction is observed in a region of the only direct radiation. The dose decreases with the increasing of the deflection angle in the field of the scattered radiation. With the corund-zirconia nanoceramic collimator channel length equal to 5 mm the dose gradient is less than 10 cm because of the range increasing the scattered radiation in the collimator material.

In the figure 6 (a, b) the simulation data of the TPU microtron collimated extracted electron beam shape with the gypsum collimator channel length equal to 5 mm and 10 mm at the collimator output taking into account the internal beam divergence are shown, correspondingly.

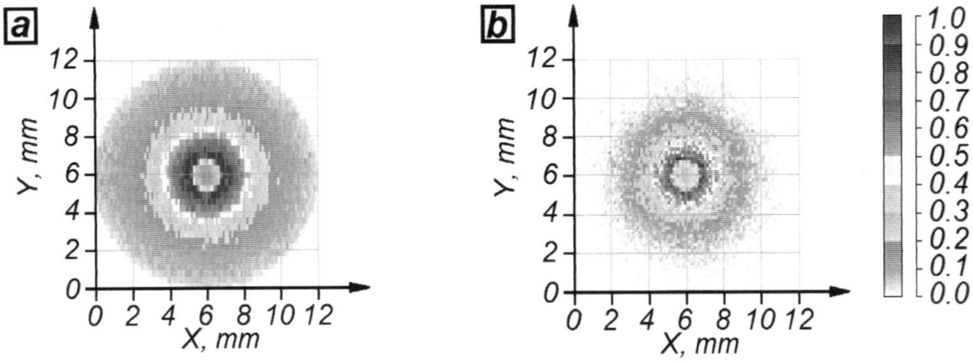

Figure 6. The TPU microtron collimated extracted electron beam shape at the gypsum collimator output: a – collimator channel length equal to 5 mm; b – collimator channel length equal to 10 mm.

The figure 6 illustrates that the shape of the electron beam at the gypsum collimator output determined by the same parameters as for corund-zirconia nanoceramic collimator (figure 5). However, the dimensions of the scattered radiation area are comparatively more due to the interaction nature of the collimator material with the electron and photon radiation.

4. Conclusion

In this paper the theoretical model of the TPU microtron extracted electron beam was calculated in the simulation program "Computer laboratory (PCLab)". The obtained results show the suitability of this program for the real electron beams analyzing and for the beam shape modulation using the different type of the collimation devices, for example this program can be used in betatron for the radiation treatment.

The obtained results show that the corund-zirconia nanoceramic exhibit more efficiency than the gypsum and can be used for the electron beam formation instead of the traditional material after the collimator-geometry optimization.

The calculation data allow estimating the electron beam size and dose distribution at the selected distance from the output window. The depth dose distribution allows estimating the radiation burden values in the electron beam propagation direction. The obtained results allow simulating the collimators for the electron beam parameters optimization necessary for the specific practical task. The next step of this research is an experimental evaluation of the obtained simulation data.

Acknowledgements

This work was financially supported by the Ministry of Education and Science of the Russian Federation in part of the science program (N 5.1485.2015).

References

[1] Gogolev A, Stuchebrov S, Wagner A, Cherepennikov Yu and Potylitsyn A 2012 *J. Phys.: Conf. Ser.* **357** Article number 012031

[2] Cherepennikov Yu and Gogolev A 2012 *7th International Forum on Strategic Technology (IFOST – 2012)* **2** pp 244-248

[3] Information on http://www-pub.iaea.org/MTCD/publications/PDF/P1433_CD/datasets/papers/ ap_ia-12.pdf

[4] Cherepennikov Yu and Gogolev A 2014 *J. Phys.: Conf. Ser.* **517** Article number 012037

[5] Choynzonov E L and Musabaeva L I 2006 The intraoperative electron and gamma-ray teletherapy of the malignant neoplasms (Tomsk: NTL) 216 p [in Russian]

[6] De la Iglesia D, Cachau R E, Garc´ıa-Remesal M and Maojo V 2013 *Comput. Sci. Disc.* **6** Article number 014011

[7] Information on http://lamp.tu-graz.ac.at/~hadley/nanoscience/references/RIVM.pdf

[8] Mel'nikov A G, Savchenko N L, Sablina T Ju and Kul'kov S N 2006 RU 2 286 316 [in Russian]

[9] Zhao Z M, Zhang L, Song Y G, Wang W G and Liu H B 2009 *J. Phys.: Conf. Ser.* **152** 012085

[10] Christ S, Schnabel M, Vorndran E, Groll J and Gbureck U 2015 *Materials Letters* **139** pp 165

[11] Information on http://arxiv.org/ftp/arxiv/papers/1503/1503.02785.pdf

[12] Information on http://www.ptcommunity.com/system/files/pdf/ptj3910704.pdf

[13] Bespalov V I 2015 Computer laboratory (Version 9.5) (TPU: Tomsk) 115 p [in Russian]

Research of nickel nanoparticles toxicity with use of Aquatic Organisms

T Morgaleva, Yu Morgalev, I Gosteva, S Morgalev
Centre "Biotest-Nano", Tomsk State University, 36, Lenina ave., Tomsk, 634050, Russia

E-mail: tg.morgaleva@gmail.com

Abstract. The effect of nanoparticles with the particle size Δ_{50}=5 nm on the test function of aquatic organisms was analyzed by means of biotesting methods with the use of a complex of test-organisms representing general trophic levels. The dependence of an infusoria *Paramecium caudatum* chemoattractant-elicited response, unicellular algae *Chlorella vulgaris Beijer* growth rate, *Daphnia magna Straus* mortality and trophic activity and *Danio rerio* fish kill due to nNi disperse system concentration, is estimated. It is determined that the release of chlorella into cultivated environment including nNi as a feed for daphnias raises the death rate of entomostracans. The minimal concentration, whereby an organism response to the effect of nNi is registered, depends on the type of test organism and the analysed test function. $L(E)C_{20}$ is determined for all the organisms used in bioassays. $L(E)C_{50}$ is estimated for *Paramecium caudatum* ($L(E)C_{50}$ = 0.0049 mg/l), for *Chlorella vulgaris Beijer* ($L(E)C_{50}$ = 0.529 mg/l), for *Daphnia m. S* ($L(E)C_{50}$ > 100 mg/l) and for fish *Danio rerio* ($L(E)C_{50}$ > 100 mg/l). According to the Globally Harmonized System hazard substance evaluation criteria and Commission Directive 93/67/EEC, nNi belongs to the "acute toxicity 1" category of toxic substances.

1. Introduction

Currently, metallic engineered nanoparticles (NPs) are prioritized nanomaterials (NM) for investigations. Particularly important is the studying of nickel nanoparticles (nNi) that have widespread application in modern industry and among other things serve as a catalyst, owing to the number of unique physical and chemical properties (high surface energy, low melting point, high specific surface area, magnetism and others). Since nNi is a promising NM, including for the purposes of platina replacing in different catalytic processes, and in the nearest time it may account for a substantial part of NPs produced by manufacture, there emerges the important task of analyzing its environmental safety. The necessity of this task is approved by normative documents both in Russia and the OECD [1, 2]. The information available in the literature concerning this issue is extremely limited, fragmented and controversial. In the last decade, particular attention has been paid to the studying of genotoxic and cancerogenic properties of nNi for cells of homoithermic animals [3, 4]. There are data showing an impact of nNi on protein turnover of mice, including albumin fraction [5]. The authors suppose that albumins take part in nNi transportation.

At the same time, the issue of nNi safety, when it inflows into aqueous environments, practically remains unresolved. An exception is that of research by Griffitt et al. [6]. The authors determined the toxicity of nNi with a particle size of 6 nm with regard to maxillopods *Daphnia pulex*, *Ceriodaphnia*

Content from this work may be used under the terms of the Creative Commons Attribution 3.0 licence. Any further distribution of this work must maintain attribution to the author(s) and the title of the work, journal citation and DOI.
Published under licence by IOP Publishing Ltd

dubia and algae *Pseudokirchneriella subcapitata.* There are data on the toxicological effect of nNiO on growth and morphological changes of *Chlorella vulgaris Beijer*. The authors stated that nNiO triggers cells' plasmolysis and chlorella's cellular membrane destruction [7]. In a chronic experiment on *Danio rerio* full-grown fish, Kovrižnych determined the high toxicity and nNiO accumulation in fish [8].

To a greater extent, the topicality in solving the issues concerning the analyzing of the impact of nNi on the test functions of aquatic organisms results from the absence of a unified system of a safety evaluation of NM.

We developed an algorithm of evaluation of ecological toxicity (as a result of the accomplishment of grants of the Federal Target Program in 2008-2013) that is oriented mainly to simultaneously studying the test reactions of an assembly of organisms from different systematic groups depending on the possible distribution sphere, application and utilization of nanoproducts.

2. Materials and methods

The nNi disperse system (DS) with an initial concentration of 50 mg/l became available by means of a laser ablation method [9]. According to the data of a BET-analysis and transmission electronic microscopy the average particle size Δ_{50} = 5 nm.

The impact of nNi on the test functions of aquatic organisms was studied on the basis of a chemoattractant-elicited response of single-celled animals *Paramecium caudatum* [10], on the speed of growth of unicellular alga *Chlorella vulgaris Beijer* [11], on the mortality rate of entomostracans *Daphnia magna Straus* [12], on *Daphnia magna Straus* trophic activity rate [10] and on the *Danio rerio* fish kill rate [13]. We were creating DS of nNi according to the technique we developed [14]. The nNi DS toxicity was studied on the concentration line of 0.00001, 0.0001, 0.001, 0.01, 0.1, 1.0, 5.0 and 10.0 mg/l. For the redispersion, DS of nNi was ultrasonicated (30 Wt/l) for 5 minutes immediately before biotesting. The evaluation of the toxicity of the analysed DS of nNi is implemented in accordance with the criteria according to certified methodologies [11, 12].

The research design is similar to the described in [15, 16].

Statistical processing of the acquired data was carried out with the help of probit analysis. The values of $L(E)C_{10}$, as upper limits of NOEC, $L(E)C_{20}$ и $L(E)C_{50}$, were calculated. Work was conducted with the application of metrologically gauged instruments: Autolumat LB953 (Germany), «Shimadzu» RF-5301 PC spectrofluorimeter (Japan), IPS-03 photoelectric colorimeter (Russia), Biotester-2 concentration meter (Russia) and diffractometer «Shimadzu» XRD 6000 (Japan).

3. Results

As a result of undertaken studies, we have determined, that cultivated environments with a content of nNi (5 nm) in 0.00001, 0.0001, 0.001, 0.01 and 0.1 mg/l concentrations resulted in changes of test functions in just two organisms: an inhibition of the chemoattractant-elicited response of *Paramecium caudatum* and an inhibition of the growth rate of *Chlorella vulgaris Beijer.* Deviations of the functions from a control did not exceed permissible values. The toxicity index varied from I=11.6±1.8 % to I=34.8±1.5 % for *Paramecium caudatum* and from I= -2.2±0.06 % to I=7.4 ±1.44 % for *Chlorella vulgaris Beijer.* In the concentrations interval from 0.00001 mg/l to 0.1 mg/l of DS of nNi, the death rate of entomostracans *Daphnia magna Straus* and fish *Danio rerio* was not noted. The trophic activity of *Daphnia magna Straus,* when cultivating entomostracans in DS of nNi with concentrations varying from 0.00001 mg/l to 1.0 mg/l, did not practically differ from the activity of a validation batch (figure 1).

The further increase of a concentration of nNi particles exacerbated the toxic effect of DS. The exposition of *Paramecium caudatum* in DS of nNi with a concentration of 1.0 mg/l during 30 minutes led to the complete inhibition of the chemotaxis of infusoria. The value of the toxicity index comprised 100 % that indicated the high toxicity of the analyzed test system (figure 2).

DS of nNi had a less pronounced toxic effect on the cells growth of *Chlorella vulgaris Beijer* with C=1.0 mg/l. After 22 hours of incubation in DS with the noted concentration, the optical density of chlorella decreased in 1.7 times with regard to the control. Moreover, the toxicity index slightly exceeded the permissible level (I=34.5±2.26 %). The exposition of *Daphnia magna Straus* and *Danio rerio* in DS of nNi with the concentration of 1.0 mg/l did not result in mortality of entomostracans and fish kill.

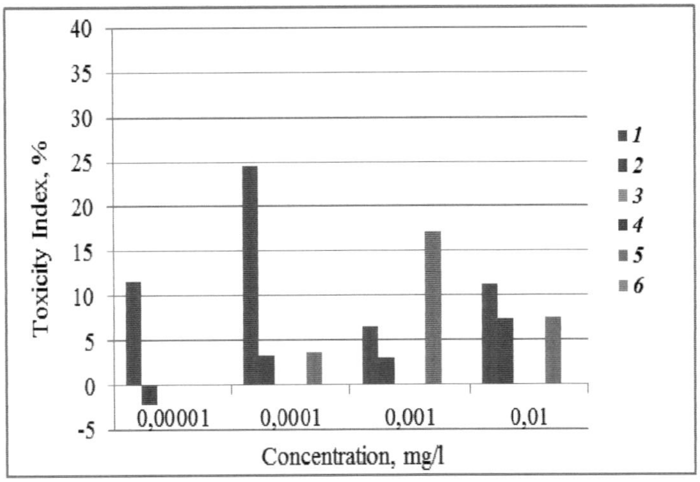

Figure 1. Toxicity Index of DS of nNi according to test functions of organisms of different systematic groups (1- chemotaxis of *Paramecium caudatum*, 2- weight gain of *Chlorella vulgaris Beijer*, 3- death rate of *Daphnia magna Straus*; 4- death rate of *Daphnia magna Straus* when inserting of chlorella as a feed; 5-trophic activity of *Daphnia magna Straus, 6* – death of *Danio rerio*)

When increasing the concentration of particles in the cultivated environment, nNi had an acute effect on all the test organisms except for fish. With nNi concentrations of 5.0 mg/l and 10.0 mg/l for *Paramecium caudatum* and *Chlorella vulgaris Beijer,* the toxicity of DS of nNi corresponded to the gradation of “High”. The exposition of daphnids in DS of nNi with the noted concentrations resulted in the death rate of entomostracans equal to 40.0±4.1 % and 43.3±6.2 % within 48 hours, and the decreasing of the trophic activity to 77.7±3.5 % and 66.9±6.2 %.

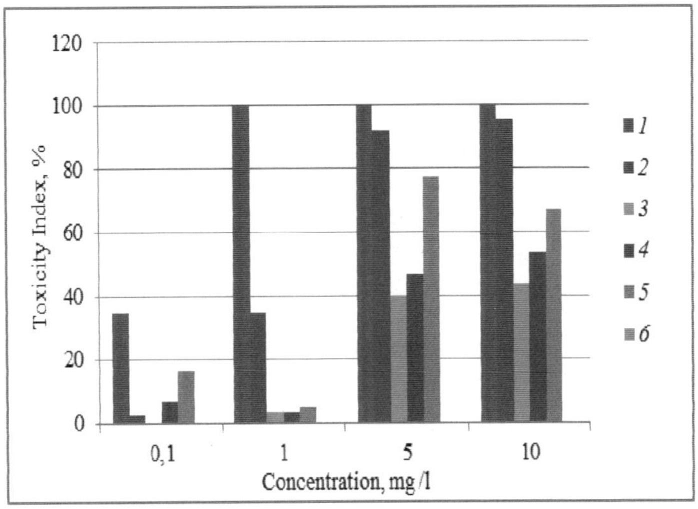

Figure 2. Toxicity index of DS of nNi according to test functions of organisms of different systematic groups (1- chemotaxis *of Paramecium caudatum*, 2- weight gain of *Chlorella vulgaris Beijer*, 3- death rate of *Daphnia magna Straus*; 4- death rate of *Daphnia magna Straus* when inserting of chlorella as a feed; 5- trophic activity of *Daphnia magna Straus, 6* - death of *Danio rerio*)

A growth of the toxic effect of DS of nNi (on the “death rate” indicator) in the series of experiments on analyzing trophic activity of entomostracans in the cultivated environment with nNi is noted. Moreover, the toxicity index increases on 6.7±1.2 % for C=5.0 mg/l and on 10.0±0.7 % for C=10.0 mg/l (figure 1). The growth of toxicity is probably connected with the effect of nNi on entomostracans by means of two ways: immediately from the environment and with contaminated algae. Insertion of chlorella in the cultivated environment as a feed for *Daphnia magna Straus* is an essential condition for the methodology of conducting experiments on analyzing of an effect of nNi on the trophic activity of *Daphnia magna Straus*. According to the data obtained by scientists of our laboratory during parallel trials (unpublished data), the concentration of nNi connected with chlorella's cells exceeded control values by more than 10000 times even after 24 hours of exposition. The ability of fresh-water algae to accumulate a significant amount of NPs and to serve as a source of their transition to consumers through the food chain was noted by the number of authors on the example of Au, TiO_2 and quantum dots [17 - 21].

The data we acquired concerning the toxicity of nNi with a particle size of 5 nm is coherent with the results of single experimental works in the literature. Griffitt et al. [6] determined the toxicity of nNi with regard to entomostracans, unicellular algae and fish. However, the figures of $L(E)C_{50}$ (3.89 mg/l, 0.35 mg/l and >10 mg/l correspondingly) given by the authors differ from the data we obtained (table 1). One can suppose that the noted differences for entomostracans and algae are connected with the application by Griffitt et al. of other representatives of systematic groups: *Daphnia pulex*, *Ceriodaphnia dubia* and *Pseudokirchneriella subcapitata*.

The data that we obtained can be compared with the results of Gong et al. and Kovrižnych et al. [7, 8], who analyzed the effect of nNiO on the test functions of aquatic organisms. The structural analysis of nNi that we tested showed that during ablation in water, nNi is acidized. Moreover, the particles with a diameter of <8 nm fully consisted of nickel oxide, and NPs with a diameter of >8 nm, a central metal core with an oxide shell is observed [22]. The same results were obtained by Sakiyama et al. [23]. During the process of Ni NPs fusion by means of the laser ablation method in water, the authors obtained shell particles with a diameter of 5÷20 nm with a core consisting of metallic Ni and an oxide layer 2nm thick.

According to Gong et al., $L(E)C_{50}$ comprised 32.28 mg/l for the first 72 hours of incubation. Moreover, the authors pointed out that nNiO triggered cells' plasmolysis and the destruction of cell membranes of *Chlorella vulgaris Beijer*. [7]. In Kovrižnych etbal. [8], the toxicity of nNiO was analyzed on full-grown fish *Danio rerio*. The effect of nNiO on *Danio rerio* over 30 days resulted in the accumulation of NPs in fishes' organisms, and as a consequence, it resulted in the high toxicity of nNiO. The value of $L(E)C_{50}$ for the 30 days of an the incubation of *Danio rerio* in a static regime comprised 45.0 mg/l, and $L(E)C_{100}$ (the minimum concentration which resulted in a 100 % death rate of fish) was 100.0 mg/l, and $L(E)C_0$ (the maximum concentration which did not resulted in deaths of fish) was equal to 6.25 mg/l for the full-grown adult of *Danio rerio*.

The studies we carried out allow us to define "limiting links" of sustainability of an ecosystem to the contamination of the nNi environment. The level of effects of NPs depends on the type of test organism and concentration of NPs. It was found that the vulnerability of separate links of the trophic level decreases in a range of *Paramecium caudatum, Chlorella vulgaris Beijer, Daphnia magna Straus* and *Danio rerio*. The ecotoxiological parameters of DS of nNi are determined for each test organism (table 1): ($L(E)C_{10}$), ($L(E)C_{20}$ and $L(E)C_{50}$.

Biological effects began to appear with nNi concentrations for protozoans *Paramecium caudatum* ($L(E)C_{20}$=0.0047 mg/l), for unicellular algae *Chlorella vulgaris Beijer* ($L(E)C_{20}$=0.067 mg/l), for entomostracans *Daphnia magna Straus* ($L(E)C_{20}$=0.237 mg/l – according to the trophic activity and $L(E)C_{20}$=15.21 mg/l – according to the death rate). As for *Danio rerio,* nNi is not active in terms of biological activity in the analyzed concentrations.

On the basis of the data obtained as a result of ecotoxicological analysis on test systems with an enhanced range of test functions, an integral estimation of the toxicity of nNi with the partical size of 5 nm was conducted.

The ecotoxicological evaluation of the obtained results of the sample toxicity was conducted taking into account the principle of the most expressed reaction. This principle was presented in normative documents [24]. It was specified that the tested sample of nNi with the particle size of 5 nm refers to dangerous substances with a high toxicity degree of "acute toxicity 1".

Table 1. The values of the ecotoxicity of DS of nNi with a size of NPs of 5 nm

Environ-ment	Test organism	Test reaction	$L(E)C_{50,}$ (mg/l)	$L(E)C_{20,}$ (mg/l)	$L(E)C_{10,}$ (mg/l)	Acute toxicity class
DS nNi	*Paramecium caudatum*	Chemoattractant-elicited	0.0049	0.00047	0.00014	1
DS nNi	*Chlorella vulgaris Beijer*	Weight gain	0.529	0.067	0.0033	1
DS nNi	*Daphnia magna Straus*	Death rate	>100	15.21	1.877	None
DS Ni + *Chlorella*	*Daphnia magna Straus*	Death rate	88.170	2.494	0.390	3
DS nNi	*Daphnia magna Straus*	Trophic activity	40.984	0.237	0.016	3
DS nNi	*Danio Rerio*	Fish kill	>100	>100	>100	None
Conclusion			Acute toxicity class 1			

4. Conclusion

It was specified that nNi (5 nm), that was obtained with the method of laser ablation, is a dangerous substance with a high toxicity degree of "acute toxicity 1".

It was shown that nNi has a selective toxicity and is toxic not for all of the analyzed test organisms. It was specified that the permissible/safe level of tested NPs does not exceed a concentration of 1.0 mg/l. The increase in the concentration of nNi leads to significant toxic damages of test organisms (except for fish). With concentrations of DS of nNi 5.0 mg/l and 10.0 mg/l, significant disorders of organisms' functions are noted. These disorders were found among representatives of phytoplancton (decrease of mass growth of unicellular algae) and zooplankton (a negative chemotaxis of infusorias, an inhibition of the trophic activity and increasing of the death rate of entomostracans). Phyto- and zooplanctonic community of hydrosphere is at risk. That leads to disorders of trophic and metabolic interrelations, injures to the integrity of natural biocoenosis and ability in terms of rehabilitation.

5. Acknowledgments

This study (research grant No 8.2.57.2015) was supported by Tomsk State University Academic D.I. Mendeleev Fund Program in 2015.
Work was conducted with the application of the Tomsk regional common use centre technical equipment acquired thanks to a grant of the Russian Ministry of the Agreement No.14.594.21.0001 (RFMEFI59414X0001).

References

[1] The concept of toxicological studies, risk assessment methodology, identification and quantification of nanomaterials 2007 Approved by the Resolution of the Chief State Inspector of the Russian Federation dated October 31 **79**

[2] OECD 2007 Current Developments in Delegations And Other International Organisations on The Safety of Manufactured Nanomaterials Tour de Table 3rd Meeting of the Working Party on Manufactured Nanomaterials
[3] Magaye R, Zhao J, Bowman L and Ding M 2012 *Experimental and Therapeutic Medicine* **4** 551-61
[4] Ahamed M, Alhadlaq H 2014 *OncoTargets and Therapy* **7** 269–280
[5] Atlaskirova A, Baradulina E, Borodulin Ya 2014 *Bulletin of Medical Internet conferences* **4** (5) 596-600
[6] Griffitt R, Luo J, Gao J, Bonzongo J and Barber D 2008 *Environmental Toxicology and Chemistry* **27** (9) 1972-78
[7] Gong N, Shao K, Feng W, Lin Z, Liang C, Sun Y 2011 *Chemosphere* **8** (4) 510-16
[8] Kovrižnych J, Sotnikova R, Zeljenkova D, Rollerova E, Szabova E 2014 *Interdiscip Toxicology* **7** (1) 23-6
[9] Svetlichnyi V, Lapin I 2013 *Russian Physics Journal* **56** (5) 581-87
[10] Morgalev Yu, Khoch N, Morgaleva T, Dunaevsky G and Morgalev S 2010 *Methodological Guide* Tomsk
[11] Morgalev Yu, Morgaleva T and Grigoriev Yu 2010 *FR.1.39. 2010.09103* (Moskow:Federal Reestr) p 45
[12] Morgalev Yu, Morgaleva T and Grigoriev Yu 2010 *FR.1.39.2010.09102* (Moskow:Federal Register) p 45
[13] The safety assessment of nanomaterials in vitro and in model systems in vivo: Guidelines 2009 Federal Center of Hygiene and Epidemiology
[14] Morgalev S, Morgaleva T, Morgalev Yu, Gosteva I 2015 *Journal Advanced Materials Research* **1085** 424-30
[15] Morgalev Yu, Morgaleva T, Gosteva I, Morgalev S, Kulizhskiy S, Astafurova T 2015 submitted to *Journal Advanced Materials Research*
[16] Gosteva I, Morgalev Yu, Morgaleva T, Morgalev S 2015 submitted to *Journal Advanced Materials Research*
[17] Morgalev Yu, Khoch N, Morgaleva T, Gulik E, Borilo G, Bulatova U, Morgalev S and Ponyavina E 2010 *Nanotechnologies in Russia* **5** 11-12
[18] Osborne-Koch M 2009 *Environmental Toxicology* 44 p
[19] Renault S, Baudrimont M, Mesmer-Dudons N, Gonzalez Pm, Brisson A 2008 *Gold Bulletin* **41** 116-126
[20] Bouldin J, Ingle T, Sengupta A, Alexander R, Hannigan R, Buchanan R 2008 *Environmental Toxicology and Chemistry* **27** (9) 1958–63
[21] Fabrega J, Luoma S, Tyler Ch, Galloway T, Lead J *Environment International* **37** 517–531
[22] Mahfouz R, Cadete Santos Aires F, Brenier A, Jacquier B, Bertolini J *Applied Surface Science* **254** 5181–90
[23] Sakiyama K, Koga K, Seto T, Hirasawa M, Orii T 2004 *J. Phys. Chem. B.* **108** (2) 523-29
[24] Globally Harmonized System of Classification and Labelling of Chemicals (GHS) The third revised edition 2009 UN

Basalt fiber manufacturing technology and the possibility of its use in dentistry

E Karavaeva[1], A Rogozhnikov[1], V Nikitin[2], Yu Cherepennikov[3] and A Lysakov[3]

[1] Perm State Medical University named after Academician E.A. Wagner, 26 Peter and Paul Street, Perm, 614990, Russia
[2] Perm National Research Polytechnic University, 29 Komsomolskiy prospect, Perm, 614990, Russia
[3] National Research Tomsk Polytechnic University, 30 Lenin Avenue, Tomsk, 634050, Russia

E-mail: sortior@mail.ru

Abstract. The article touches upon the technology of basalt fiber manufacturing and prospects of its use in dental practice. Two kinds of construction using basalt fiber have been proposed. The first one is a splinting construction for mobile teeth and the second one is the reinforced base for removable plate-denture. The work presents the results of the investigation of physical and mechanical properties of the constructions based on basalt fiber. It also describes the aspects of biomechanical modeling of such constructions in the ANSYS software package. The results of the investigation have proved that applying constructions using basalt fiber is highly promising for prosthetic dentistry practice.

1. Introduction

Basalt is magmatic rocks, highly resistant according to their chemical and mineral composition. Its geological reserves are practically unlimited in the world and range from 25 to 38% of the total area occupied by all magmatic rocks. The reserves of basalt are considered to be inexhaustible as it has been estimated that they are refilled at the rate of 1 million m^3 per year due to volcanic activity [1]. Basalt is used as a raw material for the production of road metal, cast stone material, powdering material and basalt fiber, which, in turn, is used in building, power industry, mechanical engineering and aircraft as well as chemical, nuclear and other industries. Such a wide use of basalt and basalt fiber can be explained by its low cost and good physical and mechanical properties, resistance to high temperatures, acids and alkali. Taking into account the relative inactivity of basalt fiber, it could be used in medicine. However, the analysis of scientific literature has shown no evidence of its use in medical science. In dental practice we can suggest two possibilities of basalt fiber use: periodontal splinting and reinforcing removable laminar dentures.

2. Technology of basalt fiber manufacturing

The basalt fiber production process begins with fragmentation of coarse basalt aggregate into finer grains sized 5-12 mm. Dimensions of the grains can be controlled using different methods [2, 3]. After selecting metallic inclusions by magnetic separation, basalt raw material in the form of riprap stones or chips is griddled, washed and dried under the conditions of natural airing or in a special drier. After that, the basalt chips are put in a reservoir installed above a melting furnace. Melting furnaces may be

Content from this work may be used under the terms of the Creative Commons Attribution 3.0 licence. Any further distribution of this work must maintain attribution to the author(s) and the title of the work, journal citation and DOI.
Published under licence by IOP Publishing Ltd

electrical, gas or equipped with fuel oil burners. In a melting furnace, basalt raw material is liquefied at the temperature of (1500 ± 50) °C and, when homogeneous, enters a feeder to be then formed into complex continuous basalt fibers. The melted material from a furnace gravitates into the feeder with platinum drain valves located at the bottom. Through those, the melted material flows to a platinum-rhodium electric-traced stream feeder assembly, where a strand of elementary continuous fibers is formed by extension and reeled to a pulling mechanism. It applies an oil agent and after that all fibers are gathered together into one complex thread or fiber. A complex fiber passes through a special thread-distributing apparatus and is rolled on a removable bobbin, which is in due time removed and replaced by a new one. A winding apparatus forms continuous basalt fiber consisting of 314 elementary basalt threads. Full bobbins are held under normal conditions for one day to be then taken for rewinding to receive a roving with the necessary number of compositions.

3. Basalt fiber splint construction

Splinting is constriction of teeth with a thin thread to decrease and prevent further increase in their mobility. This method is one of the ways to increase the effectiveness of periodontal disease treatment. According to the data of the World Health Organization, the incidence of periodontal diseases amounts to 80-100% [4]. Splinting is necessary for 40-50% of such patients. There are many materials for splinting but all of them have some flaws such as: separation of fibers while cutting and modeling, poor adaptation to teeth, difficulty of use while working [5, 6]. We suggest using a basalt fiber splint, which has no such disadvantages [7]. To prove the rationality of such a splint in dental practice, we carried out a biomechanical modeling of splint construction made of basalt fiber. Using the ANSYS software package we counted the intensity of resultant Von Mises stress in the lower jaw, teeth and basalt fiber. The mechanical properties of teeth, lower jaw bone tissue and basalt fiber were taken from the literature and are presented in Table 1. According to the results of the biomechanical modeling, the maximal intensity of stress occurring in case of basalt fiber splinting reaches 6.763×10^7 Pa (Fig. 1) which is well within the strength limit. That is why we can recommend basalt fiber as a periodontal splinting material.

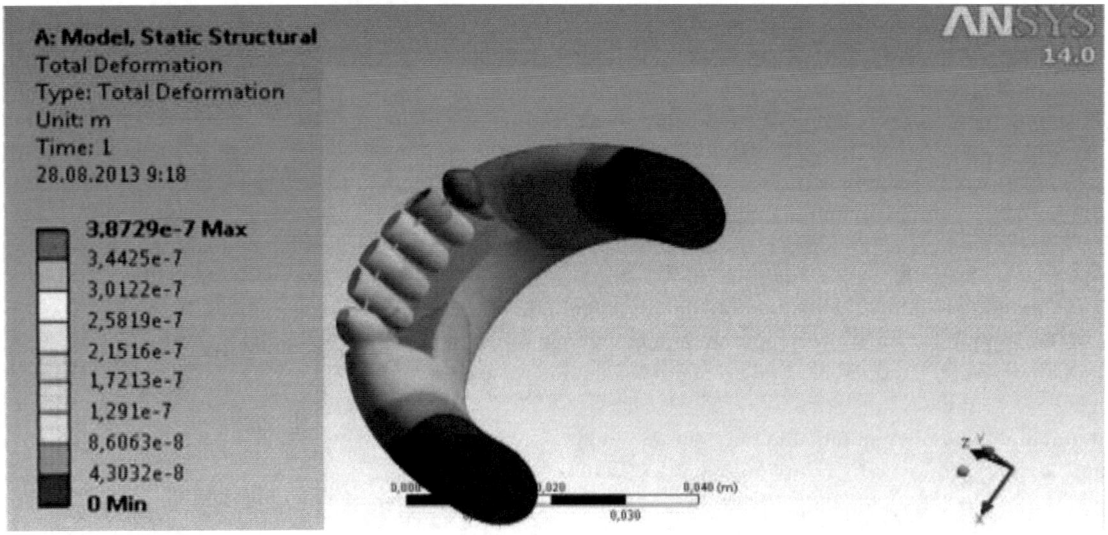

Figure 1. Intensity of resultant Von Mises stress in the lower jaw, teeth and basalt fiber

Table 1. Mechanical properties of teeth, lower jaw bone tissue and basalt

№	Parameter	Material		
		Teeth	Lower jaw bone tissue	Basalt
1.	Young modulus E, GPa	$1.8 \cdot 10^4$	10	87.1±2.3
2.	Poisson ratio ν	0.3	0.3	0.4
3.	Limit of strength σst, MPa	11.5	100	2245.3±126.8

4. Reinforcing removable laminar dentures

The need for orthopedic treatment with removable dentures is very high and amounts to 33-58% according to V.N. Kopeikin and co-authors (1994). Due to a marked demographic shift towards population ageing observed in recent years, it has become more urgent reaching 65+% [8]. Nowadays 90% of dentures are made of acrylic plastics because of their low price and easy production. However, these plastics have some disadvantages, one of which is considered to be their insufficient mechanical resistance. This is proved by prosthetic basis fractures ranging from 3 to 15% of breakages during the 1st year of use or 19% during the whole 3-year period of dispensary observation [9]. In such a case prosthetic basis fractures cause severe discomfort and additional material expenditures to a patient and bad reputation to a doctor. The existing method of reinforcing with gold-foil is not only expensive but also does not allow doctors to avoid primary fractures. We suggest using the method of acrylic plastic reinforcing with basalt fiber at the stage of removable laminar denture production. This will make it possible to avoid denture fractures during the entire period of their use. To prove the rationality of using this method, we conducted physical-mechanical investigations according to all-Union State Standard 51889-2002 "Polymeric materials for tooth denture basis. Technical demands. Test methods". 8 samples of acrylic plastic and 8 more of acrylic plastic reinforced with basalt fiber were made for the investigation. Using the Instron 5965 testing machine (Fig. 2), we investigated the dependence of strain-movement (Fig. 3, 4) with the subsequent calculation of the strength limit and elasticity modulus. As a result of the investigation, the strength limit and elasticity modulus of non-reinforced samples amounted to 67.1±0.6 GPA and 2010±9.3 MPa. In the reinforced ones, these indices were higher: 71.9±0.7 GPa and 2021±8.9 MPa respectively (Fig. 5). That is why we have made a conclusion that the use of basalt fiber to reinforce removable dentures improves their strength characteristics and can prevent their fractures.

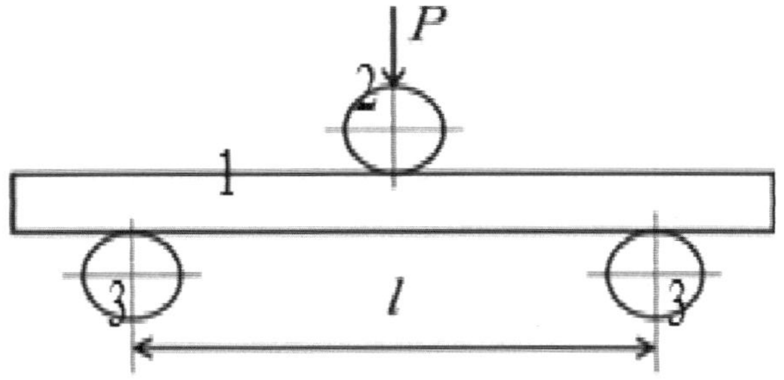

Figure 2. Instron 5965 testing the three-point bending: sample (1), loading plunger (2), supporting cylinders (3).

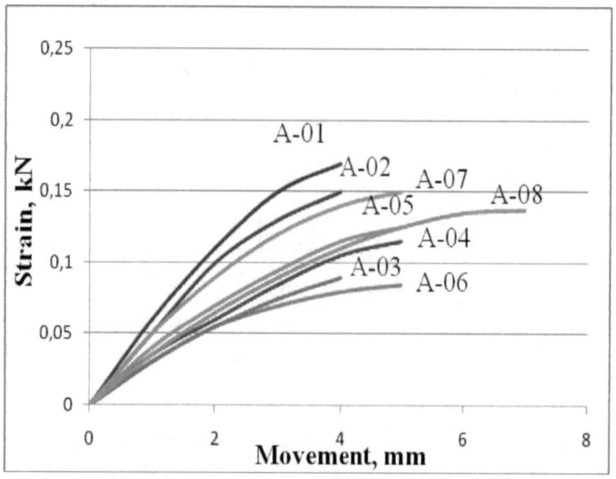

Figure 3. Dependences of strain-movement obtained while testing the three-point bending of acrylic plastic samples

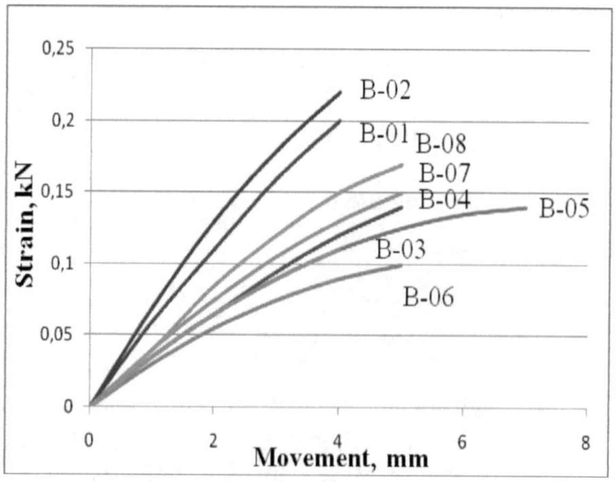

Figure 4. Dependences of strain-movement obtained while testing the three-point bending of acrylic plastic samples reinforced with basalt fiber

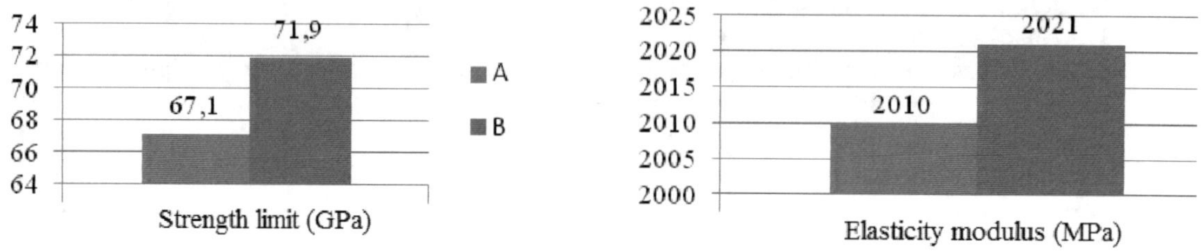

Figure 5. Comparison of the strength limit and elasticity modulus of the bases of acrylic plastic (A) and acrylic plastic reinforced with basalt fiber (B).

5. Conclusion

Having analyzed the data obtained, we have come to the conclusion that basalt fiber is a promising material for dental practice. A splint made of basalt fiber both meets all the necessary requirements and has a high strength reserve. Moreover, a basalt fiber splint is much cheaper than its competitors: splints made of polyethylene and glass fiber. And finally, bases reinforced with basalt fiber have high physical and mechanical characteristics, make it possible to avoid denture fractures and prolong the period of its use by a patient.

Acknowledgments

This work was partly supported by the Competitiveness Enhancement Program of Tomsk Polytechnic University.

References

[1] Singha K 2012 *Int. J. of Textile Sc.* **1** 19

[2] Stuchebrov S, Verigin D, Lukyanenko Y, Siniagina M and Wagner A 2014 *Adv. Mat. Res.* **880** 168

[3] Stuchebrov S, Batranin A, Verigin D and Wagner A 2014 *J. of Phys.: Conf. Ser.* **517** Article number 012046

[4] Grudyanov A, Dmitrieva J and Maksimovskiy Yu 1998 *Periodontology* **3** 5 (in Russian)

[5] Burt B 2005 Position paper: epidemiology of periodontal diseases *J. of Periodont.* **8** 19

[6] Yudina N, Azarenko V and Terekhova N 2006 *Temporary splinting in the complex treatment of periodontal diseases* (Moscow: BelMAPO) (in Russian)
[7] Rogozhnkov G.I., Ergaliev R.T., Karavaeva E.M., Rogozhnikov A.G., Chetvertnyh V.A., Astashina N.B. and Alikina N.B., Patent RU 130496 U1 (2012)
[8] Ulanova O. and Bednenko P 2000 *Proc. Int. Dental Congress Present-Day Questions of Modern Dentistry (Samara)* pp. 10-13
[9] Information on http://www.sanfarm.ru/polomka-bazisov-semnih-zubnih-protezov (in Russian)

Evaluation of the bone tissue mechanical parameters after induced alimentary Cu-deficiency followed by supplementary injection of Cu nanoparticles in rats

O Sachenkov[1], L Kharislamova[2], N Shamsutdinova[3], E Kirillova[3], Yu Konoplev[1]
[1]Kazan Federal University, 18, Kremlyovskaya St., Kazan, 420008 Russia
[2]Kazan Scientific Center, Russian Academy of Sciences, 2/31, Lobachevskogo St., Kazan, 420111 Russia
[3]Kazan State Medical University, 49, Butlerova Str., Kazan, 420055 Russia

E-mail: 4works@bk.ru

Abstract. The paper studies the evaluation of mechanical properties of bone tissue with Cu-deficiency in the body. The studies have been conducted on the bones of rats subjected to a complete Cu-deficient diet and the Cu-deficient diet, followed by rehabilitation diet. Within the frames of research testing on bending of animals' femoral bones was conducted, the scheme for flexure of the axes "move-force" midsection was drawn. The Young's modulus, ultimate normal and tangential stress limits were defined. The animals under research were divided into three groups: a) with Cu-deficiency, b) Cu-deficiency within 5 weeks with the following rehabilitation, c) Cu-deficiency within 8 weeks with the following rehabilitation. Depending on time and Cu-deficiency in the body the analysis of the obtained mechanical characteristics was performed. The effect of Cu-deficiency on bone tissues' rigidity and strength was detected. It was also shown that Cu-deficiency within 5 weeks and further rehabilitation the mechanic properties of the bone tissues are restored and the hardening analogue takes place. The recovery does not take place at the Cu-deficiency within 8 weeks, what the authors connect with the tissue texture transformation.

1. Introduction

It is known that copper is involved in maintaining adaptive mechanisms of the human body [1]. This is due to the active participation in the processes of antioxidant defense. The main factors governing the copper absorption are the concentration of copper in the diet and the organism's supply of these microelement. With a high copper content in the diet, absorption is reduced and endogenous losses increase. Copper absorption processes are competitively interconnected with zinc, cadmium, and iron. An imbalance in the copper metabolism leads to several diseases.

More than half of the copper content in the body is contained in the musculoskeletal system. Copper impact on the bone metabolism appears as participating in the synthesis of the conjunctive tissue of bone and copper deficiency leads to disruption of bone mineralization. At present people with copper deficiency are proven to suffer from osteoporosis [2, 3]. Osteoporosis is a metabolic bone disease characterized by low bone mineral density and increased risk of low-energy fractures [4-7].

The current stage of nanotechnology development allows stimulating of the living organisms with nanomaterials. That is based on the nanoparticles physical-chemical characteristics and related effects, such as the ability to penetrate into the various body organs and tissues and stimulate metabolic processes in biotic doses [8-10]. In this study the results of defining bone tissue mechanical characteristics according to copper deficiency and its various timing recovery (introduction of nanoparticles of copper in the body) in the body are shown.

Content from this work may be used under the terms of the Creative Commons Attribution 3.0 licence. Any further distribution of this work must maintain attribution to the author(s) and the title of the work, journal citation and DOI.
Published under licence by IOP Publishing Ltd

2. Materials and Methods

We used Wistar rats, which were divided into those subjected to the Cu-deficient diet (Diet) only, Diet and a Cu-restorative injections (supplementation - supp.), as well as a control group (see table 1). Laboratory animals in Groups 2 and 3 were intramuscularly administered with aqueous solutions of copper nanoparticles at a dose of 2.0 mg / kg body weight (supp.). Nanoscale copper particles with a spherical shape (103 ± 2 nm) were used in the experiment. Animal studies have been approved by the local ethics committee Kazan State Medical University, Russia.

To make analysis convenient the objects under research were divided into three groups: 1 group was subjected only to diet I; 2 group was subjected during 5 weeks to diet I, followed by the recovery diet II; 3 group was subjected during 8 weeks to diet I, followed by the recovery diet II. In several charts below the shorthand " Diet"/" Supp." will be used, i.e. 5/4 means that it was "diet" for 5 week and "supplementation" for 5 week.

Table 1. The table of diets of rats under research.

groups	Diet (weeks)	Supp. (weeks)	Total (weeks)	Number of animals	Max Weight (gramm)	Male/Female (number)
group 1	4	0	6	7	139	3/4
	5	0	7	7	146	4/3
	6	0	8	7	135	3/4
	8	0	10	7	235	3/4
group 2	5	4	11	7	202	4/3
	5	6	13	7	260	3/4
	5	8	15	7	391	3/4
	5	10	17	7	407	4/3
group 3	8	2	12	7	195	3/4
	8	4	14	7	213	4/3
	8	6	16	7	260	3/4
Control group			6	10	185	5/5

In the experimental study femoral bones of rats were taken, weighed, their density estimated, geometrical parameters measured, that followed by a bending test. For the tests in bending experimental equipment was designed and assembled, its circuit is shown in figure 1: extracted bone with the proximal part (position 1) was placed in the glasses using Wood's alloy (position 2), after which the centering of retaining ring was made (position 3), which was attached to the upper press traverse (position 5), the scales were set on a rigid support (position 4) which was attached to the lower fixed traverse (position 5). While testing the upper traverse was moved with velocity 0.2 mm/min and through the retaining ring (position 3) this move was transmitted to the bone (position 1), measurements of the applied force, movement and time being made at that. After the samples' deformation the bone walls measurement was performed.

The average length of the studied samples between the bindings of the proximal sections is 12 mm, diameter is 2 mm, thickness is 0.5 mm. figure 2 shows a typical diagram obtained in bend tests (characteristic values for strength and movements are given).

Plot OA is that of proportionality, at this section the function is close to a linear one; at this stage of loading the bone undergoes elastic deformation; section AB is that of nonlinearity (it is not long); at this stage there occurs crushing of the bone beams, similar to plastic deformation [11]; BC is the appearance of the first crack, is characterized by the fall of the applied force; CD is a crack growth (depending on samples the number of such sections might be up to 3); DE is the appearance of a large crack and subsequent destruction.

To determine the Young's modulus and the modulus of the ultimate stress the beam pattern for the ring section was applied; the boundary conditions, the load value and measurements are shown in figure 3. In this case, the conditions of load are as follows:

$$\sigma_{max} = \frac{|M_{max}|}{W} \leq [\sigma], \tag{1}$$

where σ_{max} - maximum normal load in the section, M_{max} - maximum moment, W - the moment of resistance at the bend, $[\sigma]$ - admitted normal load for material.

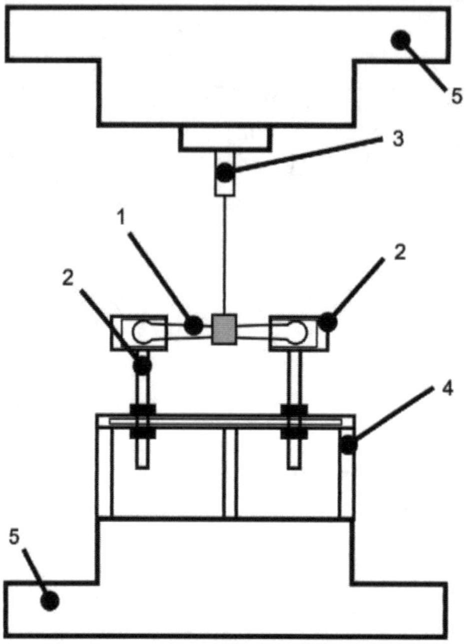

Figure 1. Scheme of the testing stand.

For the given scheme of load the movements are as follows.

$$y_{max} = \frac{Fl^3}{48EJ}, \tag{2}$$

where F - force, L – the sample length, E – the Young modulus, J – the section inertia moment.

For the linear section of the diagram (section OA in figure 2), applying the formula (1), (2), we obtain:

$$E = \frac{Fl^3}{48Jy}, \tag{3}$$

To estimate the admitted normal and tangential stresses we use the formula:

$$[\sigma] = \frac{l \cdot |F_{max}|}{2W} \tag{4}$$

$$[\tau] = \frac{2|F_{max}|S_x^*}{J \cdot d} \tag{5}$$

where S_x^* is a static moment of the intercepted section relative to Ox.

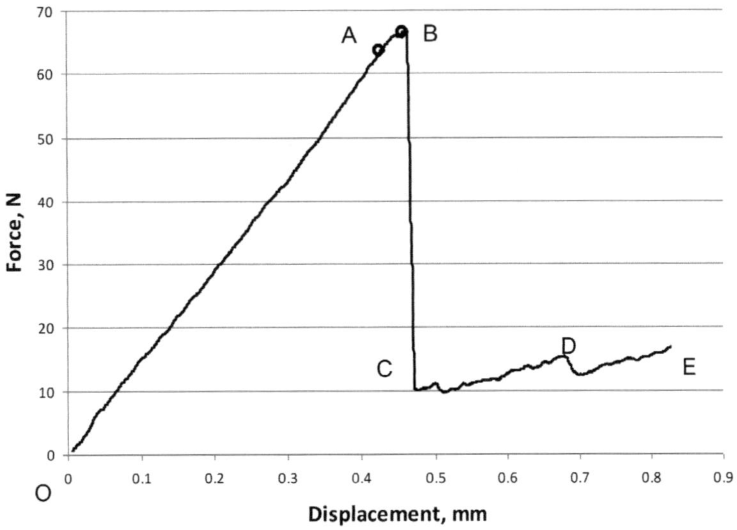

Figure 2. The characteristic diagram at the bending test: OA is the section of elastic deformation, AB is the section of nonlinearity, B is cracking formation, CD is the development of a crack, D is destruction.

The calculation of $[\sigma]$ - admitted normal stresses and $[\tau]$- admitted tangential stresses for material (in point B, figure 2) was performed along with evaluation of a crack growth at various Cu-deficiencies (section CD in figure 2).

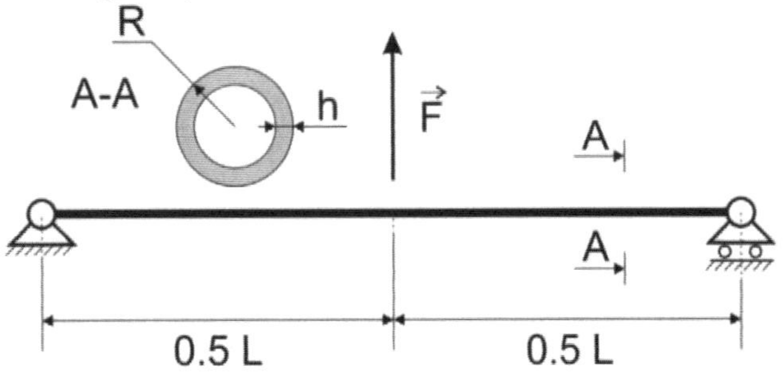

Figure 3. Calculation scheme. F - force (N), L - length of bone, R - radius of bone, h - thickness of bone

3. Results

The obtained results were subjected to statistical analysis, and then used for calculating the Young's modulus, the ultimate tangential and normal stresses. For the convenience of analysis at the graphs the calculated mechanical parameters are divided into groups, the line indicates similar characteristics for the control group, the calculated values of which do not contradict researches in the field [12-14].

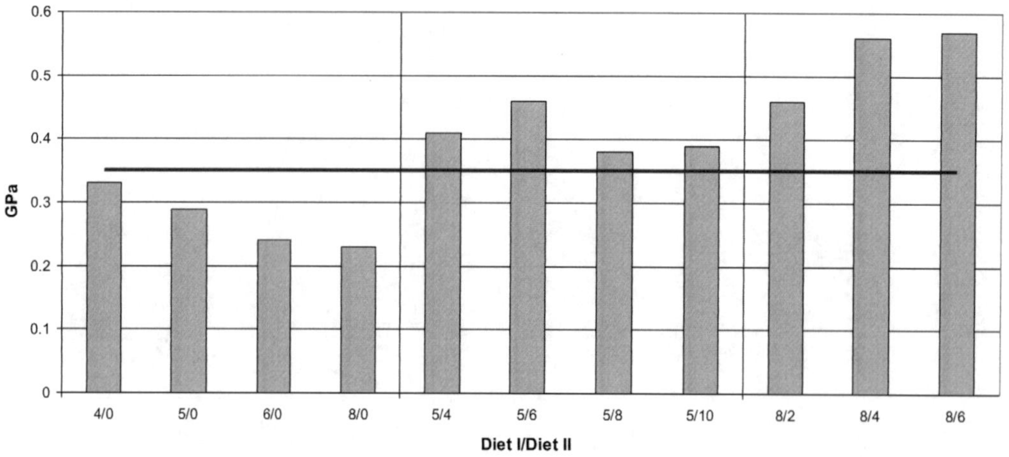

Figure 4. The average Young's modulus, horizontal line – control group

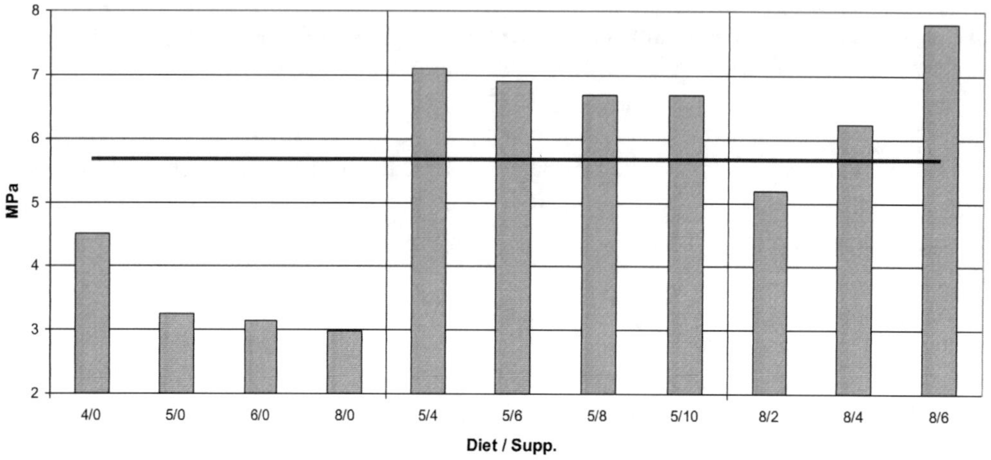

Figure 5. The average admitted tangential stresses, horizontal line –control group

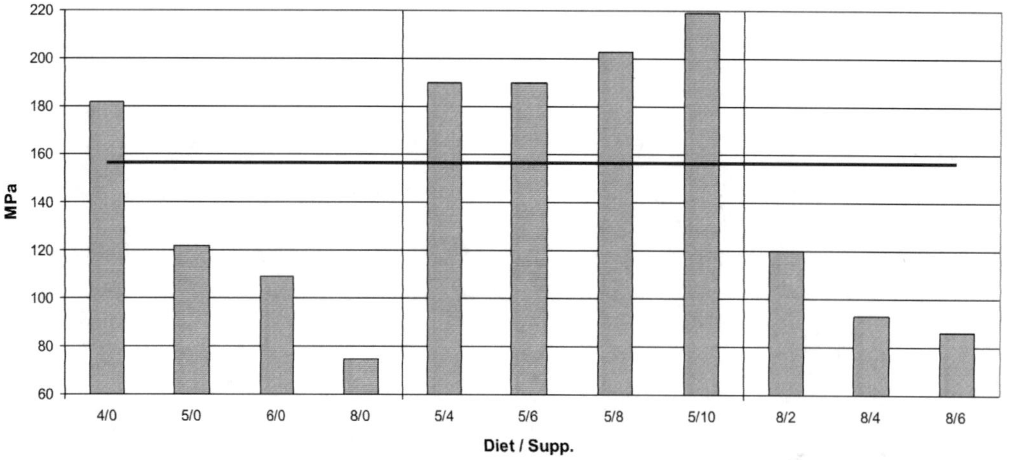

Figure 6. The average admitted normal stresses, horizontal line –control group

Group 1: with the Cu deficiency term increase in bone tissue the stiffness decreases (decrease in Young's modulus, see figure 4), the ultimate values of the normal and tangential stresses are also reduced with increasing Cu deficiency term (see figures 5 and 6), (hardness, shear strength, bending and stretching decrease).

Group 2: with increasing recovery period bone tissue stiffness tends to the average stiffness of the control group (see. figure 4), the ultimate value of the tangential stresses is also in the range of shear stresses of the control group, at the maximum term of recovery increases to 10%. The value of the ultimate normal stresses increases with the growth of Cu nanoparticles recovery period up to 20% (see figures 5 and 6), (stiffness recovers, the shear strength slightly increases and tensile strength still more increases).

Group 3: with the recovery term increase the bone tissue stiffness increases sharply compared to that of the control group (see figure 4), the value of the limiting shear stresses also increases compared to the value of the limiting shear stresses in the control group, at a maximum recovery term it increases by 30%. The value of the ultimate normal stress decreases sharply with increasing Cu nanoparticles recovery term, up to 40% (see figures 5 and 6) the stiffness and shear strength greatly increases, tensile strength is reduced.

4. Conclusion

The study has assessed the effect of Cu deficient diets followed by supplementary introduction of Cu nanoparticles on the mechanical characteristics of bone tissue in rats. It was found that copper deficiency leads to a drastic reduction in the stiffness of bone tissue and strength. In case of copper deficiency in the body within 5 weeks and subsequent supplementary recovery stiffness of the bone tissue is restored, the ultimate shear stresses increase to 10% of normal, allowable normal stresses increase by 20%, thus the hardening of bone happens. In case of 8 weeks copper deficiency and subsequent supplementary recovery the bone tissue stiffness is greatly increased (up to 40%), the allowable tangential stresses are also greatly increased (up to 30%), but the normal allowable stresses drop (to 40%), which may be explained by the irreversible bone tissue restructuring [15, 16].

Acknowledgements

This work was funded by the subsidy allocated to Kazan Federal University for the state assignment in the sphere of scientific activities.

References

[1] Klotz L, Kronche K, Buchzyk D 2003 *J Nutr.* **133** 1448.
[2] Kawada E, Moridaira K, Itoh K, Hoshino A, Tamura J and Morita T. 2006 *Ann Nutr Metab.* **50(5)** 420.
[3] Opsahl W, Zeronian H, Ellison M, Lewis D, Rucker R B, Riggins R S 1982. *J Nutr.* **112(4)** 708.
[4] Dickenson R P, Hutton W C and Stott J R R 1981, *J Bone Joint Surg Br.* **63-B(2)** 233.
[5] Hardinge M G, 1949 *Surgery gynecology and obstetrics.* **89** 439.
[6] Homminga J, McCreadie B R, Weinans H and Huiskes R. 2003 *Journal of Biomechanics* **36(10)** 1461.
[7] Khandare A L, Suresh P, Kumar P U, Lakshmaiah N, Manjula N and Rao G S. 2005 *Calcified Tissue International* **77(4)** 233.
[8] Timasheva A B, Miroshnikov S V, Notova S V and Lebedev S V *Mikroelementy v meditsine* **15(2)** 29
[9] Yausheva Ye V, Zelepukhin A G, Ryabov N I, Kvan O V, Ramenskiy V A, Zaveryukha A Kh and Sirazetdinov F Kh 2013 *Sovremennyye problemy nauki i obrazovaniya* **5(49)**
[10] Rakhmetova A A Izucheniye biologicheskoy aktivnosti nanochastits medi, razlichayushchikhsya po dispersnosti i fazovomu sostavu *Dissertation*
[11] Akulich A Yu, Akulich Yu V and Denisov A S 2010, *Russian Journal of Biomechanics* **14(4)** 7.
[12] Brown T D and Ferguson A B Jr. 1980, *Acta Orthop. Scand* **51(3)** 429.
[13] McElhaney J H, Fogle J L, Melvin J W, Haynes R R, Roberts V L and Alem N M 1970 *J Biomech.* **3(5)** 495.
[14] Rho J Y, Roy M E, Tsui T Y and Pharr G M 1999 *J. Biomed. Mater. Res.* **45** 45.
[15] Akulich Yu V, Akulich A Yu, Denisov A S, Shaymanov P S and Shulyat'yev A F 2014 *Russian Journal of Biomechanics* **18(2)** 135.
[16] Figurska M 2007 *Russian Journal of Biomechanics* **11(3)** 26

IOP Conf. Series: Materials Science and Engineering **98** (2015) 012016 doi:10.1088/1757-899X/98/1/012016

Alternative low frequency magnetic field theranostics: recent advances, safety and hazards

Y Golovin[1-3], N Klyachko[2,3], A Majouga[2,3], D Golovin[1] and S Gribanovsky[1]
[1]Nanocenter,G.R. Derzhavin Tambov State University, 33, Internatcionalnaya st., Tambov, 392000, Russia
[2]Chemistry Department, M.V. Lomonosov Moscow State University, 1, Leninsky Gory., Moscow, 119991, Russia
[3] National Research Technological University "MISIS", 4, Leninsky ave., Moscow, 119991, Russia

E-mail: golovin@tsu.tmb.ru

Abstract. The paper presents a brief review and comparative analysis of low frequency (non-heating) and radio-frequency electromagnetic nanomedicine technologies. The former are shown to have a considerable advantage over the latter ones: a higher flexibility and penetrating ability, easier to dose and control, easier to localize, as well as safer and less costly. This makes their employment promising for building a new technological platform for low frequency magnetic theranostics with a wider range of options, i.e. possessing a wider multimodality than traditional radio-frequency methods.

1. Introduction

Natural and technogenic electromagnetic fields are important factors of life on Earth (Figure 1).Their influence on many processes in living matter is very prominent [1-10]. The Earth magnetic field (MF) with the flux density of $B_E \approx 50\mu T$ as well as its low frequency variations with a $10^{-4} - 10^{-2}$ B_E amplitude, mainly induced by the increased Solar activity seems to be not only safe but also essential or, at least, useful for the biosphere function and evolution. It is known that in a number of cases after complete shielding or cancellation of external MFs depression of some functions of living organisms [11, 12] including humans [13] was observed. Investigations of adverse impacts of the Earth MF cancellation on humans [14], carried out for decades, suggest bringing a special term 'The Magnetic Field Deficiency Syndrome' as applied to the 'developed countries' population who spend much time in magnetically shielded office and apartment buildings. Thus, for example, Russiansanitary regulations generally allow for no greater than twofold decrease of the Earth's natural MF inside man-made constructions [15].

Content from this work may be used under the terms of the Creative Commons Attribution 3.0 licence. Any further distribution of this work must maintain attribution to the author(s) and the title of the work, journal citation and DOI.
Published under licence by IOP Publishing Ltd

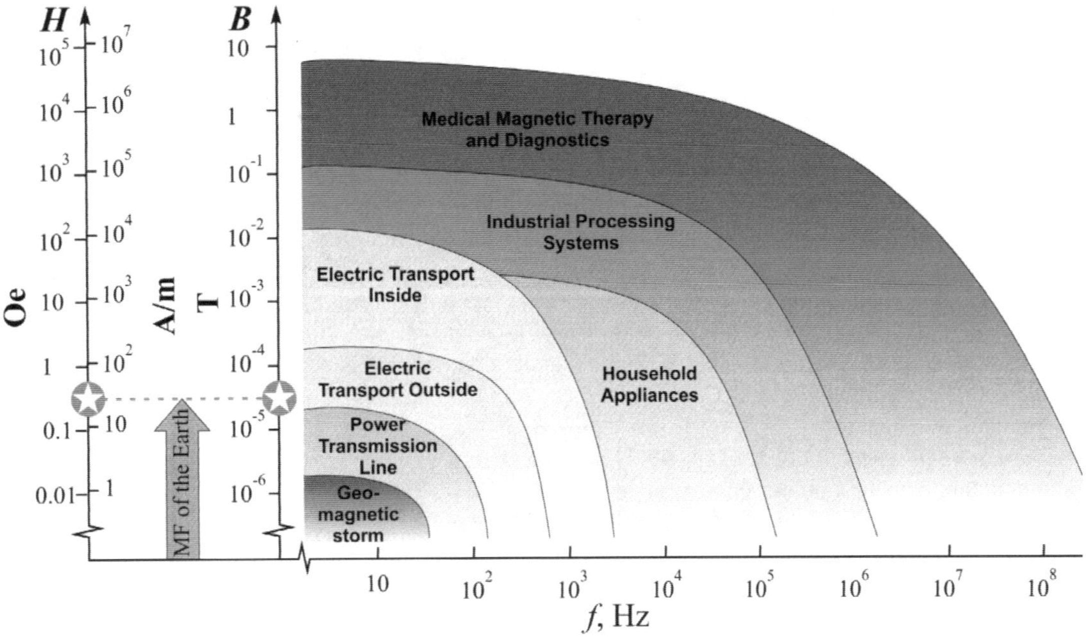

Figure 1. The map of natural and artificial magnetic fields (MF).

The recent century numerous artificial MF sources were added to the natural ones, among them: power lines, telecommunication facilities, electric transport, electric furnaces and other electric industrial equipment, medical and household electro/radio devices. The aforementioned and similar man-made sources of permanent and alternating MF produce fields comparable or even stronger than natural ones. Technogenic electromagnetic fields may be considered as certain factors of the environmental pollution. Some experts in hygiene and ecology even consider them the major pollution agents, especially in urban areas where the artificial electromagnetic background is much more intense than that in the rural areas and the ratio between the exposition period and general life time approaches 100%.

The data provided by a number of independent research groups prove that technogenic MFs may have a pronounced influence on health of industrial personnel and general public [1, 2, 8-10]. In this context, statutory documents issued by various state and international authorities limit the maximum allowed induction of alternating magnetic field (AMF) B for various groups of population depending on the field frequency f and exposition time, i.e. time spent under the influence of the field. Thus, according to the sanitary regulations of the International Commission on Non-ionizing Radiation Protection (ICNIRP)[16] $B{\cdot}f$ product at work places is limited to a constant 25 mT·Hz within the range from tens to hundreds hertz and to 2000 mT·Hz within the range of hundreds kilohertz, whereas for general public this constants are 5 mT·Hz and 920 mT·Hz respectively.

Fields with a flux density exceeding the allowed maximum values by about two orders of magnitude are considered hazardous. These guidelines vary significantly in different countries and international organizations, but everywhere an analogous dependence $B(f)$ is observed. In other words, it is generally accepted that an increase in f leads to the rise in negative effect of AMF on human organism and on biosphere. To some extent it can be explained by the fact that high values of $B{\cdot}f$ lead to the increased ability of AMF to heat any substances including biological tissues. As a result, nonspecific thermal effects can be added to the specific "magnetic" ones. Basically electromagnetic field dissipates in any initially unexcited medium which results in heating the medium. However, in easily generated experimental fields with the induction $\lesssim$ 1T, any prominent heating of biological objects is registered only in AMF with frequency values higher than certain characteristic f_T. Since dissipation of AMF energy occurs during every cycle of the field oscillations, then, with other factors being equal, the amount of heat produced in tissues under the influence of radiofrequency MF (RF MF) at a rough approximation is proportional to f. Depending on the organism's size, type of tissue, thermal exchange conditions and other factors, the frequency of 10^4- 10^5 Hz can be taken as f_T value for homoeothermic animals. Accordingly, AMF with $f < f_T$ is commonly referred to as 'non-heating' field, while AMF with $f > f_T$ is referred to as 'heating' field. Non-specific thermal influence mechanisms being added to the specific 'magnetic' ones trigger complementary risks such as possible increase in biochemical and catalytic reactions rates, biological liquids thinning, secondary structure alteration and denaturation of proteins, phase transitions and permeability increase in membranes, cell apoptosis and necrosis and etc. That is why for safety reasons, it is demanded that the higher AMF frequency is, the lower field induction should be.

Natural or artificial magnetic structures present in the organism can lower the threshold of the registered AMF effect considerably. Study and recording of such effects becomes more and more important and timely since employment of synthetic magnetic nanoparticles (MNPs) stimulated by AMF keeps growing in biomedical applications.

Recent decades witnessed a sharp increase in use of functionalized MNPs (f-MNPs) both in biomedical studies and clinical applications for various methods of biostructure visualization, diagnostics and therapy [17-19]. In particular, they allow increasing the image contrast in magnetic resonance imaging (MRI), whereas in the targeted drug delivery strategy, they support targeted transport, controlled drug release, malignant cells ablation by means of RF magnetic hyperthermia (Figure 2) and other therapeutic treatments [20-24].

All aforementioned techniques utilize f-MNPs as the mediators enhancing localization and magnitude of RF AMF effect on object structure targeted elements. Beside the discussed threats and dosage and control difficulties of overheat, aforementioned technique has another drawback. Heat transfer through tissues results in heat spreading over a long distance from heat generation point source once MNPs are considered in all biomedical applications. During typical exposition times $t_e \approx 10^2$ - 10^3 s, the thermal wave passes through $R_T = (\chi\ t_e)^{1/2} \approx 3$ - 10 mm from MNP [25,26], where $\chi \approx 10^{-7}$ m^2s is a typical value of living tissue thermal diffusivity constant. Since χ varies within one order of magnitude for all tissues, no RF magnetic hyperthermia technique enhancement could localize thermal field stronger than within 1 mm region. It completely denies its capability to be local or selective on the cellular and even more on molecular scale.

Matching the parameters of AMF used in clinical practice with the limits set by sanitary regulations shows that the former are considerably higher. This stimulates the search for and development of safer electromagnetic methods for diagnostics and therapy.

Considering these circumstances and other limitations of the traditional methods for diagnostics and therapy in RF MF, several new approaches have been recently suggested. It is proposed to use lower frequency fields to stimulate MNPs (Figure 2), since the maximum allowable fields here are much lower than those for RF MF. Various applications of LF MF with f-MNPs can lay the physical foundation for the new generation of means for safer magnetic theranostics with enhanced capabilities and modality.

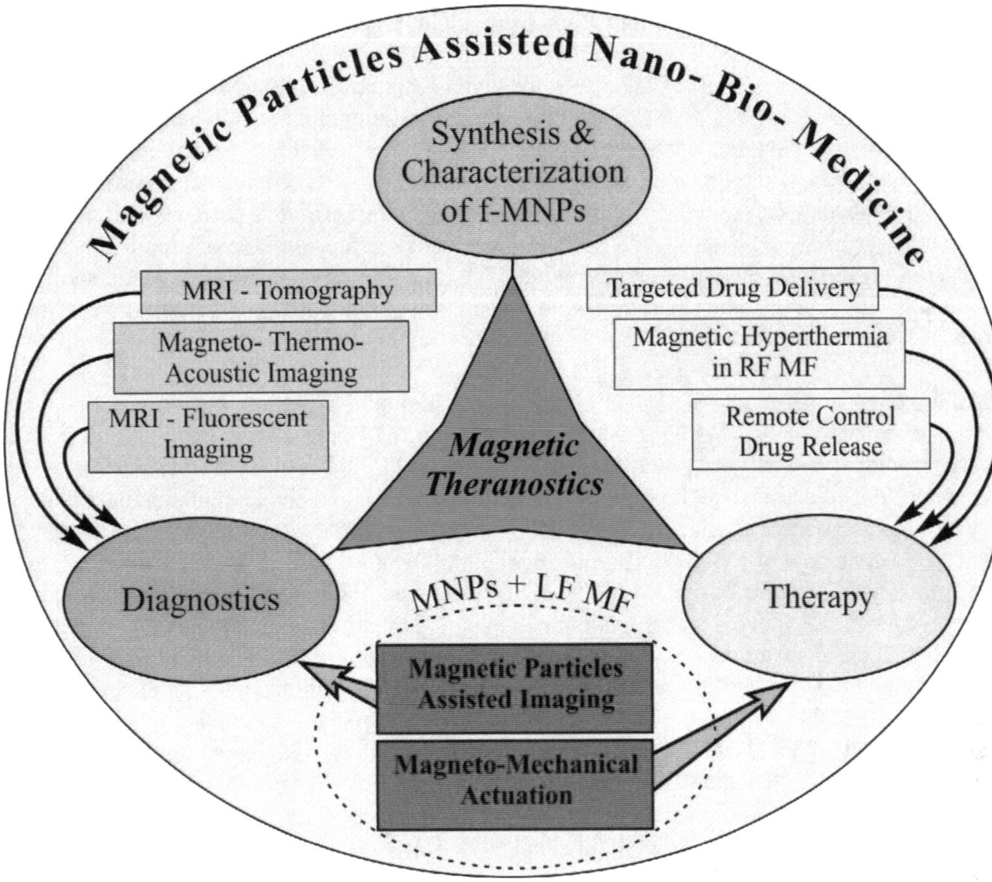

Figure 2. Innovative nanotechnology of theranostic assisted by f-MNPs and low frequency non-heating magnetic field (LF MF) vs traditional biomedical technologies

The paper presents a brief review on the principles, advantages and problems of these approaches for biomedical applications both *in vitro* and *in vivo*.

2. Multimodal visualization and diagnostic techniques

For the past several decades magnetic resonance imaging (MRI) remains one of the most widely used approaches to acquiring and visualizing information on the inner tissues in clinical practice. The main idea of the approach is a studying the spatial peculiarities of magnetic moment relaxation of hydrogen nuclei in a strong uniform MF on thoroughly designed RF pulse forcing spin system transition into an exited state. An image contrast originates from the difference in environment of hydrogen nuclei as their concentration is high and more or less uniform throughout the tissues. Technically, the object under study is placed into a strong and extremely uniform steady MF with a relative inhomogenity around 10^{-6} root mean square or even better to achieve a particularly high spatial resolution leading both to energy levels splitting depending on the nucleus spin projection on the field direction and to spin rotation with a well defined frequency known as Larmor precession. Additional RFMF short pulse on this frequency induces a significant rearrangement of the nuclei spin system followed by a gradual relaxation of the system to the equilibrium state. Specifics of this relaxation process provide the mechanism of contrasting tissues and its regions distinguished by state or functionality such as tumors localization.

MRI approach is extremely versatile because of high complexity of relaxation processes within the nuclei spin system that provides a wide variety in selecting the excitation pulse parameters, type and processing algorithms of the recorded response depending on a particular object under study and the study aim. In some cases, the image contrast is enhanced by agents, which strongly affect relaxation processes in the nuclei spin system. Such agents could be both individual paramagnetic ions or their compounds and superparamagnetic f-MNPs. The most widely used ion is gadolinium, but it is quite toxic for living organisms even being enveloped in chelate complexes, so alternatives are sought for in many ways. In particular, other paramagnetic ions, oxide nanoparticles, nanoparticles based on zeolites or polymers and functionalized by various atoms and ligands have been proposed and tested [27].

To combine capabilities of MRI and other approaches to malignant tumor visualization and treatment, one seeks to design MNPs that could stand as an MRI contrasting agent and therapeutic or another diagnostic procedure agent simultaneously. In particular, the authors of [28] propose a specific design of a multilayered nanocrystal structure made of lanthanides on the carbon nanotube matrix that can be used both as an MRI contrasting agent and a fluorescent label carrier to visualize tumor during its surgical treatment. In [29], MNPs combining MRI contrasting agent properties with a photodynamically controlled drug release carrier are designed. Those MNPs are based on graphene nanosheets functionalized by i) iron nanoparticles to provide superparamagnetic properties, ii) photosensitive fluorescent molecules that release atomic oxygen or free radicals upon light quantum absorption to affect the targeted cell and iii) hydrophilic ligands to facilitate MNPs dispersion in water solutions.

The other group of internal organ visualization techniques is based on thermoacoustic waves generation, i.e. fast pulsed localized heat generation causing acoustic waves generation. It is proposed to use f-MNPs in the regime similar to hyperthermia as a heat generation medium [30]. To achieve theoretical resolution about 1 mm one should use pulse rate about 1 MHz, MF frequency about 10 MHz and field intensity about 10 mT [31].

In the recent decade, an alternative method of MNP based visualization called magnetic particle imaging (MPI) has been developed [32]. It requires neither highly uniform MF nor satisfaction of the resonance condition and uses the magnetic moment of MNP itself as a main target of MF manipulations. MNP detection occurs due to high non-linearity of Langevin magnetization curve of superparamagnetic particles, so MNP ensemble being excited on arbitrary frequency, usually about several kHz, emits an RF signal on multiple frequencies. The parameters of this signal are directly related to media viscosity, and registration of MNP tethering to some large objects is also possible. On the other hand, it is possible to acquire information on MNPs themselves or their aggregation if they are located in the well known media for MNP characterization.

Such response signal is emitted only if external MF traverses zero point or at least approaches close enough to it, so the spatial resolution can be provided by applying an additional high gradient steady MF. Due to the ratio of MNP magnetic moment value μ to the one of proton being in the range of millions, MPI sensitivity could be better than MRI one. The MPI operation speed is much higher than for MRI too, since f-MNPs magnetic moment relaxation is provided by intermolecular viscous forces resulting in a relaxation time being in the range of microseconds, whereas the proton magnetic moment relaxation in MRI case utilizes magnetic mechanisms resulting in relaxation time about one second without a contrasting agent and a few tenth of a second with it. Another MPI advantage when MNPs are used as an agent for targeted drug delivery or other therapy techniques is the usage of those therapeutic MNPs for imaging, so that one can definitely obtain actual information on therapeutic MNPs localization and some data on environment properties of those MNPs. Thus, MPI is able to exceed other visualization techniques in some medical applications in the near future as summarized in the following table.

Table. Comparison of different medical imaging methods [33].

Perfomance	Methods			
	X-Ray computer tomography (CT)	Magnetic resonance imaging (MRI)	Positron emission tomography (PET)	Magnetic particle imaging (MPI)
Spatial resolution, mm	0.5	1	4	< 1[a]
Sensitivity	low	low	high	high
Measurement time, s	1	10-1000	100	<0.1[b]
Ionizing radiation	yes	no	yes	no

[a] Expectations based on calculations by Weizenecker et al. [34].
[b] As measured by Weizenecker et al. [35].

3. Targeted drug delivery and cancer treatment by using f-MNPs in non-heating LF MF

In the set of papers on the targeted drug delivery and remote control of the therapeutic agent (TA) carriers AMF utilization for mechanical activation of f-MNPs was suggested [36-43]. In [44,45], the results of activity control of enzyme macromolecules through f-MNPs activated by uniform AMF with the frequency from a few tens to several hundred hertz were reported. Any heating was naturally absent at this sufficiently low AMF frequency (as monitored by sensors continuously). Instead of this, macromolecule deformation forced by f-MNPs brought into oscillating rotational motion by MF was the dominating mechanism.

In [46], the quantitative magneto-mechanical models of such single molecule actuation were described (Figure 3). According to the reported computational data, AMF with the frequency up to 1000 Hz and flux density up to 0.5 T easily obtainable in the laboratory environment can result in various kinds of deformations in the macromolecules being part of f-MNP aggregates. These deformations are sufficient to change the topology of macromolecules and, therefore, to control their biochemical properties. There are four types of such deformations: tension, compression, shear, and torsion. Not only macromolecules linking MNPs in aggregate, but also those in the polymeric coating of each f-MNP ever appearing between the particles undergo them. In [47], the estimation of the effect of f-MNPs in the external uniform LF MF on vesicle and live cell bilayer membrane structure was given. It was shown that in case of the vesicle MNP influence could result in their disruption or permeability increase. The effect may be applicable in targeted delivery of vesicle-encapsulated TA. With respect to cell membrane, f-MNPs can exert the force sufficient to cause specific responses from cell receptors as well as to disturb routine function of ionic channels and to change membrane structure, leading to metabolism disruption. This opens great opportunities for controlled and selective killing of malignant cells in cancer therapy. General regularities of such biostructure stimulation were also discovered, and physical approach to optimization of AMF application mode, including field parameters, size and shape of f-MNPs, was substantiated. The authors also made estimations that show several times increase in the rate of TA macromolecules washing-out from polymeric coating under the external low-frequency uniform AMF influence. The observed increase can also be useful in targeted TA delivery applications. The effect is conditioned by the force of hydrodynamic resistance of the medium to the particle's rotational oscillations. In spite of f-MNP oscillations, this force is always directed from the center of the particle.

It is necessary to emphasize that all above mentioned technologies based on magneto-mechanical actuation have strict frequency limit determined by f-MNP hydrodynamic radius and viscosity of surrounding media. Depending on the media type (water, physiological liquids, intracellular media etc.) and f-MNP geometry, this limit lies within the range from several hertz to several kilohertz, but anyway, it is several orders of magnitude less than one of hyperthermia. Therefore, to implement the nanomechanical conception of control on biological objects, the frequency providing for the minimal AMF power dissipation must be chosen, while in the hyperthermia applications, on the contrary, dissipation enhancement is of primary importance.

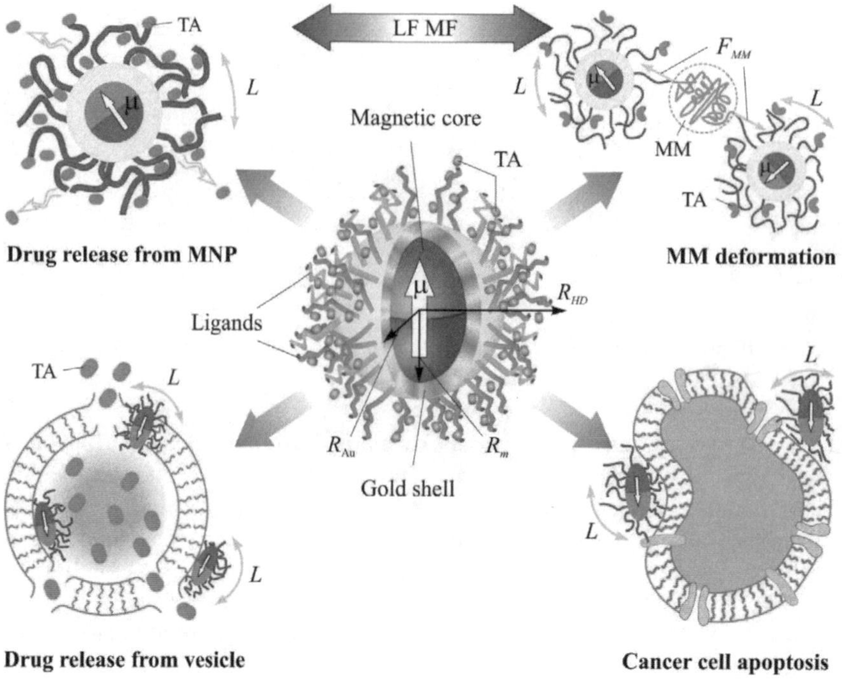

Figure 3. The main routes of f-MNPs utilization in the low frequency magnetic field (LF MF) for targeted therapeutic agent (TA) delivery. Here μ is the magnetic moment of f-MNP, *L* is the torque applied to f-MNP, MM is the therapeutic agent macromolecule, and F_{MM} is magneto-mechanical force causing macromolecule deformation.

4. Conclusion

An innovative area of magnetic-bio-medical science employing f-MNPs activated by uniform non-heating LF MF is very promising for practical applications. This platform allows developing extremely versatile biomedical technologies to fulfill various functions: i) f-MNPs characterization, ii) diagnostics and visualization of the object's inner structure and f-MNPs distribution within, iii) all the major processes of targeted delivery and on-demand remote control of TA. These technologies can be used both separately and in combination with RF magnetic hyperthermia with magneto-nanomechanical and thermal effects induced in any proportion to each other. Moreover, magneto-nanomechanical actuation of transmembrane structures of malignant cells (receptors, ionic channels) can selectively trigger their apoptosis with further ablation without using any toxic drugs (non-drug therapy).

MPI in LF MF has a higher speed of operation which allows recording the processes dynamics with the speed higher than 30 frames/sec, a more rapid response and better space resolution than typical 1 mm resolution of MRI and the necessary equipment is significantly less expensive. The magnetic-nanomechanical approach in targeted drug delivery by means of LF MF also has a number of advantages in comparison with technically resembling RF hyperthermia: it can provide selectivity and localization at the nanometer-, instead of millimeter scale, its effective implementation involves safer LF MFs which need less powerful energy sources and have higher tissue penetrating abilities. Sanitary regulations allow using LF MF of much higher values than those of RF MF (approximately in inverse proportion to its frequency). The LF MF exposure dosage can be easily controlled helping to avoid overdosage and hazardous side effects. Finally, combination of methods for f-MNPs characterization, visualization, diagnostics and therapy by means of non-heating LF MF allows developing more versatile and less costly means for theranostics carried out within the same device.

Acknowledgments
The authors are grateful to I. Gazaryan for helpful remarks on the manuscript. The work was supported by grants of Ministry of Education and Science of the Russian Federation, project #K1-2014-022 (theory and calculations), and Russian Science Foundation, project # 14-13-00731 (experimental part).

References

[1] Stavroulakis P (ed) 2003 *Biological effects of electromagnetic fields. Mechanisms, Modelling, Biological Effects, Therapeutic Effects, International Standards, Exposure Criteria* (Berlin, Heidelberg: Springer-Verlag) 793 p
[2] Rosch P and Markov M (eds) 2004 *Bioelectromagnetic Medicine* (New York: Marcel Dekker) 880 p
[3] Polk C and Postow E (eds) 1995 *Handbook of biological effects of electromagnetic fields* (Boca Raton: CRS Press) 618 p
[4] Barnes F and Greenebaum B (eds) 2007 *Handbook of Biological Effects of Electromagnetic Fields, Third Ed* (Boca Raton: CRS Press) 960 p
[5] Steiner U and Ulrich T 1989 *Chem. Rev.* **89**(1) 51
[6] Grissom C 1995 *Chem. Rev.* **95** 3
[7] Buchachenko A 2014 *Russ. Chem. Rev.***83**(1) 1
[8] Report of the independent Advisory Group on non-ionizing Radiation 2012 *Health Effects from Radiofrequency Electromagnetic Fields* (doc Health Protect Agency) 333 p
[9] Lin J (ed) 2005 *Advances in Electromagnetic Fields in Living Systems* 4 (New York: Springer Science & Business Media) 218 p
[10] Giulani L and Soffritti M (eds) 2010 *Non-Thermal Effects and Mechanisms of Interaction Between Electromagnetic Fields and Living Matter* Ramazzini Institute Eur. J. Oncol. Library 5 (Bologna, Italy) 403 p
[11] Osipenko M, Mezhevikina L, Krasts I, Yashin V, Novikov V and Fesenko E 2008 *Biophisics* **53**(4) 317
[12] Jia B, Xie L, Zheng Q, Yang P, Zhang W, Ding C, Qian A and Shang P 2014 *PLoS ONE* **9**(8) e105604
[13] Mo W, Liu Y, Bartlett P and He R 2014 *Science China Life Sciences* **57**(4) 448
[14] Nakagawa K 1976 *Jap. Med. J.* **2745** 24
[15] *SanPiN 2.2.4.1191-03* 2003 *Jelektromagnitnye polja v proizvodstvennyh uslovijah* (in Rus)
[16] ICNIRP Guidelines 2010 *Health Physics* **99**(6) 818
[17] Schleich N, Danhier F, Préat V 2015 *J. of Contr. Rel.***198** 35
[18] Nguyen D and Kim K Korean 2014 *J. Chem. Eng.***31**(8) 1289
[19] Reddy L, Arias J, Nicolas J and Couvreur P 2012 *Chem. Rev.***112** 5818
[20] E.-K. Lim, T. Kim, S. Paik, S. Haam, Y.-M. Huh, and K. Lee. *Chem. Rev.* **115** (1) 327
[21] Cohen Y and Shoushan S 2013 *Current Opinion in Biotechnology* **24** 1
[22] Thomas R, Park I and Jeong Y 2013 *Int. J. Mol. Sci.* **14** 15910-15930
[23] Lee J, Kim J and Cheon J 2013 Mol. Cells **35** 274-284
[24] Liao S, Liu C, Bastakoti B, Suzuki N, Chang Y, Yamauchi Y, Lin F, Wu K *2015 Int. J. of Nanomedicine* **10** 3315
[25] Gupta A, Kane R and Borca-Tasciuc D 2010 *J. Appl. Phys.***108** 064901
[26] Andrä W, d'Ambly C, Hergt R, Hilger I and Kaiser W 1999 *J. of Magnetism and Magnetic Materials***194**(1–3) 197
[27] Estelrich J, Sanchez-Martin M and Busquets M 2015 *Int. J. of Nanomedicine* **10** 1727
[28] Chen B, Zhang H, Du N, Zhang B, Wu Y, Shi D and Yang D 2012 *J. Coll. and Interface Sci.* **367** 61
[29] Gollavelli G and Ling Y 2014 *Biomaterials* **35** 4499-4507
[30] Nie L, Xing D, Zhou Q, Yang D, and Guo H 2008 *Med. Phys.* **35**(9) 4026

[31] Piao D, Towner R, Smith N and Chen W 2013 *Med. Phys.* 40 (6) 063301
[32] Gleich B and Weizenecker J 2005 *Nature* **435**1214
[33] Buzug T, Bringout G, Erbe M, Gräfe K, Graeser M, Grüttner M, Halkola A, Sattel T, Tenner W, Wojtczyk H, Haegele J, Vogt F, Barkhausen J and Lüdtke-Buzug K 2012 *Z. Med Phys.* **22**(4) 323
[34] Weizenecker J, Borgert J and Gleich B 2007 *Physics in Medicine and Biology* **52** 6363.
[35] Weizenecker J, Gleich B, Rahmer J, Dahnke H and Borgert J 2009 *Physics in Medicine and Biology* **54** L1
[36] Pankhurst Q, Thanh N, Jones S and Dobson J 2009 *J. Phys. D: Appl. Phys.***42** 224001 (15)
[37] Dobson J 2008 *Nature Nanotechnology* **3** 139143
[38] Hu B, El Haj A and Dobson J 2013 *Int. J. Mol. Sci.* **14** 19276
[39] Nappini S, Bonini M, Ridi F and Baglioni P 2011 *Soft Matter* **7** 4801
[40] Nappini S, Bonini M, Bombelli F, Pineider F, Sangregorio C, Baglioni P, Norden B 2011 *Soft Matter* **7** 1025
[41] Banchelli M, Nappini S, Montis C, Bonini M, Canton P, Bertia D and Baglioni P 2014 *Phys. Chem. Chem. Phys* **16** 10023
[42] Kim D, Rozhkova E, Ulasov I, Bader S, Rajh T, Lesniak M and Novosad V 2010 *Nature Materials* **9** 165
[43] Vitol E, Novosad V and Rozhkova E 2012 *IEEE Transactions on Magnetics* **48**(11) 3269
[44] Golovin Y, Klyachko N, Golovin D., Efremova M, Samodurov A, Sokolski-Papkov M, Kabanov A 2013 *Technical Physics Letters* **39**(3) 240
[45] Klyachko N, Sokolsky-Papkov M, Pothayee N, Efremova M, Gulin D, Pothayee N, Kuznetsov A, Majouga A, Riffle J, Golovin Y and Kabanov A 2012 *Angewandte Chemie. Int. Ed.* **51** 12016
[46] Golovin Y, Gribanovskii S, Golovin D, Klyachko N and Kabanov A 2014 *Physics of the Solid State* **56**(7) 1342
[47] Golovin Y, Klyachko N, Gribanovskii S, Golovin D, Samodurov A, Majouga A, Sokolsky-Papkov M and Kabanov A 2015 *Technical Physics Letters* **41**(5) 453

IOP Conf. Series: Materials Science and Engineering **98** (2015) 012017 doi:10.1088/1757-899X/98/1/012017

The toxicity of rifampicin polylactic acid nanoparticles against *Mycobacterium bovis BCG* and human macrophage THP-1 cell line

M Erokhina[1,3], E Rybalkina[1,4], G Barsegyan[2], G Onishchenko[3] and L Lepekha[1]

[1] Department of Pathomorphology, FSBSI "Central TB Research Institute", 2, Yauzskaya alleya, Moscow, Russia

[2] Russian Research Center for Molecular Diagnostics and Therapy, Sympheropolsky Bulvar, 8, Moscow, Russia

[3] Faculty of Biology, Lomonosov Moscow State University, 1-12, Leninskie Gory, Moscow, Russia

[4] Department of Tumor Cell Genetics, FSBSI "N.N.Blokhin Russian Cancer Research Center", Kashirskoe shosse, 24, Moscow, Russia

E-mail: masha.erokhina@gmail.com

Abstract. Tuberculosis is rapidly becoming a major health problem. The rise in tuberculosis incidence stimulates efforts to develop more effective delivery systems for the existing antituberculous drugs while decreasing the side effects. The nanotechnology may provide novel drug delivery tools allowing controlled drug release. Rifampicin is one of the main antituberculous drugs, characterized by high toxicity, and Poly (L-lactic acid) (PLLA) is a biodegradable polymer used for the preparation of encapsulated drugs. The aim of our work was to evaluate the toxicity of rifampicin-PLLA nanoparticles against *Mycobacterium bovis BCG* using human macrophage THP-1 cell line. Our data demonstrate that rifampicin-PLLA is effective against *M. bovis BCG* in the infected macrophages. The drug is inducing the dysfunction of mitochondria and apoptosis in the macrophages and is acting as a potential substrate of Pgp thereby modulating cell chemosensitivity. The severity of the toxic effects of the rifampicin-PLLA nanoparticles is increasing in a dose-dependent manner. We suggest that free rifampicin induces death of *M. bovis BCG* after PLLA degradation and diffusion from phago-lysosomes to cytoplasm causing mitochondria dysfunction and affecting the Pgp activity.

1. Introduction

Tuberculosis is caused by *Mycobacterium tuberculosis* that predominantly resides macrophages. In the past two decades tuberculosis has again developed into a major health problem, with 1.7 million deaths every year and 2 billion latently infected [1, 2]. Among the problems with the application of the existing antituberculous drugs are toxicity, rapid degradation and limited bioavailability [3]. The increase of tuberculosis incidences stimulates development of new antibacterial drugs as well as more effective delivery systems for the existing drugs while decreasing their side effects. The progress in nanotechnology and particular encapsulation of the drugs inside the slow biodegradable, polymer microparticles (>1000 nm, PMPs) and nanoparticles (PNPs) may advance the development of novel drug delivery systems and controlled drug release [2, 3, 4]. Highly porous polymeric materials are

Content from this work may be used under the terms of the Creative Commons Attribution 3.0 licence. Any further distribution of this work must maintain attribution to the author(s) and the title of the work, journal citation and DOI.
Published under licence by IOP Publishing Ltd

often used as drug carriers. Poly(L-lactic acid) (PLLA) and poly(lactide-co-glycolide acid) (PLGA) are biodegradable polymers used for the preparation of encapsulated drugs [5, 6, 7].

Rifampicin is one of the main antituberculous drugs, characterized by low water solubility and high toxicity. In bacterial cells rifampicin binds with RNA polymerase and thus prevents the elongation of RNA molecules [8]. In mammalian cells rifampicin initiates a complex of pathological changes ([9, 10]. In epithelial cells mitochondria are a potential target for rifampicin: the damage to mitochondria causes their dysfunction leading to oxidative stress [11]. In macrophages PMPs and PNPs predominantly localize in phago-lysosomes, where the polymer degrades, releasing the drug [12]. Compared with the accumulation of free rifampicin, rifampicine-PMPs and -PNPs can give up to a 20-fold higher concentration in macrophages [13, 14]. Released rifampicin can affect the viability of intracellular mycobacteria as well as the functional activity and structure of cellular organelles in the infected and uninfected macrophages. Thus the aim of our work was to study the toxicity of rifampicin-PLLA NPs against *Mycobacterium bovis BCG* in infected macrophages, and to determine the effects of rifampicin-PLLA NPs on cell viability, type of the induced cell death, and the role of mitochondria in this process.

2. Material and methods

2.1 Nanoparticle preparation

Rifampicin (100.8 mg, Sigma-Aldrich) and PLLA (700.1 mg, M_w ·60,000, Sigma-Aldrich) were dissolved in 10 ml chloroform with heating to 40 degrees and stirring. Then the solution was stirred 30 min at room temperature. For the water phase 180 mg of PVA (polyvinyl alhocol) was dissolved in 36 ml of water (pre-filtered using a 0.2 mm filter). The two solutions were mixed, stirring vigorously with a magnetic stirrer for 1h at room temperature, and homogenized (24,000 rpm, 3 times x 1 min) to obtain the primary emulsion. The resulting mixture of the solvent is removed by evaporation of the solution under vacuum for 1 h. Then the emulsion was filtered through a glass filter (pore size 40-110 μm). To the resulting colloidal solution was added 201.6 mg of D-mannitol, frozen in liquid nitrogen, freeze dried and stored at 4°C. The average particle size, measured using the method of autocorrelation spectroscopy of submicron laser spectrometer Coulter N4MD (USA), is 300 ± 71 nm. The loading of rifampicin in the PLLA-NPs is 8.5% (loading is given as % by weight (w/w)).

2.2 Cell lines

Both human cell lines, THP-1 (derived from human monocytic leukaemia) and KG1 (derived from human myelogenous leukaemia), were obtained from Russian Cell Culture Collection and were cultured in RPMI-1640 medium supplemented with 10% fetal calf serum. The cultures were maintained at 37°C in a humidified atmosphere (95% air/5% CO2) as a stationary suspension culture. To differentiate suspension THP-1 cells in macrophage10^{-7} M phorbol ester PMA (phorbol 12-myristate 13-acetate, Sigma) was used. After treatment for 24 h, the cells were induced to adhere to the growth substratum and to differentiate for 24-72h.

2.3 MTT-assay

The cell survival in the presence of increasing rifampicin-PLLA and rifampicin (Sigma) concentrations (from 10 to 250 μg/ml) was determined by monitoring the mitochondrial reduction activity according to standard protocol [15].

2.4 Staining of mitochondria
Mitochondria in cells were stained with 300 nM Mitotracker Orange (Molecular Probes) for 15 min at 37°C, then cell were fixed 3.7% formaldehyde in phosphate-buffered saline solution (PBS). DAPI (3 μg/ml, Molecular Probes) were added to label cell nuclei. Images were analyzed with Leica TCS SPE confocal laser microscope (Germany). The population of the active mitochondria without cytotoxic effects was stained with 10 μg/ml rhodamine 123 (Rh123) for 20 min at 37°C. The living cells was immediately observed with an Axiovert 200M (Carl Zeiss, Germany).

2.5 Immunostaining of cytochrom c
For immunocytochemical staining of cytochrome *c*, the cells were fixed for 10 min in 3.7% formaldehyde, washed in PBS (pH 7.2-7.4), treated with 0.1% Triton X-100 for 3-5 min. Then the cells were incubated in 1% BSA solution in PBS for 30 min, washed 3x10 min in PBS. Mouse monoclonal antibodies against cytochrome *c* (Sigma) were used as the first antibodies. The secondary antibodies were conjugated with Alexa 488 (Sigma). DAPI (3 μg/ml, Molecular Probes) were added to label cell nuclei. Images were analyzed with Leica TCS SPE confocal laser microscope (Germany).

2.6 Efflux of rhodamine 123
P-glycoprotein activity was determined by efflux of Rhodamine 123 (Rh123, Sigma), a fluorescent dye that is a substrate for P-gp, as previously described [16]. Briefly, 100 μl of purified cell suspension was incubated with 5 μg Rh123 for 15 min. After washing, cells were incubated in Rh123-free medium in the presence or absence of 2 mM verapamil (Sigma), a P-gp inhibitor, for 30 min at 37°C. Finally, cells were washed and up to 10,000 events were detected and analyzed with a FACScan flow cytometer (Becton Dickinson) equipped with an argon-ion laser with a wavelength setting of 488 nm and with 530/30 nm band pass (FL1).

2.7 Bacterial culture
Mycobacterium bovis BCG expressing green fluorescent protein *GFP* (*M. bovis BCG-GFP*) were grown on Middlebrook's 7H9 broth medium (Difco) supplemented with 10% OADC (V/V), glycerol 0.02% and 0.05% of Tween 80 v/v until exponential phase at 37°C.

2.8 Staining of dead bacteria
THP-1 cells were seeded on glass cover slips in 24-well plates at a density of 500 000 cells per 1 ml per well and incubated with 100 nM PMA for 48 hours in a 5% CO2 incubator. M. bovis BCG-GFP in the exponential grown phase washed twice in PBS (pH 7.4), and resuspended in serum-free RPMI medium. A 10-minute pulse in a water-bath sonicator was followed by passage through a 23-gauge needle to disrupt bacterial clumps. Before infection, residual bacterial aggregates were removed by low-speed centrifugation (120 g) for 2 minutes. Single-bacteria suspension was verified by fluorescence microscopy. Infection proceeded with BCG in the ratio of 1:10 bacteria per macrophage, to achieve about 1–10 bacteria inside macrophages after 4 hours uptake. The cells were washed twice with PBS to remove extracellular bacteria. After a 2-hour chase in medium without bacteria, rifampicine-PLLA NPs were added to the cells at 50 μg and 100 μg to each well – a total volume of 1 ml per well. After 72 hours of incubation, cells were washed with PBS, were permeabilized by 0,1% Triton for 1 min, stained by 2 μg/ml PI 5 min, washed PBS and fixed 3,7 % formaldehyde in PBS 10 min. Antibiotic was replaced freshly every day.

3. Results and Discussion
Due to the fact that the accumulation of rifampicine-PMPs and -PNPs in macrophages can be considerable, we first compared the rifampicin-PLLA NP effect on cell viability with the pure rifampicin toxicity (figure 1). Our data shows that rifampicin-PLLA NPs does not reduce the cell viability even at high concentrations (250 μg/ml) while free rifampicin induces cell death starting from

150 μg/ml. The macrophage THP-1 cells are more rifampicin-resistant compared to the suspension THP-1 cells in which 25 μg/ml of rifampicin induced apoptosis [17].

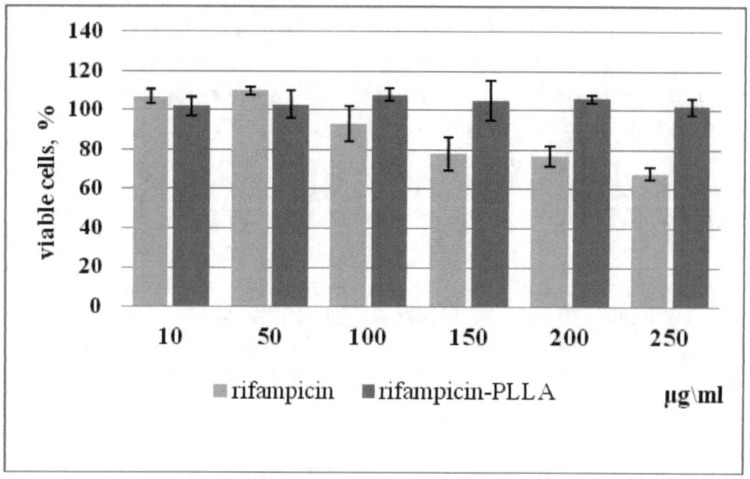

Figure 1. Rifampicin-PLLA NP effect on the viability of macrophage THP-1 cells. Data were expressed as the percent survival of the control (100% the viable cells).

To study rifampicine-PLLA effect on mitochondria, the cells were exposed to 50, 100 and 250 μg/ml of rifampicin-PLLA for 72 h. Rh123 staining showed that functional-active mitochondria formed reticular-like structure in the control cells. The similar reticular-like structures were formed upon treatment rifampicin-PLLA NPs (figure 2).

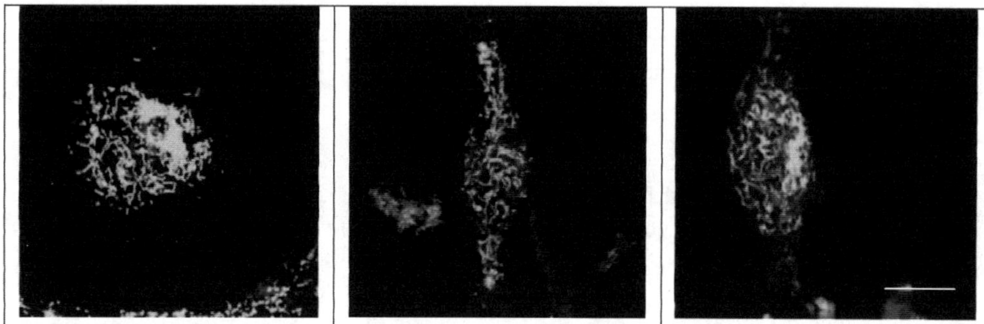

a) b) c)

Figure 2. Mitochondrial reticular-like structure in THP-1 cells exposed to rifampicin-PLLA NPs for 72 h. a) control; b) 50 μg/ml of rifampicin-PLLA NPs; c) 100 μg/ml of rifampicin-PLLA NPs.; Rh123-staining. Bar 10 μm.

Increasing the concentrations of rifampicin-PLLA up to 250 μg/ml allowed identifying the small population of cells with perinuclear localization of mitochondria in the area of the cell centre. Analysis by Mitotracker Orange revealed the disappearance of the reticular-like the mitochondria and their redistribution to the cell centre (7.9 ± 0.3% versus 1.3 ± 0.6% cells in controls) (figure 3).

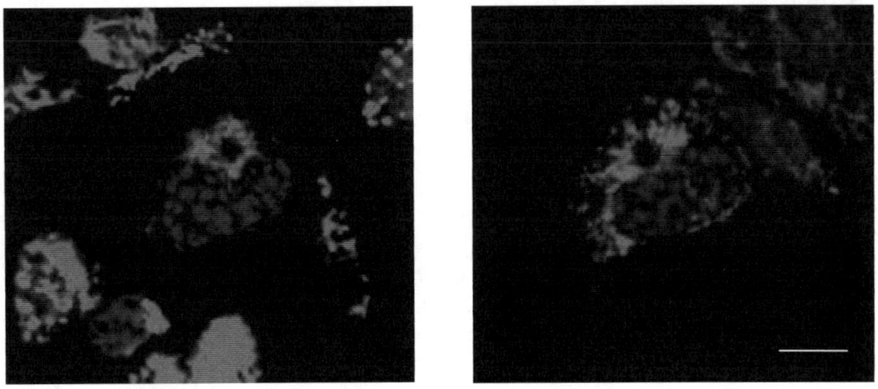

a) b)

Figure 3. The disappearance of mitochondrial reticular-like structures under the rifampicin-PLLA NP exposure for 72 h in THP-1 cells. a) control; b) 250 μg/ml of rifampicin-PLLA NPs.; Staining by Mitotracker Orange. Bar 10 μm.

Induction of apoptotic death could be caused by the changes in the mitochondrial membrane permeability and by releasing of a proapoptotic factor cytochrome *c* from the mitochondria [18, 19]. Therefore, we studied localization of the cytochrome *c* in the untreated and treated macrophages. Under the influence of rifampicin-PLLA, a diffuse distribution of the cytochrome *c* which is specific for apoptotic cells is observed in cells characterized by perinuclear localization of mitochondria (figure 4).

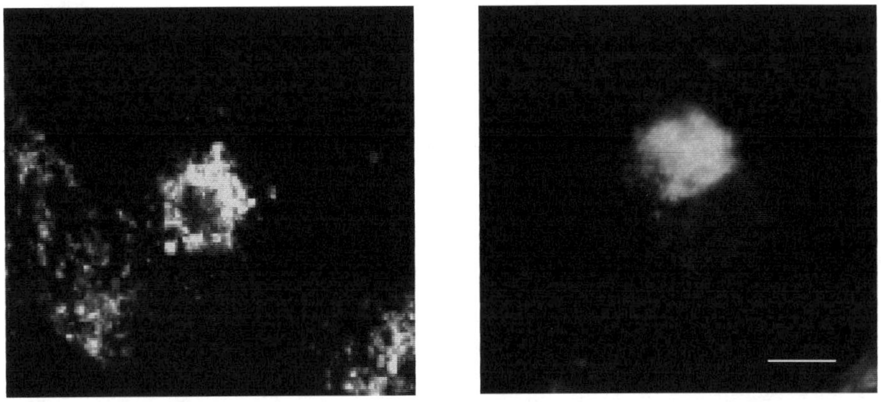

a) b)

Figure 4. Localization of cytochrome *c* under rifampicin-PLLA NP exposition for 72 h in THP-1 cells. a) control, cytochrome *c* localizes inside mitochondria; b) 250 μg/ml of rifampicin-PLLA NPs, diffuse distribution of the cytochrome *c* in perinuclear zone. Staining with first antibodies against cytochrome *c* and the secondary antibodies conjugated with Alexa 488. Bar 10 μm.

In control cells cytochrome *c* had a typical localization inside the mitochondria, including the mitochondria in the perinuclear zones around cell centre (figure 4a). Upon rifampicin-PLLA treatment, cells characterized by perinuclear localization of mitochondria displayed a diffuse distribution of the cytochrome *c* specific for the apoptotic cells (figure 4). Thus, the treatment with rifampicin-PLLA induced the release of cytochrome *c* from the mitochondria to the cytosol, implicating mitochondria in activation of apoptosis. It has been reported that damage to mitochondria and their dysfunction lead to the changes in distribution of mitochondria, development of oxidative stress and trigger the mitochondrial pathway of apoptosis [20]. We suggest that the increased frequency of the cells with perinuclear localization of mitochondria indicates that rifampicin-PLLA NPs induce dysfunction of mitochondria. The release of the cytochrome *c* from the mitochondria to the cytosol is an early event of apoptotic death that leads to activation of effector caspases and degradation of cell structures. We observed the diffuse distribution of the cytochrome *c* in the mitochondria area under the influence of rifampicin-PLLA. It may indicate that membrane organelles such as Golgi apparatus, endoplasmic reticulum, lysosomes and cytoskeleton elements maintain their integrity and localization for a certain period after the release of the cytochrome *c*. We suggest that the treated macrophages reveal specific pathways of the caspase activation and the degradation of the cell structures.

P-glycoprotein (Pgp) is a multridug transporter that belongs to the ATP-Binding Cassette (ABC) family, and is a product of the Multidrug Resistance-1 (MDR-1) gene. Pgp acts as a transporter molecule pumping a wide variety of endogenous substances and drugs from the cytoplasm to the extracellular environment [21].

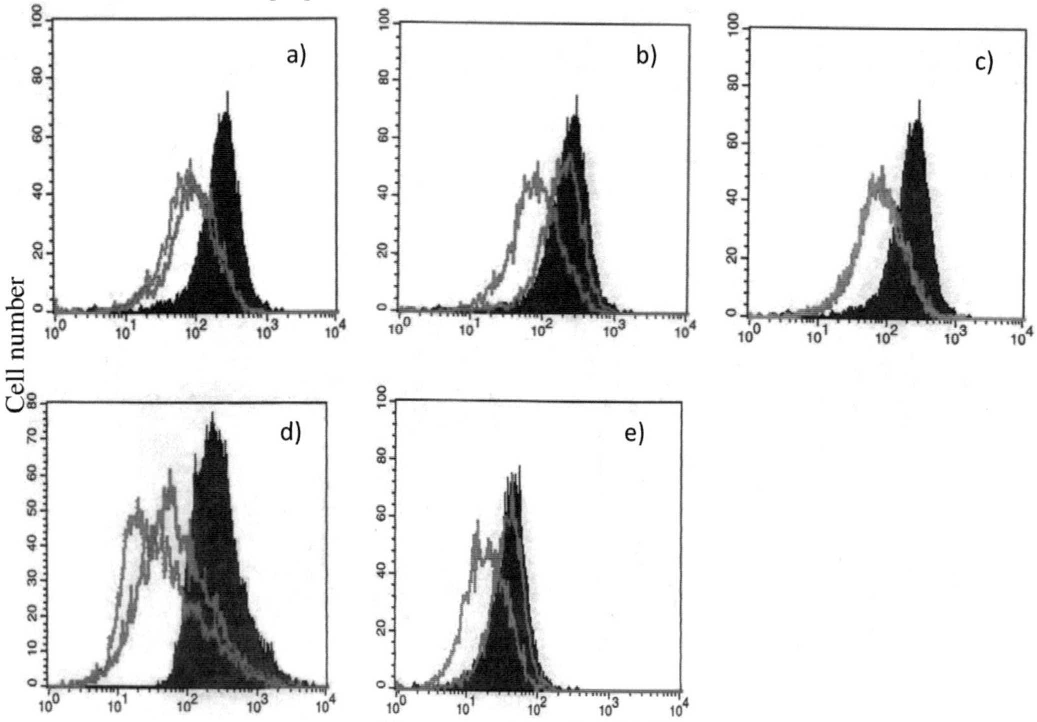

Figure 5. Effects of different drugs on P-glycoprotein activity in KG1 cells. a) 50 μg/ml of rifampicin-PLLA NPs; b) 100 μg/ml of rifampicin-PLLA NPs; c) 100 μg/ml of pure PLLA NPs; d) 10 μg/ml of rifampicin; e) 2 mM of vinblastine (a typical blocker of Pgp). FACScan analysis of the fluorescence intensity; blue areas fluorescence of cells stained with Rh123; green curves - fluorescence of cells stained with Rh123 and incubated in dye-free medium; red curves - fluorescence of cells stained with Rh123 and incubated in the dye-free medium in the presence of drugs.

Rh123 and several other fluorescent dyes are substrates for the Pgp-mediated transport, and Rh123-efflux is the functional test for Pgp-activity. We tested the effect of rifampicin-PLLA NPs (50 and 100 μg/ml) on Rh123 efflux in human leukemic KG1 cell line characterized by high Pgp expression [22]. It has been shown that rifampicin-PLLA NPs affect the Rh123-efflux in a concentration-dependent manner: 50 μg/ml of rifampicin-PLLA has no impact on Rh123 efflux, whereas 100 μg/ml completely blocks the Rh123 efflux (figure 5a and 5b).

We have found that PLLA-NPs alone (100 μg/ml), don't have effect on Rh123 efflux (figure 5c). In contrast, 10 μg/ml of free rifampicin (comparable to rifampicin concentration in the NP composition) inhibits the efflux (figure 5d). Thus, the rifampicin PLLA-NPs have a stronger effect on the functional activity of Pgp as compared to both, rifampicin and PLLA NPs. We propose that effect is reached due to the higher intracellular concentration of free rifampicin after polymer degradation. These data indicate that rifampicin acts as a potential substrate of Pgp and therefore may modulate cell chemosensitivity.

Analysis of rifampicin-PLLA NP effect on viability of *M. bovis BCG-GFP* in the infected macrophage THP-1 cells showed the appearance of dead *M. bovis BCG* (PI-staining is effective only for bacteria with damaged permeability of bacterial cell wall) inside the macrophages (figure 6).

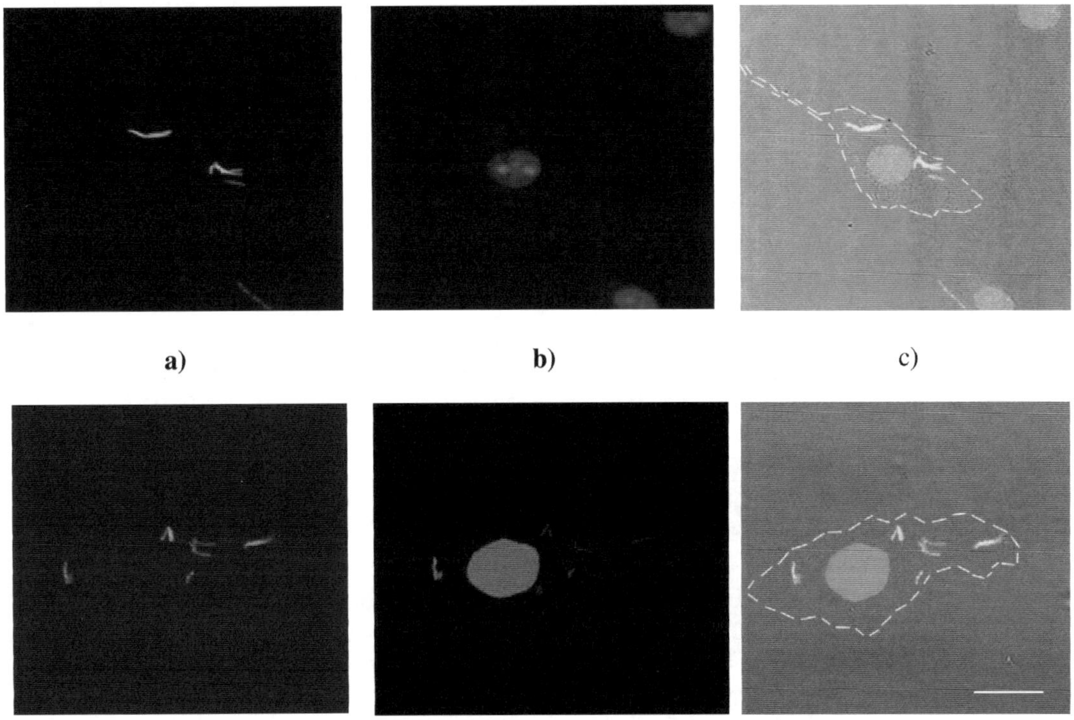

a) **b)** c)

d) e) f)

Figure 6. Influence of rifampicin-PLLA NPs on viability of *M. bovis BCG-GFP* in infected macrophage THP-1 cells. a)-c) – control; d)-f) - 50 μg/ml rifampicin-PLLA NPs; a) and d) intracellular *M. bovis BCG-GFP*; b) and e) PI-staining; c) and f) DIC, merge. Confocal laser microscopy. Bar 10 μm.

4. Conclusion

Our data demonstrate that rifampicin-PLLA NPs: (i) are non-toxic for a significant part of macrophage population; (ii) can induce dysfunction of mitochondria and lead to apoptosis only in a small part of the cell population; (iii) are effective against *M. bovis BCG* in the infected macrophages; (iv) act as potential substrate of Pgp and therefore may modulate cell chemosensitivity. We suggest that after degradation of polymer in phago-lysosomes, the free rifampicin induces death of *M. bovis BCG* and diffuses from phago-lysosomes to cytoplasm causing the dysfunction of mitochondria and modulating the Pgp activity. The increasing concentration of rifampicin-PLLA NPs increases the severity of the toxic effects on the macrophage cells.

Acknowledgments

The authors thank Dr. Valentina Factor for her very helpful suggestions in the preparation of this manuscript. This work was supported by the *Russian Science Foundation* (project №14-50-00029).

References

[1] Young D B, Perkins M D, Duncan K and Barry C E, 2008 *J. Clin. Invest.* **118** 1255
[2] Dube D, Agrawal G P, Vyas S P 2012 *Drug Discov. Today* **17(13–14)** 760
[3] Sosnik A, Carcaboso A M, Glisoni R J, Moretton M A and Chiappetta D A 2010 *Adv. Drug Deliv. Rev.* **62** 547
[4] Shegokar R, Al Shaal L, Mitri K *J. Pharm. Pharmaceut. Sci.* 2011 **14 (1)** 100
[5] Mao S, Guo C, Shi Y and Li L C 2012 *Expert Opinion Drug Delivery* **9** 1161
[6] Mao S, Guo C, Shi Y and Li L C 2012 *Expert Opinion Drug Delivery* **9** 1209
[7] Sasaki T, Matsuura H and Tanaka K 2014 ISRN *Polymer Science* **2014** 1
[8] Campbell E A, Korzheva N, Mustaev A, Murakami K, Nair S, Goldfarb A and Darst S A *Cell* 2001 **104** 901
[9] Bottcher T, Gerber J, Wellmer A, Smirnov A, Fakhrjanali F, Mix E, Pilz J, Zettl U K and Nau R 2000 *J. Infect. Dis.* **181** 2095
[10] Bi W, Zhu L, Jing X, Liang Y and Tao E 2013 *Neurol. Sci.* **34** 137
[11] Erokhina M, Kurynina A, Onishchenko G 2013 *Biochemistry (Mosc)* **78(10)** 1473
[12] Kalluru R, Fenaroli F, Westmoreland D, Ulanova L, Maleki A, Roos N, Madsen PM, Koster G, Egge-Jacobsen W, Wilson S, Roberg-Larsen H, Khuller G K, Singh A, Nyström B and Griffiths G 2013 *J. Cell Sci.* **126** 3043
[13] Anisimova Y, Gelperina S, Peloquin C and Heifets L B 2000 *J. Nanopart. Res.* **2** 165
[14] Hirota K, Hasegawa T, Nakajima T, Inagawa H, Kohchi C, Soma G, Makino K and Terada H 2010 *J. Control. Release* **142** 339
[15] Carmichael J, De Graff W, Gazdar A F, Minna J and Mitchell J 1987 *Cancer Res.* **47** 936
[16] Stromskaya T, Rybalkina E, Shtil A A, Zabotina T N, Filippova N A and Stavrovskaya A A 1998 *Br. J. Cancer* **77** 1718
[17] Erokhina M B, Aleksandrova E A, Prokopenko A V, Lepekha L N and Onishchenko G E 2009 *Tuberkulez Bolezni Legkikh* **11** 49
[18] Bossy-Wetzel E, Goldberg M, Allen T, Barber M J, Green D R and Newmeyer D D 1999 *J. Cell Biol.* **147** 809
[19] Scorrano L, Ashiya M, Buttle K, Weiler S, Oakes S A, Mannella C A and Korsmeyer S J 2002 *Developmental Cell*, **2**, 55
[20] Izyumov D S, Domnina L V, Nepryakhina O K, Avetisyan A V, Golyshev S A, Ivanova O Y, Korotetskaya M V, Lyamzaev K G, Pletjushkina O Y, Popova E N, Chernyak B V 2010 *Biochemistry (Mosc)* **75** 123
[21] Stavrovskaya A A 2000 *Biochemistry (Mosc)* **65(1)** 95
[22] Bailly J, Muller C, Jaffrézou JP, Demur C, Gassar G, Bordier C and Laurent G 1995 *Leukemia* **9(5)** 799

Study of ecologo-biological reactions of common flax to finely dispersed metallurgical wastes

O Zakharova [1], A Gusev [1,2], E Skripnikova[3], M Skripnikova[3], Yu Krutyakov [4,5], A Kudrinsky[4,5], I Mikhailov[2], S Senatova[2], C Chuprunov[2], D Kuznetsov[2]

[1]G.R. Derzhavin Tambov State University, 33, Internatsionalnaya street, Tambov, 392000, Russia
[2]National Research Technological University "MISIS", 2, Leninsky ave., Moscow, 119049, Russia
[3]Michurinsk State Agrarian University, 101, Internatsionalnaya street, Michurinsk Tambov reg., 393760, Russia
[4]M.V. Lomonosov Moscow State University, Department of Chemistry, 1-3 Lenin Hills, Moscow, 119991, Russia
[5]National Research Center "Kurchatov Institute", 1, Academician Kurchatov Square, Moscow, 123182, Russia

E-mail: olgazakharova1@mail.ru

Abstract. Study was carried out on the influence of metallurgic industrial sludge on morphometric and biochemical indicators as well as productivity of common flax under laboratory and field conditions. In laboratory settings negative influence on seed germinating ability and positive influence on sprouts biomass production in water medium were observed. In sand medium suppression of biological productivity under the influence of sludge together with photosynthetic system II (FS II) activity stimulation were registered. Biochemical study showed peroxidase activity decrease in laboratory, while activity of polyphenol oxidase, superoxide dismutase and catalase were given a mild boost under the influence of sludge. In the field trial, positive influence of sludge on flax photosynthetic apparatus was shown. Positive influence of sludge on vegetation and yield indicators was observed. The analysis of heavy metals content showed excess over maximum allowable concentration (MAC) of copper and zinc in control plants, it may point to the background soil pollution. In the plants from the trial groups receiving 0.5 and 2 ton/ha heavy metals content below the control values was registered. Application of 4 ton/ha led to the maximum content of copper and zinc in the plants among the trial groups. The analysis of soils from the test plots indicated no excess over maximum allowable concentrations of heavy metals. Thus, further study of possibilities of using metallurgic industrial sludge as a soil stimulator in flax cultivation at the application rate of 0.5 t/ha seems promising.

1. Introduction

The problem of production and recycling of finely dispersed industry-related wastes is one of the crucial scientific issues of the 21st century lacking thorough study. In Russian Federation about 7 bln ton of industrial wastes is produced yearly, while no more than 28% (2 bln ton) of the volume is recycled. Metallurgical wastes comprise major part of all the unusable wastes.

Content from this work may be used under the terms of the Creative Commons Attribution 3.0 licence. Any further distribution of this work must maintain attribution to the author(s) and the title of the work, journal citation and DOI.
Published under licence by IOP Publishing Ltd

At the present time in RF there are no effective industrial technologies for utilizing finely dispersed ashes and sludges from electrometallurgical and blast-furnace processes containing high amounts of iron and heavy metals such as zinc, arsenic, lead, cadmium, nickel, copper, etc. In spite of high concentration of iron in such wastes amounting up to 40-50%, the presence of heavy metals, especially zinc, in concentrations higher than 0.5%, makes it impossible to use them for agglomeration and blast-furnace production with further iron and steel smelting. As a result, most of the metallurgical companies in Russia utilize such wastes by dumping in disposal areas [1]. The present work suggests a prospective way to utilize the stocks of finely dispersed metallurgical wastes accumulated by the traditional storage practices in Russia. The novelty of the suggested approach is in the attempt to simultaneously solve two problems: ecologically safe bioutilization of metal-containing sludge and development of effective and cheap agricultural fertilizers. For instance, iron is found in a number of plant enzymes, it is also involved in chlorophyll synthesis, in breathing and metabolic processes; zinc plays an important role in protein, carbohydrate and phosphoric metabolism as well as in vitamins and growth substances (auxins) synthesis. In patent literature a method is suggested for ecologically safe bioconversion of finely dispersed metallurgical wastes containing heavy metals [2], but it considers employment of sludge as a component for artificial soil, which inacceptable for a large number of popular industrial crops cultivated in the open.

2. Research target and methods

Research target. For the research a sample of finely dispersed metallurgical waste represented by blast-furnace gas cleaner sludge was used. Blast-furnace sludges contain a number of elements, beside iron, such as zinc, copper, calcium, aluminum, lead, cadmium, etc. Zink content is one of the main factors limiting recycling of such sludges. A sample of metallurgical sludge (blast-furnace gas cleaner dust) was collected from blast-furnace gas cleaner discharge water, it was filtered by a vacuum filter and air dried at room temperature for 48 hours. The dried sludge was homogenized by means of a mechanic triturating machine (Fritsch Pulverisette 2, no on-bottom weight applied, mixing time 15 min.). The prepared sample of dried sludge was stored in an air-tight polyethylene bag at room temperature, the sample's humidity was no higher than 0.8 %. Detailed chemical examination of the sample was carried out according to standard methods, the analysis results are summarized in table 1.

Table 1. Chemical composition of a metallurgical sludge sample

Analyte denomination	Measurement result, %	Specified uncertainty, %	Measurement procedure denomination
Hygroscopic moisture	0.68	±0.20	GOST 23 581.1-79 [3]
Metallic iron	<1	–	GOST 26 482-90 [4]
Total iron	38.4	±9.1	PND F 16.3.24-2000 [5]
Manganese	0.086	±0.029	PND F 16.3.24-2000 [5]
Copper	<0.025	-	PND F 16.3.24-2000 [5]
Oil derivatives	0,028	±0.011	PND F 16.1.41-04 [6]
Nickel	<0.05	-	PND F 16.3.24-2000 [5]
Aluminium oxide	1.38	±0.39	GOST 23 581.17-81 [7]
Iron oxide (II)	3.44	±0.17	GOST R 53657-2009 [8]
Calcium oxide	4.08	±0.59	GOST 23 581.16-81 [9]
Silicon oxide (IV)	5.87	±0.59	GOST 23 581.15-81 [10]
Magnesium oxide	1.67	±0.29	GOST 23 581.16-81 [9]
Losses on ignition	39.3	±0.7	GOST 23 581.13-79 [11]
Sulfur	0.42	±0.1	GOST 23 581.20-81 [12]
Phosphorus	0.043	±0.098	GOST 23 581.19-91 [13]
Chrome	0.117	±0.018	PND F 16.3.24-2000 [5]
Zinc	8.30	±0.72	PND F 16.3.24-2000 [5]

The sludge chosen for the research is characterized by total iron content of 38.4% w/w. While zinc content is the highest among heavy metals amounting to 8.3% w/w.

Phase composition examination was carried out by means of qualitative X-ray phase analysis using a desktop X-ray difractometer Difray 401 (ZAO 'Scientific Instruments', St. Petersburg). A chromium anode tube (Cr-Kα) was used as an X-rays source; with operating current of 4mA and voltage of 25 kV. Focus control was carried out according to Bragg-Brentano X-ray optics. The obtained diffraction pattern is represented in figure 1.

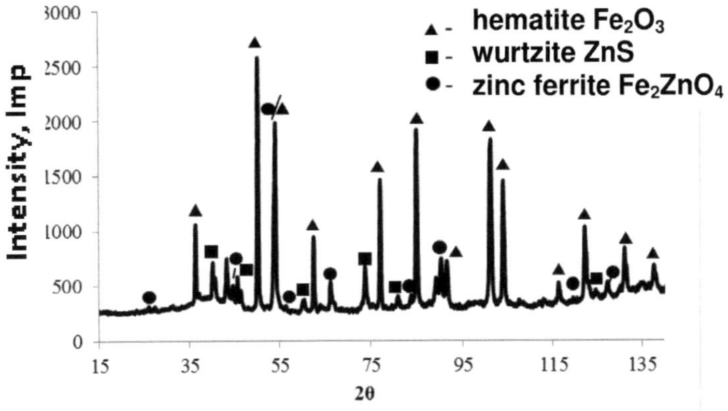

Figure 1. An X-ray diffraction pattern of a metallurgical sludge sample

Quantitative phase study was not carried out due to low amount of phases (except hematite Fe_2O_3, wurtzite ZnS and zinc ferrite Fe_2ZnO_4) in the sample. Nevertheless, the obtained diffractometric data were processed (software for X-ray phase analysis Match! v.2.0) and evaluation of content ratio of the main phases was carried out, the results are presented in table 2.

Table 2. Content ratio of the main phases in a metallurgical sludge sample

Phases	Content, %
Fe_2ZnO_4	12
Fe_2O_3	84
ZnS	4

As one can see from Figure1 and Table 2, the major part of iron in the sample is represented by Fe_2O_3 and Fe_2ZnO_4, a quantity of zinc is represented as wurtzite ZnS phase.

The data on size distribution of metallurgical sludge particles (grain size distribution analysis) were obtained by means of laser diffraction analysis carried out using a Fritsch Analysette 22 NanoTec device in submicro- and micro-range analysis mode. The analysis results in the form of a histogram and a grain size distribution curve are represented in figure 2.

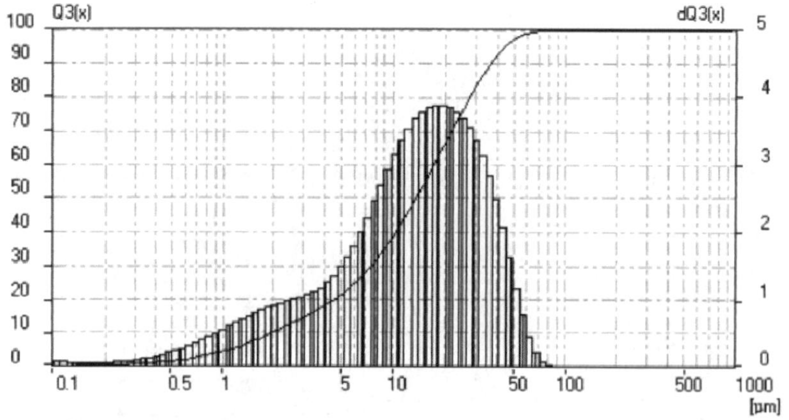

Figure 2. Grain size distribution analysis of a metallurgical sludge sample

It was observed that the sludge particles sizes range from 0.1 to 100 µm. Grain size distribution is close to bimodal distribution by its character with distribution median d50 value of about 14 µm.

Morphology of the particle aggregates in the studied sample was examined by means of scanning electron microscopy (HitachiTM-1000 equipment) and the images are represented below in figure 3.

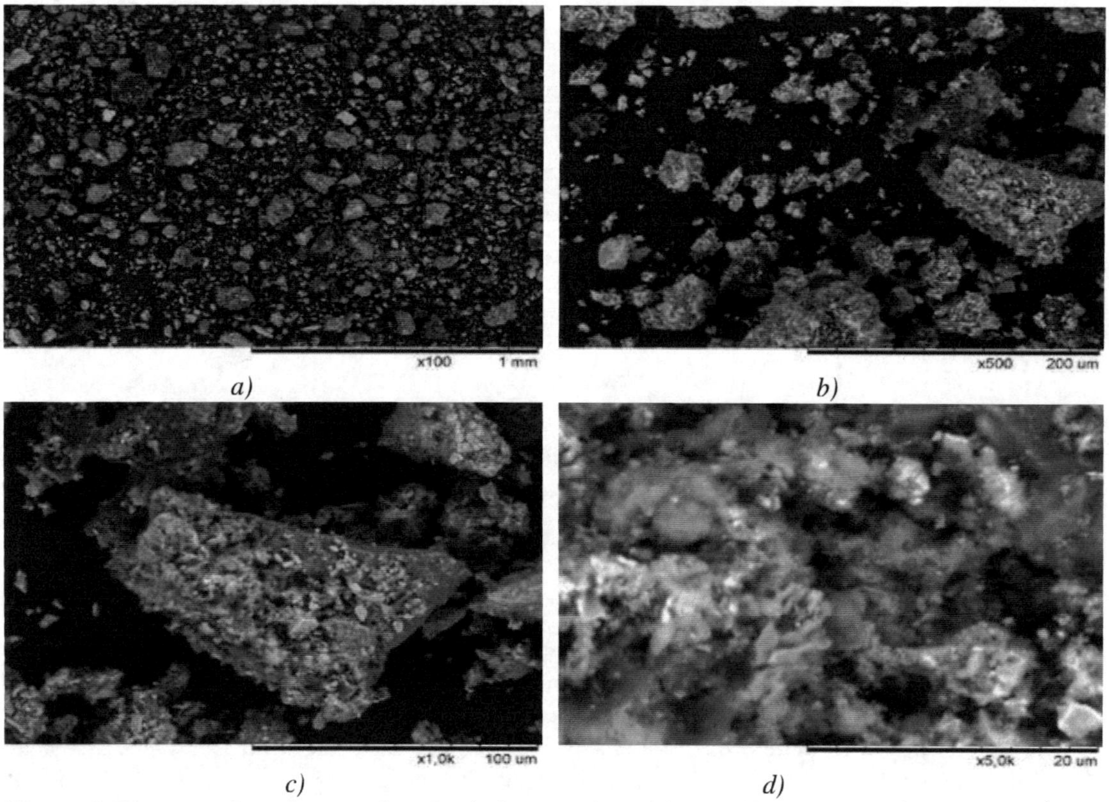

a) *b)* *c)* *d)*

Figure 3. Electron microphotography of a sludge sample (with magnification: a – 100, b – 500, c – 1000, d – 5000)

As one can see from the presented images, a metallurgical sludge sample features finely dispersed powder consisting both of agglomerates and separate particles with irregular morphology (their shape is mostly irregular) and varying dispersiveness. The particle size varies within the range of values from below 1 μm to about 80 μm (for the largest agglomerates).

Test object. Considering possible accumulation of heavy metals in plants, common flax (*Línum usitatissimum L., 1753*), being a widely cultivated commercial crop, was chosen as a test object.

Methods of research. Germination is a critical stage of plant ontogenesis, that is why this stage attracted special attention during the study. Laboratory experiments to evaluate sludge influence on seeds germination were set in water media and in sand soil.

For the experiment in water media sludge suspensions in distilled water (10…0.001%) were prepared. For this purpose batches of metallurgical sludge (100…0.01g = 10…0.001%) were weighed using analytical scales ViBRA HT (Shinko Denshi, Japan, (precision ±0.0001 g)), poured into prepared reservoir containing 1l of water and stirred for 20 sec. with a glass stick. After that the suspension was dispersed by ultrasound with power of 300W and frequency of 23,740 kHz.

For germinations the seeds were placed in humid cameras on paper filters moistened with the liquid (10…0.001%) corresponding the variant. Seeds on filters were placed in Petrie dishes. Each variant of the experiment was produced in 4 replications. In each replication 100 seeds were used. The experiment was divided into two periods. Thus, totally 800 seeds were germinated for each variant. Germination was carried out at 20°C.

The seeds were observed every 24 hours, filters with seeds were sprayed with distilled water.

Germinating energy and germinating ability were measured according to GOST 12038-84 [14] regulations on the 3rd and 7th day accordingly. After that the aboveground and underground parts of the plants were measured, dried for 1 hour at 105°C and weighed.

To discover the peculiarities of seed germination and seedlings development in solid medium a substrate was prepared consisting of sand and sludge in the ratio of 9 weight parts of sand to 1 weight part of sludge (equal to the maximum sludge concentration in water). Triple-washed fluvial sand was used as control substrate.

500 ml plastic containers were used for the experiment. Each one contained either 440g of fluvial sand (control) or 400g of sand and 40g of sludge (experiment), the components of the experimental substrate were mixed thoroughly and moistened till complete wetting. After that the seeds were sown on the surface and the containers were covered with lids to create humid camera conditions. The containers were initially placed in darkness, they were placed in the light when the seedlings appeared. Biological productivity of the plantings was evaluated on the 7th and 15th day. For this purpose samples of plants with the substrate were taken with a probe of 1 sm^2 in diameter. The substrate was washed off, excess of water was removed from the plants by filter paper and the plants weight was measured (first wet weight was measured and then, after drying in boxes at 105°C, dry weight was measured).

Besides, activity of antioxidant system enzymes, such as peroxidase, polyphenol oxidase, superoxide dismutase and catalase was analyzed in seedlings grown in solid medium [15-17].

For the study of metallurgical sludge influence on biological and industrial productivity of flax in the field conditions the following doses of sludge were used: 0.5 t/ha, 2 t/ha and 4 t/ha. During the field experiment the following indicators were measured: main photosynthetic characteristics of flax plantings [18], main biometric and yield characteristics [19]. After the experiment was accomplished, analysis of heavy metals content in the plant tissues and in the soil was carried out [20-24].

3. Results and discussion

Water medium. It was ascertained that sludge has negative influence on germinating energy and germinating ability of flax seeds (up to 20% decrease was observed) (figure 4).

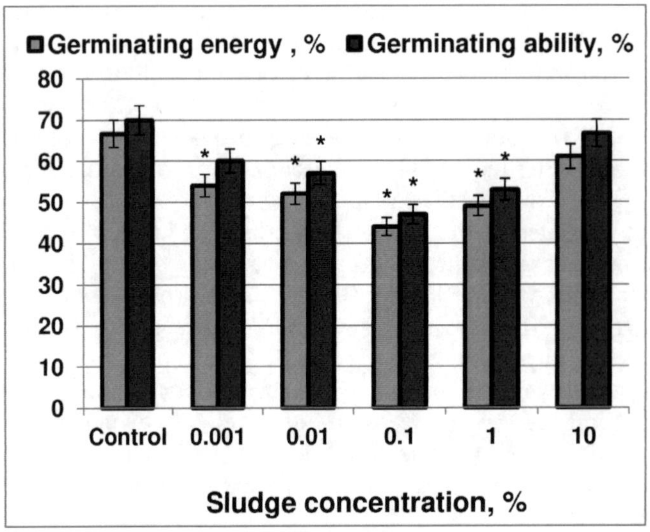

Figure 4. Influence of metallurgical sludge in water medium on germinating energy and germinating ability of common flax

Besides, decrease of the root length was observed by 30-40% in the whole, though the value of average weight of the seedlings root part displayed no difference from the control group, while at 0.01% concentration consideable, more than twofold, increase in root weight was registered (figure5). Average stalk length underwent no considerable alteration under the influence of sludge, while average stalk weight exceeded the control values by 1.5-2 times at some concentrations.

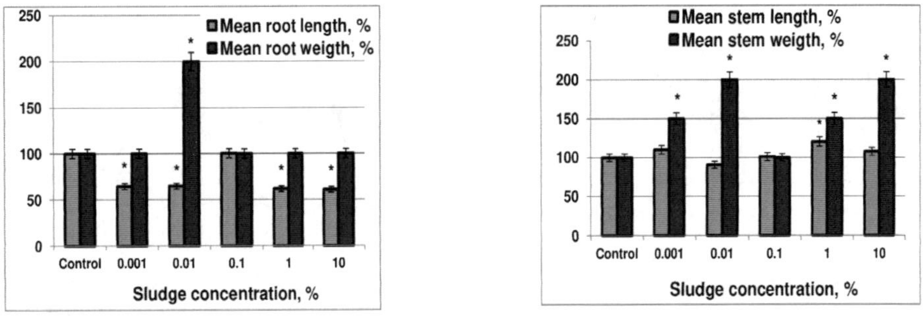

Figure 5. Influence of metallurgical sludge in water medium on morphometric characteristics of common flax seedlings

Sand medium. Comparison of biological productivity of the plantings in sand medium on the 7th day showed that high doses of sludge (10% by weight) have negative influence on the studied characteristic (figure 6 (a)). On the 15th day after germination the characteristics of slow chlorophyll fluorescence induction differed in the test and control variants, growing power coefficient being higher in the experimental variant [18]. It was accompanied by increase in the plants' desiccation resistance [25]. After 3-day incomplete drying the experimental seedlings totally recovered their turgor within an hour, while only 80% of the control plants recovered their turgor within 24 hours after being placed in humid cameras.

At the same time, sludge application to the sand substrate stimulates photosynthetic system II (PS II) activity which may lead to the plantings bioproductivity increase (figure 6 (b)).

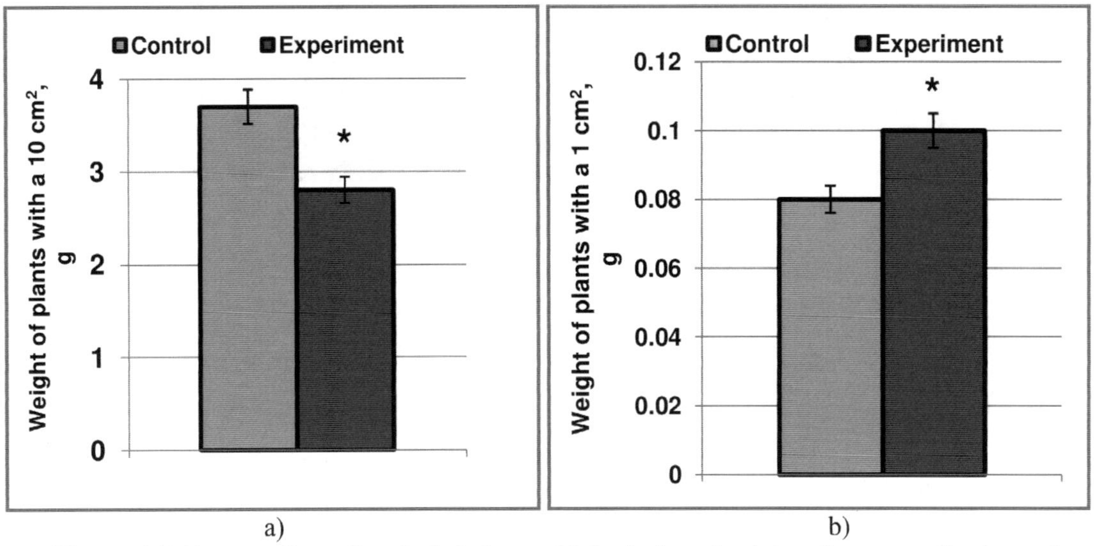

Figure 6. Influence of metallurgical sludge on biological productivity of common flax in sand medium: a) 7th day; b) 15th day

Seed germination is accompanied by a variety of biochemical processes and performance of oxidation-reduction reaction enzymes is the most important one. In the process of seed germination active lipids oxidation can lead to decrease in germination rate and in growth-regulating activity of seedlings. The study showed that in flax peroxidase activity was decreased in the experimental plants, while polyphenol oxidase, superoxide dismutase and catalase activity slightly increased under the influence of sludge (table 3). Association of enzyme activity in roots with seedlings growth activity proves the crucial role of these enzymes in formation of protective mechanisms, taking place in a plant undergoing influence from external factors which can stimulate the activity of oxidation processes in a cell.

Table 3. Influence of metallurgical sludge on enzyme activity in seedlings roots on the 7th day

Variant	Peroxidase, umol/min g d.w.	Polyphenol oxidase, rel. act. unit/min	Superoxide dismutase, rel. act. unit	Catalase, umol/min g
Control	264.0±16.8	13.4±1.7	15.8±1.1	0.79±0.02
Experiment	249.0±9.9	13.8±1.5	16.4±1.0	0.84±0.01

Field experiment: Assessment of the impact of finely dispersed sludge wastes on photosynthetic apparatus of common flax grown in the field showed that flax plants receiving 0.5 t/ha of sludge formed the largest assimilating surface amounting to 20.76×10^3 m^2/ha. The minimal assimilating surface area, amounting to 18.07×10^3 m^2/ha was registered in plants from the control group (figure 7). Photosynthetic characteristics (PSC) varied accordingly (table 4).

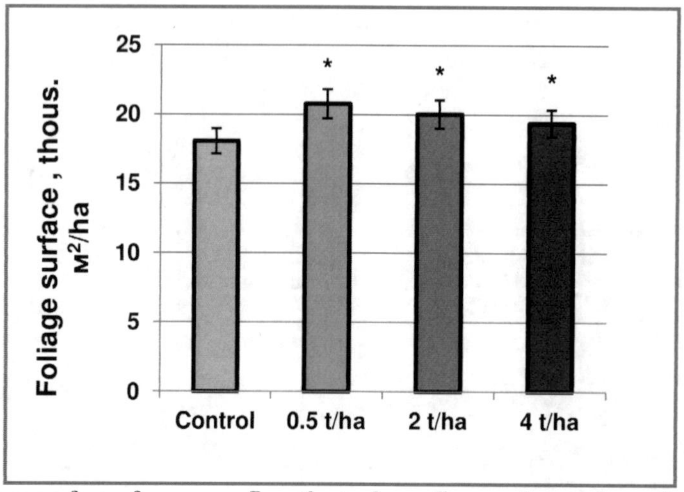

Figure 7. Foliage surface of common flax plants depending on the amount of applied product

Table 4. Main photosynthetic characteristics of common flax plantings (flowering period)

Variant	PSC, 10^3 m^2 day/ha	Dry matter yield, t/ha	Net primary productivity, g/(m^2day)
Control	371±8	2.65±0.11	7.43±0.19
0.5 t/ha	438±9	3.71±0.18	8.33±0.21
2 t/ha	407±10	3.38±0.14	8.12±0.16
4 t/ha	379±9	2.95±0.11	7.86±0.17
$LCD_{.05}$	12.6	0.27	0.36

Increase in foliage surface affects photosynthetic potential changing it from 371×10^3 m^2day/ha on the control plot to 438×10^3 m^2day/ha on the plot with 0.5 t/ha amount of introduced slurry. Increase in photosynthetic potential has been observed at the plots with 2 t/ha or even 4 t/ha sludge.

Maximal photosynthesis net yield 8.33 g/m^2day is recorded with 0.5 sludge, while minimal yield 7.43 g/m^2day is recorded on the control plot.

On the plots with 0.5 t/ha and 2 t/ha sludge growth acceleration has been observed at all stages. Seedlings appear 3-4 days earlier than on the control plot, and blooming and seed ripening is intensified (figure 8).

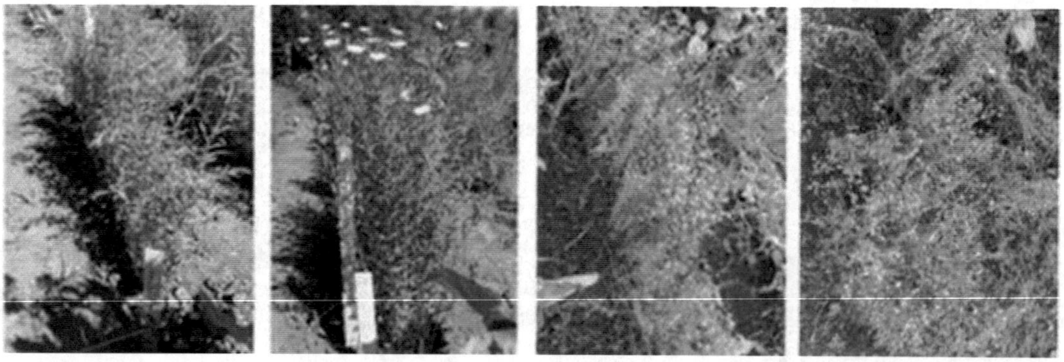

a) b) c) d)

Figure 8. Flowering and bearing of common flax: (a), (b) – control, (c), (d) – 2 t/ha

Sludge application leads to plant height increase by 17% for 4 t/ha solution amount and 25% for 0.5 t/ha solution amount (figure 9).

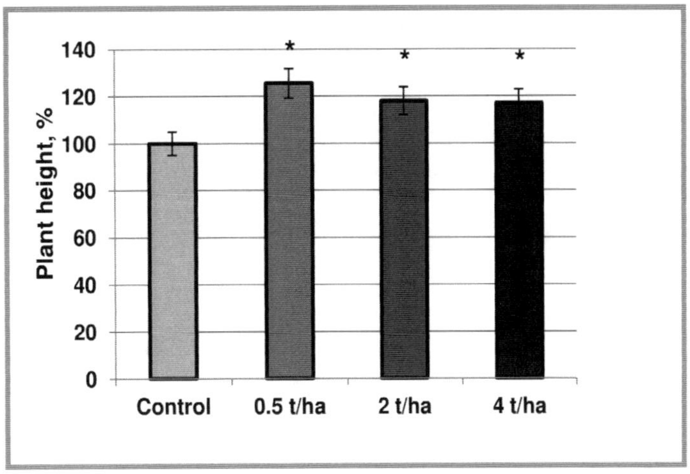

Figure 9. Common flax plant height depending on the amount of applied sludge

Flax yield is constituted by total number of plants per unit area, the quantity of seeds per plant and seed weight. Flax yield component analysis shows different impact of the sludge depending on applied amount and specific yield component (figure 10).

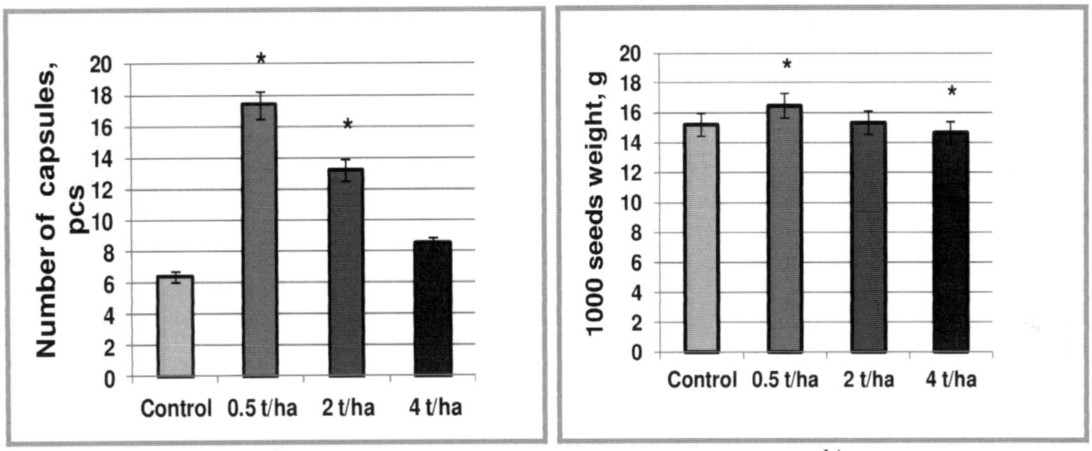

a) b)

Figure 10. Flax crop yield depending on the amount of applied sludge: a) seed heads per 1 flax plant ($LCD_{.05}$ = 2.32); b) average weight of 1000 seeds ($LCD_{.05}$ = 0.45)

The largest number of seed capsules per plant has been recorded on the lot with 0.5 t/ha sludge and equals 17.4, while the lowest number equals 6.4 and was found on the control plot (figure 10a). The heaviest seeds have been collected on 0.5 t/ha sludge lot also. Mass of 1000 of those seeds equals 16.46 g, while the minimal mass equals 14.7 and is recorded on 4 t/ha plot (figure 10b).

As could be seen from the above, to increase flax yield and plant quality, i.e. its height, the optimal amount of introduced sludge is 0.5 t/ha. Besides, reliable positive impact of higher sludge amount on flax growth has been seen also, but particular attention should be paid to ecological safety.

Accounting for high heavy metals concentration in the sludge, control and test plants have been tested on content of heavy metals which are known to be in the sludge and regulated by Russian sanitary standard SanRaN 42-123-4089-86 [26] (table 5). Content of the heavy metals which MAC in soil is regulated by [27] has been monitored also (table 6).

Table 5. Heavy metal content in mg/kg depending on sludge amount in t/ha.

Element	Roots				Stem				Seeds				MAC*
	0	*0.5*	*2*	*4*	*0*	*0.5*	*2*	*4*	*0*	*0.5*	*2*	*4*	
Pb	0.31	0.39	0.41	0.43	0.16	0.18	0.21	0.19	0.21	0.30	0.28	0.43	0.5
Cd	0.053	0,051	0.064	0.075	0.039	0.045	0.043	0.051	0.043	0.053	0.059	0.086	0.1
Cu	7.90	11.52	13.91	9.62	4.72	4.04	4.79	2.51	30.3	9.57	11.66	50.2	10
Zn	7.10	25.46	30.63	22.50	26.1	20.7	27.90	10.84	140.0	34.8	51.54	73.0	50
Fe	388.0	411.3	1132.0	1240.0	91.6	91.3	79.7	200.1	89.6	15.0	26.9	223.5	-

* SanRaN RF 42-123-4089-86 [26]

As could be seen from table 5, roots and seeds heavy metals content exceed MAC on zinc and copper, moreover MAC is exceeded in seeds more than three times in control that could be an evidence of soil background contamination. In case of plants from the 0.5 and 2 t/ha sludge amount plots, a minor excess was observed, but 4 t/ha amount result in maximal copper and zinc content among tested groups. Maximal iron content is recorded at plant roots from 4 t/ha lot.

Table 6. Heavy metal content in mg/kg in the soil after harvesting.

Element	Control	0.5 t/ha slurry introduced	2 t/ha slurry introduced	4 t/ha slurry introduced	MAC in soil*
Pb	0.4	0.91	0.89	0.93	6.0
Cu	0.1	0.11	0.17	0.44	3.0
Co	0.1	0.14	0.16	0.36	5.0
Zn	3.6	7.07	12.2	13.56	23.0
Cr	3.49	4.91	4.54	5.08	6.0
Ni	0.96	1.63	1.82	2.46	4.0

*HS 2.1.7.2041-06 [27]

As could be seen from table 6, MAC in soil has been never exceeded.

4. Conclusions

It is found in laboratory experiment that sludge presence suppresses flax germination and average root length in water environment, but average root mass remains the same and even increases twofold at 0.01% sludge concentration. Average stem length does not change significantly in the presence of sludge, while its average mass increases by 1.5 times at 0.001% and 1% sludge concentrations, and twofold at 0.01% and 10% sludge concentrations.

Biological productivity is found to be reduced by sludge application in sand soil, but viability and resistance to water deficiency increase with slurry introduction. Besides, it stimulates photosynthesis that could be explained by high zinc and iron content in the sludge that enhance ascorbic acid and chlorophyll synthesis thus affecting photosynthesis, plant growth, overheating resistance and protein metabolism [28]. Biochemical study educes reduction of peroxidase activity, slight increase in polyphenoloxidase, superoxidedismutase and catalase activities. It could be accounted for active oxygen including radical generation under unfavorable conditions resulting in diverse alterations of enzymes activity depending on stress factor intensity and duration along with plant susceptibility and growth stage [29].

Evaluation of highly dispersed sludge impact on flax photosynthesis in the field shows the highest assimilating surface of 20.76×10^3 m^2/ha has been recorded in case of 0.5 t/ha amount of sludge, while the lowest one of 18.07×10^3 m^2/ha has been recorded at the control plot. Photosynthetic potential and photosynthesis net production increase with introduction of 0.5 t/ha sludge accordingly from 371×10^3 to 438×10^3 m^2day/ha for the former and from 7.43 to 8.33 g/m2day for the latter. 0.5 and 2 t/ha applied amounts of sludge accelerate plant development and intensify blooming and seed ripening. Sludge application leads to increase of flax height up to 25% at 0.5 t/ha amount and yield increase at all tested amounts. Both maximal number of seed capsules and average seed mass has been registered at 0.5 t/ha sludge amount. Yield increase can be accounted for the presence of zinc and copper in the sludge. It is known that zinc is involved deeply into physiological and biochemical processes, and its deficit leads to metabolic disorder, plant productivity and quality reduction. Besides, zinc enhances plant resistance to unfavorable meteorological conditions such as excessive or deficient amounts of water or insolation, while copper reduces plant sensitivity to fungous and bacterial diseases [30].

Heavy metals content analysis discovers exceeds over MAC on zinc and copper, notably on the control lot it is found to be the highest and exceeds MAC by about three times that could be accounted for background soil contamination. In case of 0.5 and 2 t/ha of sludge applied those exceeds are insignificant, while 4 t/ha results in maximal copper and zinc content among the sludge treated plants. No MAC excess in soil examination on heavy metals content after harvesting have been found in any tested amount of introduced sludge.

Results of this work are in agreement with that of research [31], where sludge stimulating effect on rapeseed plant growth has been found. Thus, further studies of metallurgical sludge usage as a flax soil growth stimulator with introduced amount of about 0.5 t/ha seems to be of promising.

Acknowledgments

The authors gratefully acknowledge the financial support of the Ministry of Education and Science of the Russian Federation (project no. RFMEFI57814X0080) (in part of analysis of the properties of colloidal suspensions) and within the framework of Increase Competitiveness Program of NUST «MISiS».

References

[1] Bryzgalov S, Pugin K, Weisman Ya, Maximovich N and Kalinina E 2009 Effect of placement of domain metallurgical slag on the environment *Issue of articles "Waste: ecology, technology, resource» VI International Forum on Waste Management and Environmental Technology "VeysTEK 2009"* (Moscow) pp 35-39 (in Russian).

[2] Patent RF 2541642 2015 The process of bioconversion of highly clean steel industry waste containing heavy metals (in Russian).

[3] GOST RF 23 581.1-79 Iron ores, concentrates, agglomerates and pellets. Method for determination of hygroscopic moisture (in Russian).

[4] GOST RF 26 482-90 Iron ores, concentrates, agglomerates and pellets prereduced. Method for determination of metallic iron (in Russian).

[5] Natural Federal Regulations 16.3.24-2000 Quantitative chemical analysis of soil. Methods of measurement of mass fraction of metals (iron, cadmium, aluminum, magnesium, manganese, copper, nickel, calcium, chromium, zinc) in samples of industrial waste (slag, sludge metal production), atomic absorption method (in Russian).

[6] Natural Federal Regulations F 16.1.41-04 Quantitative chemical analysis of soil. Methods of measurement of the mass concentration of petroleum products in the soil samples by gravimetric method (in Russian).

[7] GOST RF 581.17-81 23 Iron ores, concentrates, agglomerates and pellets. Methods for determination of aluminum oxide (in Russian).

[8] GOST RF 53657-2009 Iron ores, concentrates, agglomerates and pellets. Method for determination of iron (II) in terms of oxide (in Russian).
[9] GOST RF 581.16-81 23 Iron ores, concentrates, agglomerates and pellets. Methods for determination of calcium oxide and magnesium oxide (in Russian).
[10] GOST RF 581.15-81 23 Iron ores, concentrates, agglomerates and pellets. Methods for determination of silica (modified N 1) (in Russian).
[11] GOST RF 581.13-79 23 Iron ores, concentrates, agglomerates and pellets. Method for determination of loss on ignition (in Russian).
[12] GOST RF 581.20-81 23 Iron ores, concentrates, agglomerates and pellets. Methods for determination of sulfur (in Russian).
[13] GOST RF 581.19-91 23 Iron ores, concentrates, agglomerates and pellets. Methods for determination of phosphorus (in Russian).
[14] GOST RF 12038-84 Agricultural seeds. Methods for determination of germination (in Russian).
[15] Ermakov A, Arasimovich V, Jarosch N, Peruvian J, Lukovnikova G and Ikonnikova M 1987 *Methods of biochemical research plants*. (Leningrad: Agropromizdat) p 44 (in Russian).
[16] Sibgatullina G, Haertdinova L, Gumerova E, Akulov A, Kostyukova Yu, Nikonorova N and Rumyantseva N 2011 *Methods for determining the redox status of cultured plant cells: a teaching aid*. (Kazan: Kazan (Volga) Federal University) p 61 (in Russian).
[17] Chupahina G 2000 *Physiological and biochemical analysis methods of plant*: *Workshop* (Kaliningrad: Kaliningrad University) p 59 (in Russian).
[18] Korneev D, 2002 *Information possibilities of the method of chlorophyll fluorescence induction*. (Kiev: Alterpres), p 188 (in Russian).
[19] Vashenko I 1991 *The workshop on the basics of agriculture* (Moscow: Education) p, 431 (in Russian).
[20] GOST RF 27998-88 Fodder plant. Methods for determination of iron (in Russian).
[21] GOST RF 26931-86 Raw materials and food products. Methods for determination of copper (in Russian).
[22] GOST RF 26934-86 Raw materials and food products. Method for determination of zinc (in Russian).
[23] Methodical Guidelines for determination of heavy metals in the soils of farmland and crop production. M.TSINAO 1992 (in Russian).
[24] GD 52.18.289-90. Guidance document. Methodical instructions. Methods of measurement of the mass fraction of mobile forms of metals (copper, lead, zinc, nickel, cadmium, cobalt, chromium, manganese) in soil samples by atomic absorption analysis (in Russian)
[25] Tretyakov N, Karnauhova T and Panichkin L1990 *Workshop on Plant Physiology*. (Moscow) Agropromizdat, 271 (in Russian).
[26] SanRaN RF 42-123-4089-86. Sanitary Rules and Norms. Maximum permissible concentrations of heavy metals and arsenic in food raw materials and food products (in Russian)
[27] Hygienic Standards 2.1.7.2041-06 Maximum permissible concentration (MPC) of chemicals in the soil (in Russian).
[28] Abdurakhmonova Z 2002 Influence kompleksonatov iron and zinc on physiological and biochemical parameters of cotton: Author. Dis. cand. biol. Sciences (Dushanbe, RF)
[29] Kutskir M 2014 Determination of the environmental safety of nanomaterials based on morphological and physiological and biochemical indices of agricultural crops. Author. Dis. cand. biol. Sciences (Ryazan, RF)
[30] Solovyov A, 1989 *Flax cultivation*. - 2nd ed., rev. and ad., (Moscow: Agropromizdat), 320 (in Russian).
[31] Gusev A, Akimova O, Zakharova O, Godymchuk A, KrutyakovYu, Klimov A, Denisov A, and Kuznetsov D 2014 *Adv Mater Res* **880** 212

Membrane materials for storing biological samples intended for comparative nanotoxicological testing

A Metelkin[1], D Kuznetsov[1], E Kolesnikov[1], K Chuprunov[1] , S Kondakov[1,2], A Osipov[1,2] , J Samsonova[1,2]
[1] National Research Technological University "MISiS", 2, Leninsky ave., Moscow, 119049, Russia
[2] M.V. Lomonosov Moscow State University, 1-3 Lenin Hills, Moscow, 119991, Russia

E-mail: ksekse@mail.ru

Abstract. The study is aimed at identifying the samples of most promising membrane materials for storing dry specimens of biological fluids (Dried Blood Spots, DBS technology). Existing sampling systems using cellulose fiber filter paper have a number of drawbacks such as uneven distribution of the sample spot, dependence of the spot spreading area on the individual biosample properties, incomplete washing-off of the sample due to partially inconvertible sorption of blood components on cellulose fibers, etc. Samples of membrane materials based on cellulose, polymers and glass fiber with applied biosamples were studied using methods of scanning electron microscopy, FT-IR spectroscopy and surface-wetting measurement. It was discovered that cellulose-based membrane materials sorb components of biological fluids inside their structure, while membranes based on glass fiber display almost no interaction with the samples and biological fluid components dry to films in the membrane pores between the structural fibers. This characteristic, together with the fact that membrane materials based on glass fiber possess sufficient strength, high wetting properties and good storage capacity, attests them as promising material for dry samples of biological fluids storage systems.

1. Introduction

In 1963 a Scottish scientist Robert Guthrie introduced into practice diagnostics of phenylketonuria in newborn infants by analyzing blood samples dried-up on sheets of porous cellulose paper [1]. Since then this method of sample preparation (DriedBloodSpots or DBS) gained wide popularity in many countries for so called neonatal screening of infants. In spite of large-scale application of the approach, based on obtaining dried blood spots for neonatal screening, by specialists in many countries, it is only in recent times that we see increase in the number of works where DBS technology is applied not only in medical diagnostics, but also in such areas as veterinary, pharmacokinetic studies, genetic diagnostics, creating specialized data banks, etc. [2, 3, 4]. Drawbacks of the existing sampling systems using cellulose fiber filter paper are: uneven distribution of the sample spot, dependence of the spot spreading area on the individual biosample properties, incomplete washing-off of the sample due to partially inconvertible sorption of blood components on cellulose fibers, etc. That is why studies aimed at discovering new materials for collecting and storing biological samples, including samples for nanotoxicological research, seem so promising.

2. Research target and methods

Content from this work may be used under the terms of the Creative Commons Attribution 3.0 licence. Any further distribution of this work must maintain attribution to the author(s) and the title of the work, journal citation and DOI.
Published under licence by IOP Publishing Ltd

Research target. The work studied characteristics of the following membranes: cellulose membrane TFN («Munktell», Germany); glass fiber membrane 8964 («Ahlstrom», USA) and glass fiber membrane GFSP 223000 («Millipore», USA), as presented in the Table 1.

Table 1. Membrane materials

Denomination	Membrane	Chemical composition
Paper 01 (P01)	CFSP223000 Lot 92072701	Cellulose fiber
Paper 02 (P02)	CFCP203000 Lot 92073601	Glass fiber
Paper 03 (P03)	MAPDS-0300 Lot 4072402 (Grade 8964)	Glass fiber
Paper 04 (P04)	TFN 179 г/см2 Lot 11156	Cellulose fiber
Paper 05 (P05)	Grade 111 102170	Cotton fiber
Paper 06 (P06)	Grade 121 102168	Cotton fiber
Paper 07 (P07)	Grade 142 102563	Cotton fiber
Paper 08 (P08)	Grade 1281 102446	Cellulose fiber
Paper 09 (P09)	Grade 6613 Lot 496651	Polymer fiber
Paper 10 (P10)	Grade Lot091061	Polymer fiber
Paper 11 (P11)	Grade 8950 Lot 106951	Glass fiber
Paper 12 (P12)	Grade 8951 Lot 106951	Glass fiber
Paper 13 (P13)	Grade 8964 Lot 100791	Glass fiber
Paper 14 (P14)	Grade 320 Lot 102674	Cotton fiber
Paper 15 (P15)	Grade 222 Lot 106671	Cotton fiber
Paper 16 (P16)	Grade 237 Lot 101461	Cotton fiber
Paper 17 (P17)	Grade 238 Lot 101666	Cotton fiber
Paper 18 (P18)	Grade 319 Lot 102464	Cotton fiber
Paper 19 (P19)	Grade 601 Lot 102915	Cotton fiber
Paper 20 (P20)	Grade 270 Lot 102741	Cotton fiber
Paper 21 (P21)	CytoSep 1660 Lot 8290501	Mix of natural and synthetic fibers
Paper 22 (P22)	CytoSep 1663 Lot 829053	Mix of natural and synthetic fibers

Research methods. In order to study the properties of membranes after biological fluid application, samples about 5 mm on a side were impregnated with blood serum by dipping the membranes into a biosample till complete saturation with subsequent air drying at room temperature.

The membrane material samples were studied by methods of scanning electron microscopy (scanning electron microscope JEOLJSM-6610LV, «JEOL», Japan), angle of wetting analysis (equipment KRUSSEASYDROP, «KRÜSS», Germany), mechanical tension testing (universal testing machine Zwick/Roellz020, «Zwick», Germany), surface composition analysis by means of FT-IR spectroscopy (FT-IR spectrometer Nicolet 380, «Thermo Scientific», USA).

3. Results and discussion

It was discovered that membranes based on cellulose materials tend to sorb components of the biological fluid inside their structure, while membranes based on glass fiber display almost no interaction with the samples and biological fluid components dry to films in the membrane pores between the structural fibers (Figure 1).

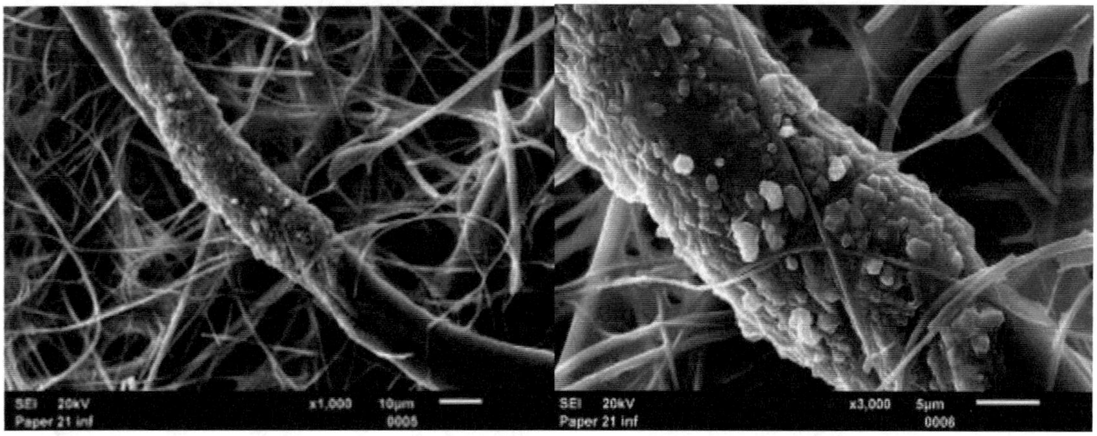

a b

Figure 1. Microphotograph of a P21 sample (mix of natural and synthetic fibers) presenting an area with the urea layer on fiberglass (a) and urea crystal in and on cellulose fiber materials (b)

In addition it has been discovered that the samples of glass fiber membranes possess the highest strength among the studied samples, therefore they can withstand higher external mechanical stresses. Also it has been shown that glass fiber membranes require thrice less time for soaking than cellulose ones, which makes them more convenient for sampling biological fluids by application to a thin stripe of membrane material. Such characteristics as capillary rise and specific volume of glass fiber membranes are also higher than those of cellulose-based samples.

4. Conclusions

It has been proved that cellulose-based membranes sorb components of biological fluids inside their structure, while membranes based on glass fiber display almost no interaction with the samples and biological fluid components dry to films in the membrane pores between the structural threads. This characteristic, together with the fact that membrane materials based on glass fiber possess sufficient strength, high wetting properties and good storage capacity, attests them as a promising material for storage systems for dry samples of biological fluids.

Acknowledgments

The authors gratefully acknowledge the financial support of the Ministry of Education and Science of the Russian Federation within the Federal Target Program "Research and development on priority directions of scientific-technological complex of Russia in 2014-2020" (project No. RFMEFI57814X0010).

References

[1] Guthrie R and Susi A 1963 A simple phenylalanine method for detecting phenylketonuria in large populations of newborn infants*Pediatrics*. **32**(3):338

[2] McDade T, Williams S and Snodgrass J 2007 What a drop can do: dried blood spots as a minimally invasive method for integrating biomarkers into population-based research*Demography* **44**(4) 899

[3] Smit P W., Elliott I, Peeling R W., Mabey D and Newton P N. 2014 An Overview of the Clinical Use of Filter Paper in the Diagnosis of Tropical Diseases. *Am. J. Trop Med Hyg*. **90**(2) 195.

[4] Meesters R and Hooff G 2013State-of-the-art dried blood spot analysis: an overview of recent advances and future trends. *Bioanalysis***5**(17) 2187.

IOP Conf. Series: Materials Science and Engineering **98** (2015) 012020 doi:10.1088/1757-899X/98/1/012020

Metal nanoparticles in DBS card materials modification

A Metelkin[1], G Frolov[1], D Kuznetsov[1], E Kolesnikov[1], KChuprunov[1], S Kondakov[1,2] , A Osipov[1,2], J Samsonova[1,2]

[1] National Research Technological University "MISIS", 2, Leninsky ave., Moscow, 119049, Russia

[2] M.V. Lomonosov Moscow State University, Department of Chemistry, 1-3 Lenin Hills, Moscow, 119991, Russia

E-mail: ksekse@mail.ru

Abstract. In the recent years the method of collecting and storing Dried Blood Spots (DBS) on special cellulose membrane (paper) has gained wide popularity. But possible damage of biosamples caused by microorganisms in case of their incomplete drying is a disadvantage of the method. It can be overcome by treating sample-collection membranes with colloidal solutions of metal nanoparticles, having antibacterial effect. The team studied antibacterial properties of nonwoven material samples with various coatings (alcohol sols of copper, aluminium, iron, titanium, silver and vanadium nanoparticles). Colloidal solutions of nanoparticles were obtained by means of electroerosion method with further low-temperature plasma condensation. Antibacterial activity of fiberglass and cellulose membrane samples with nanoparticle coatings was studied using B. cereus and plaque bacteria cultures. It was revealed that nanostructured coatings can suppress bacterial activity; in addition they can diffuse from the membrane surface into medium which leads to widening the areas of inhibiting testing cultures' growth. Thus, membrane materials treatment with alcohol-sols of metal nanoparticles can be seen as promising for conferring antibacterial properties to DBS carriers.

1. Introduction

In the recent years the method of collecting and storing Dried Blood Spots (DBS) on special cellulose membrane (paper) has gained wide popularity. But possible damage of biosamples caused by microorganisms in case of their incomplete drying is a disadvantage of the method. It can be overcome by treating the collecting membranes with metal nanoparticles having antibacterial properties. Modification of membranes surfaces with nanomaterials and polymers with antibacterial properties is among most promising methods for maintaining stability of physico-chemical parameters of membranes. Silver and zinc oxide nanoparticles range among the most active nanomaterials of this type. Silver and zinc oxide nanoparticles, as well as based on them antibacterial polymers gain wide popularity due to their prominent antibacterial properties [1]. Considerable effect is obtained because of high specific surface area of the particles providing good contact of the nanoparticles' surfaces with the environment. In addition to that, small particles sizes let them penetrate through cell membranes, thus increasing their antibacterial activity [2]. In particular, silver nanoparticles are widely used as an antibacterial agent in wound dressings, implant coatings, etc. [3]. Still, mechanism of their activity is not totally clear. Antibacterial effect is in proportion to the amount of silver ions ($Ag+$) released and to their interaction with a bacterial cell membrane [4]. High specific surface area of the particles surfaces enables greater number of atoms to interact with the environment so, antibacterial activity of silver

Content from this work may be used under the terms of the Creative Commons Attribution 3.0 licence. Any further distribution of this work must maintain attribution to the author(s) and the title of the work, journal citation and DOI.
Published under licence by IOP Publishing Ltd

nanoparticles depends on their size [5]. Ions of silver interact with peptidoglycan cell wall and with cytoplasmic capsule [6]. Besides, ions of silver prevent DNA replication by reacting with sulfhydryl groups of bacteria proteins, especially with enzymes engaged in electron transport [7]. Thus, studies aimed at creating nanostructured coatings for DBS membranes providing antibacterial effect seem very promising.

2. Research target and methods

2.1. Research target.

In the present work properties of the following membranes were studied: cellulose membrane TFN («Munktell», Germany); glass fiber membrane Grade 8964 («Ahlstrom», USA) and glass fiber membrane GFSP 223000 («Millipore», USA), see the list in Table 1.

Table 1. List of materials and coatings under study

#	Material	Coating
1	Millipore CFSP 223000	Silver alcohol-sol
2	Millipore GFCP 203000 Glassfiber coujugade pad	Aluminium alcohol-sol
3	Mista PPDS-0300 (Grade 89640)	Tantalum alcohol-sol
4	Munklel TFN 179 g/cm^2	Titanium alcohol-sol
5	Gzade 8950	Iron alcohol-sol
6	Grade 8964	Vanadium alcohol-sol
7	Grade 319	Copper alcohol-sol

2.2. Research methods.

Method for alcohol-sols of metals application onto porous material surfaces.

Colloidal solutions of silver, copper, tantalum, vanadium, titanium, iron and aluminium in ethyl alcohol were prepared using specially designed electro-pulse dispersion-condensation equipment. Porous materials were saturated with the obtained alcohol-sols after the following technique:

- after short shaking of colloidal solution a 5 ml aliquot was drawn with a single-use syringe;
- the solution was slowly and evenly applied onto the samples of nonwoven material, situated in a petri dish, up to complete saturation (when a small amount of free liquid appeared);
- the material was air dried due to alcohol evaporation at room temperature.

The samples were saturated twice following the same method.

To study the properties of membranes after application of biological fluids, membrane samples about 5 mm on a side were treated by blood serum by immersing them into biosample till full saturation, then the samples were air dried at room temperature.

Test-objects: *B. cereus* bacilli, spores of STI-1 vaccinal strain and tooth plaque bacterial culture.

Tooth plaque bacteria were obtained by isolation from teeth surfaces with an aseptic cotton wool ball and transferred to a test tube with Hottinger broth. The bacteria were cultivated in the broth for 18-20 hours at $35^{O}C$; bacteria concentration in the broth culture was 1 x 10^8 CFU/ ml.

In the experiments the team used Hottinger broth and solid medium (SM) for antibiotic sensitivity testing (AGV medium, produced by FSSIC 'Microgen', Ministry of Health and Social Development of the Russian Federation, Moscow).

Antibacterial properties of the samples were assessed by means of disk-diffusion method (zones of microbial growth inhibition determination) and by suspension method.

3. Results and discussion

Antibacterial activity of nonwoven materials (fiber felts of various composition, such as fiberglass and cellulose) with various coatings (Table 1) was studied by disk-diffusion method. Sample identification numbers were from 1 to 7, number 8 was a control sample. Tooth plaque bacteria and *B. cereus* bacilli were used as test cultures as the most resistant for disinfectology purposes. In the first experiment squares of the studied nonwoven materials as well as the control material square treated with silver-based nanodisperse system were placed in petri dishes on the surface of SM with test-culture inoculation (Figure 1).

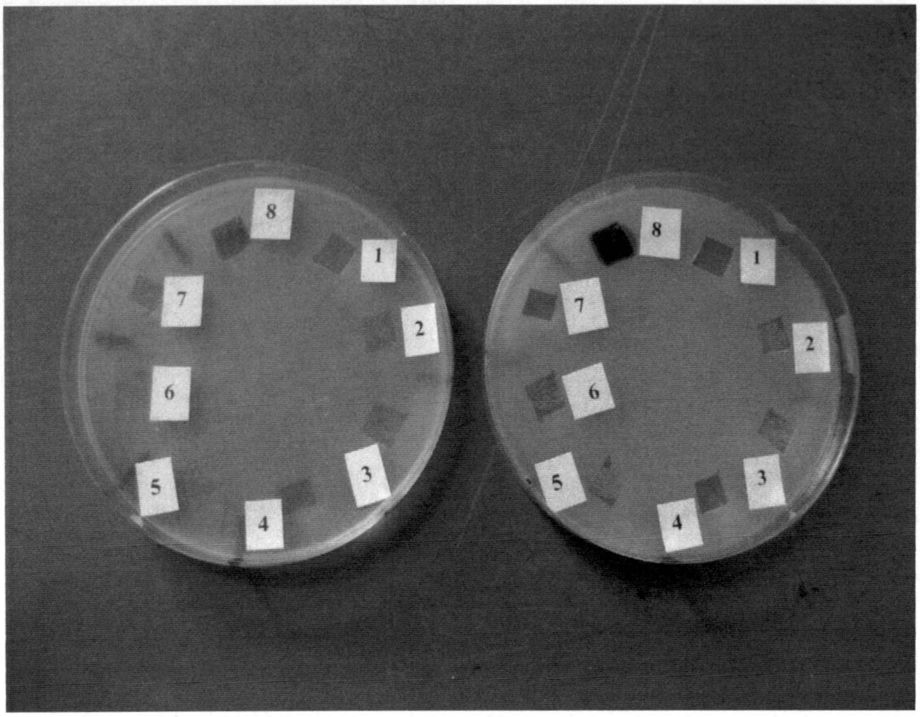

a b

Figure 1. Antibacterial activity of nonwoven materials to tooth plaque bacteria (a) and *B. cereus* (b)

It can be observed from the Figure 1 that microbial growth inhibition zones were formed only under the squares. Growth inhibition of tooth plaque bacteria was especially pronounced under samples ## 1, 2, 3, 4, 7 while under samples ## 5 and 6 bacterial growth inhibition was very slight. On the *B. cereus* bacterial lawn growth inhibition was prominent under samples ## 1, 2, 4, 6, 7 and weak under samples ## 3, 5. Growth inhibition under the control sample # 8 is observed only under the sample itself.

In the second variant of the experiment squares of the studied nonwoven materials as well as the control material square # 8 treated with (ZnO+1.2% cationic surfactant) were placed in petri dishes on the surface of SM with test-culture inoculation of tooth plaque bacteria (Figure 2). The squares were previously wetted with distilled water.

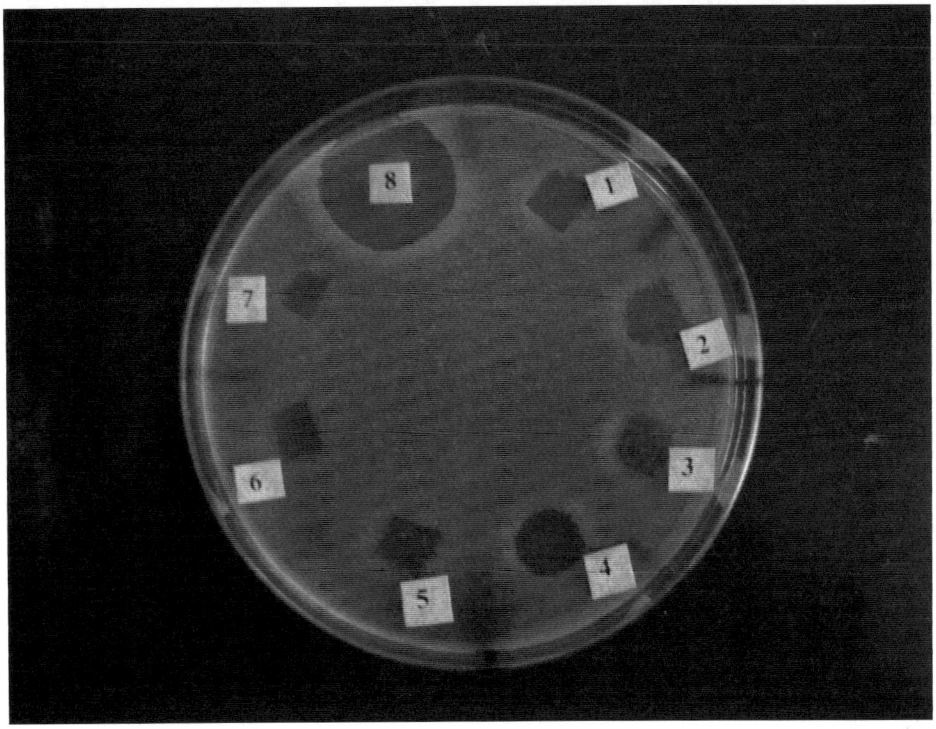

Figure 2. Antibacterial activity of prewetted nonwoven materials towards tooth plaque bacteria

Zones of tooth plaque bacteria test-culture growth inhibition were revealed, exceeding the outlines of the squares #2 (by 2 mm), #3 (by 1.5 mm), #4 (by 3 mm), # 5 (by 1 mm) as well as a prominent zone of tooth plaque bacteria growth inhibition around the contours of control square # 8.

Thus, pre-treatment of the studied materials displayed in some of them the ability of the coating to diffuse into agar and widen the zones of test-culture growth inhibition.

4. Conclusions

It was revealed that nanostructured coatings can inhibit activity of bacteria and they also can diffuse from membrane surfaces into agar thus widening test-culture growth inhibition areas. Treatment of membrane materials with alcohol sols of metal nanoparticles can be promising for adding antibacterial properties to DBS carriers.

Acknowledgments

The authors gratefully acknowledge the financial support of the Ministry of Education and Science of the Russian Federation within the Federal Target Program "Research and development on priority directions of scientific-technological complex of Russia in 2014-2020" (project No. RFMEFI57814X0010)

References

[1] Hajipour M J, Fromm K M, Ashkarran A A, de Aberasturi D J, de Larramendi I R, Rojo T, Serpooshan V, Parak W J and Mahmoudi M 2012 Antibacterial properties of nanoparticles. *Trends in Biotechnol.* **30** (10) 499–511

[2] Park H J, Park S, Roh J, Kim S, Choi K, Yi J, Kim Y and Yoon J 2013 Biofilm-inactivating activity of silver nanoparticles: a comparison with silver ions. *J. Industr. and Eng. Chem* **19** (2) 614–619

[3] Sotiriou G A and Pratsinis S E 2010 Antibacterial Activity of Nanosilver Ions and Particles. *Env. Sc. & Technol.* **44** (14) 5649-5654
[4] Peng J J Y, Botelho M G and Matinlinna J P 2012 Silver compounds used in dentistry for caries management: a review *J. of Dentistry.* **40** (7) 531–541
[5] Radzig M A., Nadtochenko V A., Koksharova O A., Kiwi J, Lipasova V A and Khmel I A 2013 Antibacterial effects of silver nanoparticles on gram-negative bacteria: Influence on the growth and biofilms formation, mechanisms of action *Coll. and Surf. B-Biointerf.* **102** 300.
[6] Chaloupka K, Malam Y and Seifalian A M. 2010 Nanosilver as a new generation of nanoproduct in biomedical applications. *Trends in Biotechnol.* **28** (11) 580-588
[7] Seth D, Choudhury S R, Pradhan S, Gupta S, Palit D, Das S, Debnath N and Goswami A 2011 Nature-inspired Novel Drug Design Paradigm Using Nanosilver: Efficacy on Multi-Drug-Resistant Clinical Isolates of Tuberculosis. *Current Microbiol.* **62** (3) 715-726

In vitro toxic effects of different types of carbon nanotubes

T O Khaliullin[1], L M Fatkhutdinova[1], R RZalyalov, E R. Kisin[2], A R Murray[2], A A Shvedova[2]

[1]Department of Hygiene, Occupational Medicine, Kazan State Medical University, 49 Butlerova str., Kazan, 420039, Russia
[2]National Institute for Occupational Safety and Health, Morgantown, WV 26505, USA

E-mail: khaliullin.40k@gmail.com

Abstract. We have carried out a comparative assessment of industrial single- and multiwalled carbon nanotubes (SWCNTs / MWCNTs) toxic effects in the cultures of macrophages (RAW 264.7) and bronchial epithelial cells (BEAS-2B). Lactate dehydrogenase (LDH) levels were measured directly in the medium. Intracellular reduced glutathione levels (GSH) were determined in culture lysates. Cell viability was determined by fluorescence method. RAW 264.7 were much more sensitive to the effects of carbon nanotubes (CNTs): MWCNT addition was not accompanied by a significant reduction in the viability of macrophages, but caused damage to cell membranes (release of LDH) with a dose and time-dependent oxidative stress. SWCNTs caused a significant decrease in viability and induction of oxidative stress. Improved darkfield microscopy revealed adsorption and accumulation of MWCNTs on the surface and inside the macrophages. Adding SWCNTs to BEAS-2B culture caused small statistically significant dose-and time-dependent decrease in viability and pronounced reduction in glutathione only at the highest concentration of nanoparticles. Bronchial epithelial cells BEAS-2B appeared to be much less susceptible to MWCNTs. The findings suggest that there are differences in the toxic action of different CNTs and there is a need for thoughtful approach to the assessment of nanomaterials toxicity.

1. Introduction

Single- and multiwalled carbon nanotubes (CNT), being ones of the most promising nanomaterials are becoming more and more common. The world is witnessing a steady increase in companies producing and using CNTs. At the same time the number of persons potentially exposed to the aerosol of carbon nanotubes at their workplaces increases as well [1, 2].

Accumulated data suggest a risk to human health upon occupational exposure to carbon nanoparticles. It has been suggested that nanosized particles are more dangerous than micro-particles due to the greater penetrating power, surface area and reactivity [6], whereas fibrous structure of carbon nanotubes may account for the pathogenicity inherent to mineral fibers (asbestos, silica) [3, 4, 5].

The aim of our study was a comparative evaluation of the industrial singlewalled and multiwalled carbon nanotubes toxicity in the cell cultures originating from pulmonary system.

Content from this work may be used under the terms of the Creative Commons Attribution 3.0 licence. Any further distribution of this work must maintain attribution to the author(s) and the title of the work, journal citation and DOI.
Published under licence by IOP Publishing Ltd

IOP Conf. Series: Materials Science and Engineering **98** (2015) 012021 doi:10.1088/1757-899X/98/1/012021

2. Materials and Methods

2.1. Particles.
Single- and multiwalled carbon nanotubes used for experiments were produced using the catalytic vapor deposition. Characteristics of both carbon nanotubes types provided by producers are shown in the Table 1.

Table 1. CNT characteristics, provided by producers

Parameters	MWCNT	SWCNT
External diameter (nm)	8-15	1-2
Internal diameter (nm)	4-8	0.8-1.4
Length (um)	2 and more	3 and more
Total amount of impurities – catalysts leftovers (%)	up to 5	4-7
Bulk density (g/cm^3)	0.03-0,05	0.15
Specific surface area (m^2/g)	300-320	1100
Thermal stability in air (°C)	up to 600	Up to 700

2.2. Dosage.
CNTs were sonicated in the presence of dipalmitoylphosphatidylcholine (DPPC) (0.01 µg/ml) to increase dispersion [7, 8] and visually were represented by tangles up to 5 µm and individual fibers. To assess the dose-dependent effects in the cultures, concentrations of 0.02, 0.2, 2.4 and 24 micrograms per cm^2 of the monolayer surface were selected (0.1, 1, 10, 100 µg/ml of medium respectively). These concentrations were used in previous studies on the comparative evaluation of the carbon nano-and microsized particles toxicity [9, 10, 11].

2.3. Cell cultures.
Two cell lines were taken for the study: RAW 264.7 transformed murine macrophages and immortalized normal human bronchial epithelium BEAS-2B. Cells were grown using MEM (Minimal Essential Media, «Gibco», USA) and DMEM (Dulbecco's modified Eagle medium, «Gibco», USA) media supplemented with 10% fetal bovine serum, 100 U/ml penicillin and 100 µg/ml streptomycin in the incubator (37 °C, 5% CO_2). Cells were seeded on 96-well plates at approximately 2,000 per well.

Lactate dehydrogenase (LDH) levels were measured directly in the medium as a marker of cell membranes damage [12]. Intracellular reduced glutathione levels (GSH) were determined in culture lysates using the ThioGlo® («Calbiochem», USA) reagent to evaluate indirectly the activity of free radical oxidation. Cell viability was determined by fluorescence method.

Part of the cells (RAW 264.7 exposed to MWCNT concentration of 0.2 µg/cm^2 after 2, 4, 6, 24 and 48 hours) was collected, washed off and analyzed using the improved dark field microscopy («CytoViva®», USA).

Statistical analysis was performed using the t-test in the Microsoft Excel 2010 software package.

3. Results

We had selected RAW 264.7 murine macrophages as they are the cells initiating and maintaining different types of immune responses in the case of contact with the foreign bodies [13]

Adding MWCNT to the RAW 264.7 culture caused no significant reduction in viability (Figure 1a), however it was accompanied by the increased concentrations of LDH in the extracellular medium after 48 hours at all doses (but not at 24 hours) (Figure 2a). Oxidative stress manifestations had a dose and time-dependent relationship (Figure 3a) - the higher the dose and exposure time (for the highest dose) the lower the levels of intracellular reduced glutathione.

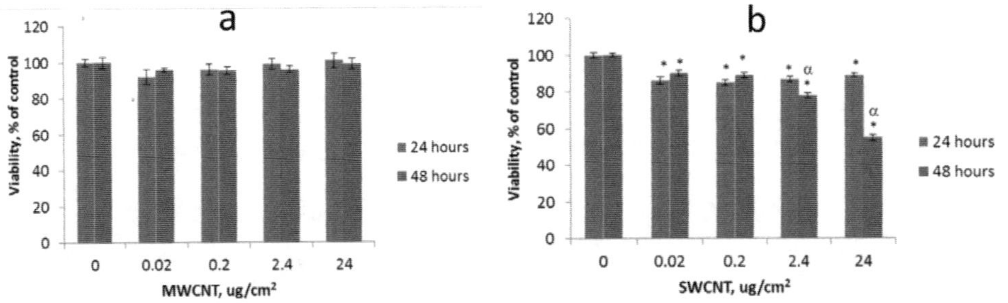

Figure 1. Viability of RAW 264.7 macrophages exposed to different types of CNTs. * - P <0.05 vs control, α - p <0.05 vs. 24 hours

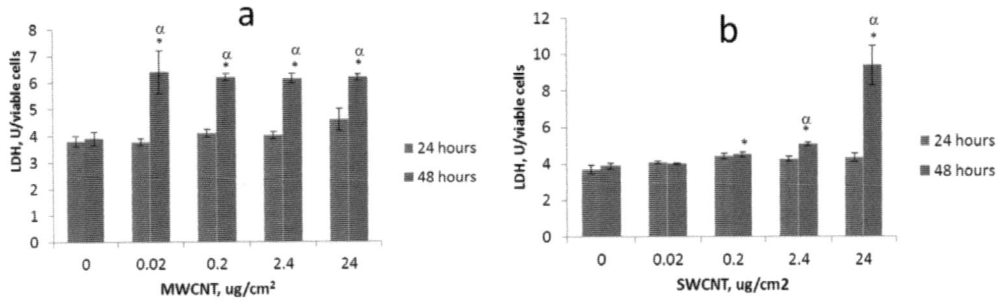

Figure 2. LDH leakage in the culture of RAW 264.7 macrophages exposed to different types of CNTs. * - P <0.05 vs control, α - p <0.05 vs. 24 hours

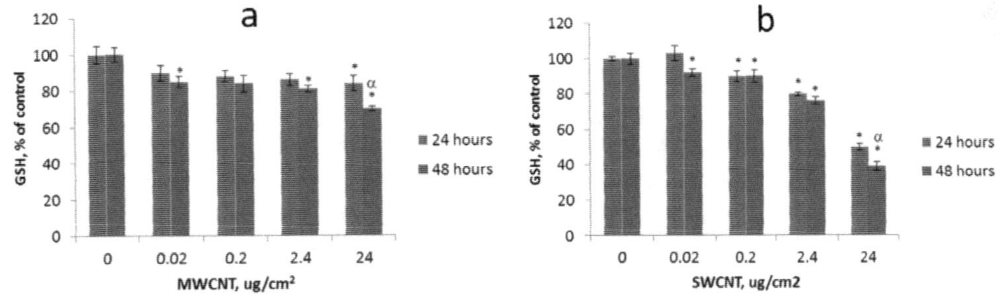

Figure3. Oxidative stress in RAW 264.7 macrophages exposed to different types of CNTs. * - P <0.05 vs control, α - p <0.,05 vs. 24 hours

Adding SWCNT to RAW 264.7 resulted in reduced viability (by 10-15%) at 24 hours after exposure, regardless of the dose. After 48 hours at 2.4 μg/cm^2 SWCNT viability decreased to 78 % and at 24 μg/cm^2- to 50% of the control cells (Figure 1b). Levels of LDH increased sharply after 48 hours at concentrations of 0.2, 2.4, and 24 μg/cm^2 (Figure 2b), which coincides with a noted reduction of viability in the same groups. Significant decrease in reduced glutathione levels was also found: the greater were the exposure dose and time, the lower GSH levels (Figure 3b).

Bronchial epithelium cells BEAS-2B, being the cells that play a major role in the evacuation of dust particles from respiratory tract, appeared a little susceptible to the introduction of MWCNT after 24 and 48 hours: there had been no significant reduction in viability compared to the control culture. The same results were obtained in the evaluation of cell damage and oxidative stress (data not shown).

Adding SWCNT to the BEAS-2B culture caused a small statistically significant dose- and time-dependent decrease in viability after 24 and 48 hours as well as marked reduction in the level of reduced glutathione at the highest concentration of 24 μg/cm^2 (data not shown).

Subsequent investigation of exposed RAW 264.7 cells by darkfield microscopy revealed adsorption and accumulation of MWCNTs on the surface and inside the cells. With increased exposure time number of CNT visually observed inside and outside the cells also increased (Figure 4).

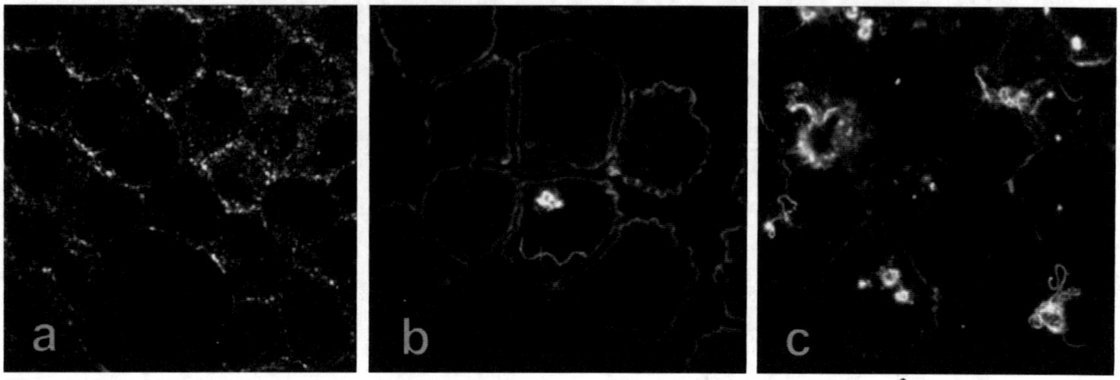

Figure 4. RAW 264.7 exposed to MWCNTs: **a** – control culture; **b** – 0.2 μg/cm^2, 2 hours; **c** – 0.2 μg/cm^2, 24 hours

4. Discussion

Obtained data suggest that the studied industrial SWCNT have a toxic effect, causing the induction of oxidative stress and cell death. At the same time MWCNTs administration, whereas not leading to a significant reduction in viability, cause damaging effect on the membrane of sensitive cells (macrophages). This complies well with the data obtained by other research groups (Hirano et al., Shimizu et al.) showing that MWCNT may associate with the membrane of macrophages and disrupt its integrity [14, 15].

Comparative studies of the SWCNT and MWCNT *in vitro* toxicity are few. Existing data have a multidirectional nature. Some researchers have found that SWCNT had greater cytotoxicity [11, 16, 17]. Some reported no acute toxic effects for both types of CNTs [18]. Others showed decreased viability when exposed to CNTs, but found no quantitative differences between single- and multiwalled nanotubes [19, 20]. Bacteria E.Coli, P.Aeruginosa and others were the most sensitive to the SWCNTs while MWCNT showed moderate toxicity [21]. According to our results, both SWCNT and MWCNT have toxic effects, but mechanisms and manifestations are different, which should be considered in the future in comparative toxicological studies.

One of the features of our study was to investigate macrophages response. Experiments *in vivo* suggest a significant role of the latter in the development of acute and chronic inflammation as well as profibrotic changes in the lungs upon exposure to CNT [9, 22, 23]. It was also shown that SWCNT are accumulated by macrophages to a much lesser extent than MWCNT [24]. Researchers studying genotoxic effects of CNTs in RAW 264.7 macrophages found signs of damage to the genetic apparatus at concentrations greater than 0.1μg/ml for SWCNT and above 1 μg/ml for MWCNT, although cytotoxic effects appeared only at the highest doses (100μg/ml) [25]. Mechanisms of adverse effects remain poorly understood and studies using macrophage cultures are relatively scarce.

Our study has shown that CNT do have a toxic effect on the RAW 264.7 macrophages, unlike bronchial epithelium cells. It should be noted that we used immortalized leukemic macrophages in the experiment, which, however, are capable of phagocytosis, lysosomal activity, express and secrete most of the biologically active substances typical of normal macrophages, including various cytokines and mediators.

The experimental data obtained for one CNT type should be extrapolated on other CNTs produced by other companies and/or in other conditions with caution. Thus, there were found different responses to administration of MWCNT of differing morphology and length into the pleural cavity of rats - from minor local inflammatory response in case of short and tangled nanotubes to severe inflammatory response and granuloma formation for long rigid particles [26]. CNTs produced by catalytic vapor deposition also differ in catalysts content. Occupational exposure at primary manufacturers involves contact with both unpurified and purified material, while secondary production and handling implies contact with the already treated (and/or functionalized) carbon nanotubes. Investigators have already pointed out the role of contaminants - catalysts in the development of oxidative stress upon the cells' contact with CNT [18, 27]. In experiments on mice, purified SWNTs were less toxic to animals than samples containing residual catalysts, such as iron and nickel [28].

5. Conclusion

The macrophages (RAW 264.7) were much more sensitive to the effects of different types of CNTs than bronchial epithelial cells (BEAS-2B). SWCNT exposure resulted in the induction of oxidative stress and cell death. Addition of MWCNT did not cause a significant reduction in viability, but had a damaging effect on the cellular membranes of macrophages. The results show that it is necessary to develop *in vitro* or *in silico* models for the evaluation of various parameters of nanopaticles toxicity. Comparison of different CNT types should be based not only on the viability/cytotoxicity assessment, but also on specific adverse effects, such as oxidative stress, cell membranes damage and cytokine profiling. Furthermore, for regulatory documents it is necessary to classify nanoparticles not only by their chemical composition, but also by their morphological characteristics and properties.

References

[1] Lux Research. The Nanotech Report. 5th ed. New York, NY: Lux Research; 2007.
[2] Global Markets and Technologies for Carbon Nanotubes report.URL: http://www.bccresearch.com/report/carbon-nantubes-markets-technologies-nan024e.html
[3] Wick P, Manser P, Limbach LK, Dettlaff-Weglikowska U, Krumeich F, Roth S, Stark WJ and Bruinink A. 2007*Toxicol Lett* **168**(2) 121
[4] Kanno J, Takagi A, Nishimura T and Hirose A 2010 *The toxicologist.* (Salt Lake City, Utah: Oxford University Press) A1397
[5] Kim JS, Song KS, Lee J K, Choi YC, Bang IS, Kang CS and Yu IJ 2012 *Arch Toxicol* **86**(4) 553
[6] Nel A, Xia T, Madler L and Li N 2006 *Science* **311**(5761) 622
[7] Davoren M, Herzog E and Casey A 2007 *Toxicol In Vitro* **21**(3) 438
[8] Herzog E, Byrne HJ, Davoren M, Davoren M, Lenz AG, Maier KL, Duschl A and Oostingh GJ 2009 *Toxicol Appl Pharmacol* **236**(3) 276
[9] Shvedova AA, Kisin ER, Mercer R Murray AR, Johnson VJ, Potapovich AI, Tyurina YY, Gorelik O, Arepalli S, Schwegler-Berry D et al. .2005 *Am J Physiol Lung Cell Mol Physiol* **289** 698
[10] Tabet L, Bussy C, Amara N Setyan A, Grodet A, Rossi MJ, Pairon JC, Boczkowski J and Lanone S 2009 *J Toxicol Environ Health* A **72**(2) 60
[11] Tian F, Cui D, Schwarz H, Estrada GG and Kobayashi H 2006 *Toxicol In Vitro* **20**(7) 1202
[12] Drent M, Cobben NAM and Henderson RF 1996 *Eur Respir J* **9** 1736
[13] Mossman BT and Churg A *Am J Respir Crit Care Med* **157** 1666

[14] Hirano S, Fujitani Y, Furuyama A and Kanno S 2010 *Toxicol Appl Pharmacol* **249**(1) 8
[15] Shimizu K, Uchiyama A, Yamashita M, Hirose A, Nishimura T, Oku N 2013 *J Toxicol Sci* **38**(1) 7
[16] Jia G, Wang H, Yan L Wang X, Pei R, Yan T, Zhao Y and Guo X 2005 *Environ Sci Technol* **39**(5) 1378
[17] Hu X, Cook S, Wang P, Hwang HM, Liu X and Williams QL 2010 *Sci Total Environ* **408**(8) 1812
[18] Pulskamp K, Diabaté S and Krug HF 2007 *Toxicol Lett* **168**(1) 58
[19] Murr LE, Garza KM, Soto KF Carrasco A, Powell TG, Ramirez DA, Guerrero PA, Lopez DA and Venzor J 3rd 2005 *Int J Environ Res Public Health.* **2**(1) 31
[20] Palomäki J, Karisola P, Pylkkänen L, Savolainen K and Alenius H 2010 *Toxicology* **267**(1-3) 125
[21] Kang S, Mauter MS and Elimelech M 2009 *Environ Sci Technol* **43**(7) 2648
[22] Porter DW, Hubbs AF, Mercer RR, Wu N, Wolfarth MG, Sriram K, Leonard S, Battelli L, Schwegler-Berry D, Friend Set al. 2010*Toxicol.* **269** 136–147
[23] Shvedova AA, Kisin E, Murray AR, Johnson VJ, Gorelik O, Arepalli S, Hubbs AF, Mercer RR, Keohavong P, Sussman N et al. 2008 *Am J Physiol Lung Cell MolPhysiol* **295**(4) 552-65
[24] Mercer RR, Hubbs AF, Scabilloni JF, Wu N, Wolfarth MG, Sriram K, Leonard S, Battelli L, Schwegler-Berry D, Friend S, et al. 2011 *Part Fibre Toxicol* **8**:21.
[25] Migliore L, Saracino D, Bonelli A, Colognato R, D'Errico MR, Magrini A, Bergamaschi A and Bergamaschi E 2010 *Environ Mol Mutagen* **51**(4) 294
[26] Poland CA, Duffin R, Kinloch I, Maynard A, Wallace WA, Seaton A, Stone V, Brown S, Macnee W and Donaldson K 2008 *Nat Nanotechnol* **3**(7) 423
[27] Kagan VE, Tyurina YY, Tyurin VA Konduru NV, Potapovich AI, Osipov AN, Kisin ER, Schwegler-Berry D, Mercer R, Castranova V and Shvedova AA. 2006 *Toxicol Lett* **165**(1) 88
[28] Lam CW, James JT, McCluskey R and Hunter RL. 2004 *Toxicol Sci* **77**(1) 126

IOP Conf. Series: Materials Science and Engineering **98** (2015) 012022 doi:10.1088/1757-899X/98/1/012022

Danger due to the translocation of nanoparticles in soil: mathematical modeling

Yu Lisyuk, S Vorobyev, L Izerskaya and Yu Morgalev

Centre "Biotest-Nano", Tomsk State University, 36, Lenina ave., Tomsk, 634506, Russia

E-mail: yu.morgalev@tsu.ru

Abstract. A necessary step was taken towards the formation of a migration model of nanoparticles (NPs) from the surface deep into the soil, taking into account the frequency of precipitation and the processes of adsorption and desorption which occur in the soils. An equation for migration of nanoparticles in soil is proposed. A method of obtaining the averaged equations for long-term migration of NPs in the soil profile has been developed. Similarly, partial differential equations may be obtained which describe more complex models, for example, including capillary phenomena, etc. The obtained equations allow the use of integral transformations in order to find solutions. The model can be used to plan natural experiments in different types of soils.

1. Introduction

Over the last two decades an increase in diversity and overall production of nanomaterials, along with exponential growth in the number of goods which contain engineered nanoparticles and the extension of their application areas has been observed. The extension of nanotechnological manufacturing leads to an inflow of substantial amounts of nanomaterials to the environment, their accumulation in biota and the abiotic environment followed by their possible transmission to humans. The soil is a natural accumulator of NPs both from industrial aerosols and in the process of turnover of nanomaterials (including recycling). The degree of mobility of nanoparticles in the soil and their ability to migrate identifies the risk of NPs to biota, modulating the bioavailability of toxic or biologically active NPs. The possibility of NPs entering the food chains of agricultural production and the transmission of nanotoxicity to an ultimate consumer is also linked to this factor.

The development of technology for protecting agricultural production chains and assessment of the risks to agricultural production in areas of possible contamination with nanoparticles (including the migration of nanoparticles to adjacent water bodies by means of catena) is not possible without an understanding of the factors determining the intensity of migration. Mathematical modeling of physical and chemical processes which occur in the soil is the basis of effective forecasting of NPs' behavior. The study of vertical and lateral migration of substances in the soil profile for nanoparticles has so far practically not been carried out. At the same time, models of vertical migration were developed through the example of contaminants [1, 2] and they can be used for the migration modeling of NPs. In the existing literature on mathematical modeling of the migration of contaminants different models have been considered. The most complex ones take into account the hydrodynamic features of moisture movement [3-7], capillary effects, multistaging of sorption and desorption processes [8] and others [9-16].

Content from this work may be used under the terms of the Creative Commons Attribution 3.0 licence. Any further distribution of this work must maintain attribution to the author(s) and the title of the work, journal citation and DOI.
Published under licence by IOP Publishing Ltd

However, all models consider the process of migration as a single one. This approach corresponds to the mode of humidification and it is far from the description of a real migration process. Thus, the influence of the factor of repeated wetting and drying on the migration process of substances in mathematical models appears to have been totally excluded from consideration. In other words, the influence of natural climatic factors is almost ignored.

In our opinion, a correct account of natural climatic factors even in the framework of a simple model of the elementary migration process is more important than an increase in complexity of the migration process model since this is the only way to achieve adequate compliance of the modeling results with the real natural process. In this work, on the basis of natural assumptions, partial differential equations were obtained. They describe the spatial and temporal distribution of nanoparticles in the soil profile from the source of NPs which appeared on the surface of the soil.

2. The model of the migration process

It is evident that a long-term process of migration is the result of overlap of a large number of elementary acts of migration which occur during each act of humidification.

In this paper we shall use the simplest physical and mathematical description of these elementary migration processes. In particular, some factors are not taken into consideration, namely, the presence of stagnant zones, possible variety of sorption centers, nonlinear motion of the humidification front, dissimilarity of humidification (within the humidified area), capillary phenomena, the hierarchy of structural organization of the soils, etc. In the future these factors can be taken into consideration if the model should require developing.

Thus we assume that on the surface of a soil profile there is a source of accrual of NPs with a surface density N. When humidified, the nanoparticles migrate down the soil profile along with the filtered moisture. Moving with the flow of the filtered moisture, the nanoparticles make chaotic diffusive motions together with the moisture and they are also adsorbed onto the soil particles. After being desorbed, the NPs again continue moving along the profile.

The main components of the flow of nanoparticles in the soil profile are convective and diffusive. The convective component prevails at the first stage. At the second stage, the diffusive component starts to prevail with the termination of the directional filtering movement of the moisture. When this stage of migration is described mathematically, the initial conditions for moving and adsorbed components of NP should be considered the distribution of them at the end of the first stage of humidification. The sum of moving and adsorbed components of NP after a period of diffusion is the initial distribution for the adsorbed component of the act of humidification $n+1$. For the $n+1$ act at the initial time the moving component equals 0 everywhere apart from the origin of coordinates.

Thus, in order to obtain a general long-term image of migration of NPs, it is necessary to consider the phenomenon of consistent overlap of individual acts of humidification. Moreover, it is evident that for simplification, the stage of diffusion spreading achieved at the stage of filtration of the distribution of moving and adsorbed components can be rejected.

We shall assume that the movement of moisture and the NPs occurs only vertically.

We shall use the following designations.

T – the macroscopic time of migration,

t – the time inside the time interval of an individual act of humidification $(T-T_0+t_{01}-t_{02})$,

t_{01} – the duration of an elementary act of humidification until the moment when the movement of the water front ceases,

t_{02} – the duration of the whole act of humidification until the moment of drying,

t_{02} - t_{01} – the duration of the mainly diffusive movement inside the humidified profile after the filtrational movement has stopped,

t – the current time inside the interval of elementary humidification,

Obviously: t_{02}, $t_{01} << T$,

Δ – the length of the free path of the NPs between collisions,

V – the movement speed of the moisture being filtered (to simplify, we assume that V does not depend on $x_I t_I$; generalizations are obvious),

u – the speed of the chaotic diffusive movement,

x – the distance from the surface of the soil profile,

c_I – the density of the NPs which move with the moisture, c_2 – the density of the adsorbed NPs,

D – diffusion coefficient c_I,

τ – the "lifetime" of the NPs in a free condition, which determines the decrease c_I inside the flow under filtration due to adsorption

$$\frac{dc_1}{d_1 t} = -\frac{c_1}{\tau}$$

γ – the value which determines the speed of change N due to destruction of NPs by moisture (or trapping soil particles with NPs by moisture),

γ_2 – the speed of desorbtion

$$\frac{dc_1}{d_1 t} = -\gamma_2 c_2$$

τ – the average time of collisionless motion of the NPs during diffusive motion.

3. The equation of migration during an elementary act of humidification

For convenience, we shall temporarily omit the dependence of c_I ,c_2 from T within one act of humidification, i.e. we shall assume

$$c_1(T+t) \equiv \bar{c}_1(t)$$
$$c_2(T+t) \equiv \bar{c}_2(t)$$

We obtain the expression for the flow of the component c_I including directional filtrational movement of the component c_I, I_I.

The flow density can be depicted as:

$$I \equiv \frac{1}{\tau_0} \int_{-\infty}^{t} e^{\frac{-(t-t')}{\tau_0}} \bar{c}_1(\acute{x},\acute{t})(V+u) * \delta(x - (\acute{x} + (V+u)(t-\acute{t})))dt'dx \int_{-\infty}^{+\infty} W(u)du \quad (1)$$

In (1) δ(z) - Dirac delta function, which equals zero everywhere except the point z=0, where it is equal to infinity,

Its integral is

$$\int_{-\infty}^{+\infty} \delta(z)dz = 1$$

t' – the moment of collision with complete chaotization of speed u; $W(u)$ – distribution function of chaotic speed u.

Obviously:

$$\int_{-\infty}^{+\infty} W(u)udu = 0$$

The formula (1) reflects the fact of summation of contributions into the flow of all the NPs which arrive at the chosen point (x, t) and they experienced the last collision at the moment t' and they have the speed $(V+u)$ after this collision.

The presence of the δ function allows us to perform the integration in (1) over x:

$$I \equiv \frac{1}{\tau_0}\int_{-\infty}^{\infty} e^{-\frac{(t-t')}{\tau_0}}\bar{c}_1(x-(t-t')_1 t')W\,(u)dudt$$

The average time between collisions is a very small value; that is why it is possible to consider t' as coinciding with t in the time dependence c from t'. The dependence from x can be depicted as:

$$\bar{c}_1(x-(V+u)(t-t')_1 t) \cong \bar{c}_1(x_1 t) - \frac{\partial \bar{c}_1(x_1 t)}{\partial x}(V+u)(t-t')$$

Then:

$$I \equiv \frac{1}{\tau_0}\int_{-\infty}^{t} e^{-\frac{t-t'}{\tau_0}}\left[c_1(x_1 t) - \frac{\partial c_1(x_1 t)}{\partial x}(V+u)(t-t')\right]dt' \int_{-\infty}^{\infty}(V+u)W(u)dudt \tag{2}$$

Performing the required integration in (2), considering (3) we obtain:

$$I \equiv \bar{c}_1 V - D\frac{\partial \bar{c}_1}{\partial x} \tag{3}$$

Where: $D = \bar{u}^2\tau$
u^2 – the average square speed of the chaotic movement.

The second summand in (3) is a diffusion component of the flow. The larger it is, the higher the spatial heterogeneity c_1. Its influence can be evaluated by comparing both summands in (3).

$$\bar{c}_1 V = D\frac{\partial \bar{c}_1}{\partial x}$$

which implies that under equilibrium conditions

$$\bar{c}_1 \sim e^{\frac{V}{D}}, \text{ or } \bar{c}_1 \sim e^{\frac{Vx}{v\Delta}}$$

Since the length of the free path Δ is a small value, then even where V and u are equal the diffusion component plays its role only when the distribution c is highly heterogeneous. However, the influence of diffusion can be affected when there is a long-term period of humidification without filtration.

In view of (3) the balance equation can be written as:

$$\frac{\partial \bar{c}_1}{\partial t} + \gamma_2\bar{c}_2 - \frac{\bar{c}_1}{\tau} = -divI$$

Or

$$\frac{\partial \bar{c}_1}{\partial t} - \gamma_2\bar{c}_2 + \frac{\bar{c}_1}{\tau} + V\frac{\partial \bar{c}_1}{\partial x} - D\frac{\partial \bar{c}_1}{\partial x^2} = 0 \tag{4}$$

The equation (4) must be supplemented by the equation for c_2:

$$\frac{\partial \bar{c}_2}{\partial t} = -\gamma_2\bar{c}_2 + \frac{\bar{c}_1}{\tau} \tag{5}$$

The first summand in (4) describes the desorption of the absorbed NPs, the second describes the speed of adsorption.

It should be noted that the equations (4) and (5) are true when the humidification front has reached the considered point x, i.e. when $x<Vt$. Otherwise $c_1, c_2 \neq f_{1,2}(x,t)$ and then in (4) and (5) we have

$$c_1, c_2, {}^1/_\tau, \gamma_2, D, V = 0$$

Thus, the equations under consideration (4) and (5) are equations with variable coefficients which are not solved in a standard way by Laplace and Fourier integral transformations, and which require a different approach.

In accordance with the aforementioned, it rational to divide the time of humidification into 2 parts: the time of directionally moving moisture during filtration (from 0 to t_{10}) and the time of the drying of the moistened profile (from t_{10} to t_{20}). The value t_{20} is indefinite to a large degree, but it can be useful for completeness of modeling studies.

Thus, in the area from 0 to t_{01} we have:

$$\frac{\partial \bar{c}_1}{\partial t} - D\frac{\partial^2 \bar{c}_1}{\partial x^2} + \frac{\bar{c}_1}{\tau} + \bar{c}_2 y_2 = 0 \quad (6)$$

$$\frac{\partial \bar{c}_2}{\partial t} + \frac{\bar{c}_1}{\tau} + \bar{c}_2 \gamma_2 = 0$$

Similarly, in the area t from t_{01} to t_{02} we write:

$$\frac{\partial \bar{c}_1}{\partial t} - D\frac{\partial^2 \bar{c}_1}{\partial x^2} + \frac{\bar{c}_1}{\tau} + \bar{c}_2 y_2 = 0$$

The initial condition for the equations (7) are the values $c_1(x, t_{01})$ and $c_2\,(x_1, t_{01})$ which have been obtained as a result of solving the equations (12).

$$\frac{\partial \bar{c}_2}{\partial t} - \frac{\bar{c}_1}{\tau} + \bar{c}_2 \gamma_2 = 0 \quad (7)$$

In this work we neglect the influence of diffusion in order to simplify it and we obtain long-term equations for the component of NP which is averaged over the external conditions.

4. The boundary conditions

We determine the boundary conditions for $c_1\,(0,t)$ (i.e. on the surface of the soil profile).

It can be assumed that:

$$c_1(0,t)=\gamma N(t) \quad (8)$$

where γ is a constant which does not depend on time. It can be assumed that $\gamma \sim 1/V$, since the higher saturability of the soil solution of NP, the lower the speed of movement through a thin but ultimate layer of soil with NPs. So far we leave aside the dependence of γ on the temperature.

On the other hand, we can write:

$$\frac{\partial N}{\partial T} = -\bar{c}_1 V \quad (9)$$

(8) and (9) give

$$N(T+t) = N_{0e}{}^{-\gamma V(T+t)} = N_0(T)e^{-\gamma Vt}$$

where T_0 is the initial value of N (when $T=0$).

The boundary condition for c_1:

$$c_1(0, T+t) = \gamma N_0(T)e^{-\gamma Vt} \tag{10}$$

Since during the drying of the soil profile c_1 disappears, transforming in accordance with the accepted model into c_2 the boundary condition (9) is true at any stage of humidification.

The boundary conditions for c_2 can be determined using balance equations (5). We depict c_2 in the form:

$$c_2(0, T+t) = c_2(0, T)\bar{c}_2(0, t)$$

The equation (5) then gives

$$\frac{\partial \bar{c}_1(0,t)}{\partial t} = -\gamma_2(c_2 + \bar{c}_2(0,t)) + \frac{c_1(0, T+t)}{\tau}$$

Taking into account the small change of c_1 and c_2 during the act of humidification, for the increment c_2 during the whole act of humidification can be recorded (including the component c_1, which transforms into c_2 at the end of the humidification cycle t_0):

$$c_2(t_0) = -t_0\gamma c_2(T) + \frac{\gamma N(T)t_0}{\tau} + \gamma N(T) \tag{11}$$

The boundary condition for $\Delta c_2(0,t)$ depicts the relation (19) without the last summand:

$$\bar{c}_2(x,t) = -\gamma c_2(T)t + \frac{\gamma N(T)}{\tau}t \tag{12}$$

In (11) the assumption that there is little desorption during humidification ($\gamma Vt_0 << 1$) is used.

Of interest are only the values of concentration c_1 and c_2 averaged over a long period of time. We average (12) over t_0 (the time of humidification or the relation $Vt_0=l_0$, of wetted depth l_0).

$$\bar{c}_2(t) = -\gamma_2 c_2(T)\nabla + \frac{\gamma N(T)\nabla}{\tau} + \gamma N$$

or

$$\frac{\bar{c}_2(t)}{\nabla} = -\gamma_2 c_2(T) + \gamma N(T)\left(\frac{1}{\tau} + \frac{1}{\nabla}\right) \tag{13}$$

Where ∇ is the average time of humidification.

The left part (13) in fact depicts a derivative

$$\frac{\partial c_2(0,T)}{\partial T}$$

Then:

$$\frac{\partial c_2(0,T)}{\partial T} = -\gamma_2 c_2(T) + \frac{\gamma N(T)}{\tau} \tag{14}$$

The integration (14) gives:

$$c_2(0,T) = Ae^{-\gamma_2 T} + \frac{\gamma N_0(\frac{1}{V}+\frac{1}{\tau})}{\gamma V-\gamma_2}(e^{-\gamma_2 T} - e^{-\gamma V T}) \quad (15)$$

The initial value for border density $c_2(0,0)$ is zero. This determines A in (15) and we finally obtain the boundary condition for c_2.

$$c_2(0,T) = \frac{\gamma N_0(\frac{1}{V}+\frac{1}{\tau})}{\gamma V-\gamma_2}(e^{-\gamma_2 T} - e^{-\gamma V T}) \quad (16)$$

This boundary condition is necessary in order to obtain a unique solution of the partial differential equation found below, which describes the spatio-temporal dynamics of migration of the NPs.

We note that according to (16) $c_2(0,T)$ has a maximum, as the function of "large" time T. Its appearance reflects the obvious fact of accumulation of the NPs in the surface region and further flushing of it with a soil solution of low concentration with gradual depletion of the surface source of NPs.

5. Averaged (climate-oriented) equations of migration

The increment of the adsorbed component c_2 is considered small during one act of humidification and can be found. The total increment Δc_2 during several acts of humidification can be considered small as well. This allows us to average the obtained c_2 (during one humidification) in accordance with the statistics of precipitation. Then, considering it the differential increment per unit time (the gap between two humidifications), the differential equation for the value $c_2(x,T)$, averaged over the statistics of humidification, can be obtained. Here we practically do the same as when deriving the boundary conditions for $c_2(0,T)$. Since the component $c_2(x,T)$ does not accumulate with an increase in time by means of its transformation into c_2, we can depict c_2 as:

$$c_2(x,T+t) = c_2(x,T) + c_2(x,t) \quad (17)$$

Where c_2 is the increment of c_2 during the investigated humidification $c_2 << c_2$ (T).

We write down these equations neglecting the small values c_2 in the first equation (6):

$$V\frac{\partial c_1}{\partial x} + \frac{\partial c_1}{\partial t} + \frac{c_1}{\tau} = \gamma_2 c_2(x,T) \quad (18)$$

$$\frac{\partial c_2}{\partial t} + \gamma_2 c_2 = \frac{c_1}{\tau} \quad (19)$$

The equation (18) is heterogeneous. Its solution can be presented as a general solution of the homogeneous equation and a specific solution of the heterogeneous equation $c_{11}+c_{12}$. The general solution c_{11} is found by separation of variables:

$$c_{11} = Ae^{-\propto t} * e^{-\left(\frac{1}{V\tau}-\frac{\propto}{V}\right)x}\left(Al^{-\propto t}e^{-\left(\frac{1}{V\tau}-\frac{\propto}{V}\right)x}\right)$$

Where A is the arbitrary constant determined from the boundary conditions when $x=0$, α is a yet unknown constant determined also from the boundary condition (10).

After the required calculations are completed, we obtain:

$$c_1(x,T+t) = \gamma N_0 e^{-\gamma V(T+t)+\left(\gamma-\frac{1}{\tau V}\right)x} + \frac{\gamma_2}{V} ex \int_0^x c_2(x',T)e^{\left(\gamma-\frac{1}{V\tau}\right)(x-x')}dx' \quad (20)$$

Where $\tilde{t} = \frac{x}{V}$

For the NPs we consider that their solvability or transition into another mobile form C_1 during one act of humidification is small.

$$\gamma V t_0,\ \gamma_2 t_0 << 1$$

Then:

$$C_1(x_1T+t) = \gamma N_0 e^{-\gamma Vt + \left(\gamma - \frac{1}{tV}\right)x} + \frac{\gamma_2}{V}\int_0^x c_2(x'_1T)e^{-\frac{1}{V\tau}(x-x')}dx' \tag{21}$$

The equation (19) including (17) can be rewritten:

$$\frac{\partial C_2}{\partial t} + \gamma_2 C_2(x_1T) = \frac{C_1}{\tau}\frac{(x_1T+t)}{t} \tag{22}$$

Where C_I is determined by the relation (20), and the boundary condition for ΔC_2 is given by the relation (11). The integration (22) taking into account the weak time dependence C_2 and C_1 within one humidification gives

$$C_2 = -\gamma_2 C_2(x,T)(t-\tilde{t}) + \frac{\gamma N_0 e^{-\gamma VT + \left(\gamma - \frac{1}{\tau V}\right)x}}{\tau}(t-\tilde{t}) + \frac{\gamma_2}{V_\tau}(t-\tilde{t})\int_0^x c_2(x',t)e^{-(x-x')\frac{1}{\tau V}}dx' \tag{23}$$

$$\tilde{t} = \frac{x}{V}$$

It is necessary to remember that the expressions for c_I and c_2 in the formulas (21) and (23) are true, when x and t correspond to the point x with the water front which has reached this point. In other words, (21) and (23) are true when

$$\frac{x}{V} < t < t_0$$

During drying of the profile, c_I is transformed into C_2 and the total increment of the related component c_2 (which equals $C_I(x,t_0)+C_2(x,t_0)$) then:

$$\bar{c}_2 = -\gamma_2 c_2\left(t_0 - \frac{x}{V}\right) + \frac{\gamma N_0}{\tau}\left(1 + \frac{t_0 - \frac{x}{V}}{\tau}\right)e^{-\gamma VT - \frac{x}{\tau u}} + \frac{\gamma_2}{\tau V}\left(1 + \frac{t_0 - \frac{x}{V}}{\tau}\right)\int_0^x c_2(x',T)e^{-\frac{(x-x')}{\tau V}}dx' \tag{24}$$

In (24) t_0 can be changed into l_0/u, where l is the depth of wetting. It is obvious that not every wetting provokes the change of the c_2 component; it is reasonable to consider in (24) only $x<t_0u$. For subsequent (24) it is convenient to depict as follows:

$$\bar{c}_2 = -\gamma_2 c_2\left(\frac{l-x}{V}\right) + \gamma N_0\left(1 + \frac{l-x}{V\tau}\right)e^{-\gamma VT - \frac{x}{V\tau}} + \frac{\gamma_2}{V}\left(1 + \frac{l-x}{\tau V}\right)\int_0^x c_2(x',T)e^{-\frac{(x-x'_1)}{\tau V}}dx'$$

Individual wetting at the pace of migration of the NPs plays a minor role. Notable changes in the distribution of the NPs can occur only after multiple humidifications (remembering that we are

IOP Conf. Series: Materials Science and Engineering **98** (2015) 012022 doi:10.1088/1757-899X/98/1/012022

considering the migration of slightly soluble and slightly destructible NPs). This imposes calculation of the average increment during multiple humidifications Δc_2. Averaging can only be performed in accordance with the function of power distribution of precipitation (or, to be more exact, the depth of wetting *l*). We determine the distribution function *w(l)*. Then the averaged value of increment c_2 is:

$$\bar{\bar{c}}_2 = \int_x^{\infty} w(l) dl \bar{c}_2 \tag{25}$$

Where *w(l)* is a normalized distribution function.

The integration in (25) is spread from x for the reasons given above. The spread of the interval of integration to ∞ is insignificant because

$$\frac{l}{\tau V} \gg 1, \quad (\bar{l} = Vl)$$

c_2 is nothing other than the average increment c_2 during one humidification.

We can write it as:

$$\bar{\bar{c}}_2 = \frac{\partial c_2(x,T)}{\partial T} \nabla$$

We obtain the differential equation:

$$\frac{\partial c_2(x,T)}{\partial T} \nabla = \int_x^{\infty} w(l) dl \left\{ -\gamma_2 c_2(x,T) \left(\frac{l-x}{V}\right) + \gamma N_0 \left(1 + \frac{l-x}{V\tau}\right) e^{-\gamma V T \frac{x}{V\tau}} + \frac{\gamma_2}{V_\tau} \left(1 + \frac{l-x}{\tau V}\right) \int_0^x c_2(x',T) e^{\frac{(x-x')}{v\tau}} dx' \right\} \tag{26}$$

The equation (26) depicts an integro-differential equation, the solution of which depends on the particular kind of *w(l)*. The solution of the equation (26) with the boundary condition (16) and zero, for example, the initial condition (*c(x,0)=0*) uniquely determines the spatial and temporal distribution of c_2.

The exponential distribution function *w(l)* is good because, in its main features, it reasonably describes the distribution function of humidification: the probability of weak humidifications is higher than strong ones. In addition to this, it is convenient for calculations.

For the exponential distribution function

$$w(l) = \frac{e^{-\frac{l}{l^-}}}{l^-} = \frac{1}{l^-} exp\left(-\frac{l}{l^-}\right)$$

the equation (26) will have the form:

$$\frac{\partial c_2(x,T)}{\partial T} \nabla = \frac{\gamma_2}{V}\left(1 + \frac{\bar{l}}{\tau V}\right) e^{-\frac{x}{\bar{l}}} \int_0^x e^{-\frac{(x-x')}{\tau V}} c_2(x'T) dx - \gamma_2 c_2(x,T) \frac{e^{-\frac{x}{\bar{l}}} * \bar{l}}{V} + e^{-\gamma V T - \frac{x}{V\tau}} \gamma N_0 \left(1 + \frac{l}{\tau V}\right) e^{-\frac{x}{\bar{l}}} \tag{27}$$

After differentiating over *x*, we obtain the partial differential equation of the second order:

$$\frac{\partial^2 c_2(x,T)}{\partial x \partial T} - \frac{1}{l} * \frac{\partial c_2}{\partial T} + \frac{1}{\tau V} * \frac{\partial c_2}{\partial T} + \frac{\gamma_2 \bar{l}}{\nabla V} * \frac{\partial c_2}{\partial x} * e^{\frac{x}{\bar{l}}} - \frac{\gamma}{V \nabla} c_2(x,T) e^{\frac{x}{\bar{l}}} = 0 \tag{28}$$

Thus, we obtained a differential equation which describes the long-term migration of NPs. This equation takes into account multiple wettings of the soil profile in accordance with statistical distribution of precipitation.

The equations (27) and (28) are spread over the whole plane x,T. The known methods of solving can be applied to them.

We must note that the time T in the equations is the total time of humidification and not the absolute time.

6. Conclusion

A necessary step has been taken towards the formation of the migration model of NPs in the soil profile, which adequately reflects actual physical and chemical processes occurring in the soil profile and the influence of natural and climatic factors on them. The method of obtaining averaged equations for the description of the long-term migration of NPs in the soil profile has been developed. In the same way partial differential equations, which describe more complex models, for example, those using diffusion, capillary phenomena and others, can be obtained. This will be the subject of our further research. The obtained equations permit the use of integral transformations in order to find solutions. The obtained model will be applied in the design of field studies on the migration of marker nanoparticles in different types of soils.

Acknowledgments

This study (research grant No 8.2.57.2015) was supported by Tomsk State University Academic D.I. Mendeleev Fund Program in 2015.

References

[1] Popov V, Kutnyakov I, Zhirnov V, Pirchenko E, Siverina A and Bobovnikova C 1994 *Pochvovedenie* (in Russian) **1** 41

[2] Abdullaev A and Shein E 1993 *Pochvovedenie* (in Russian) **8** 49

[3] Nizovtsev V, Panchenko and Garifullin R Investigation of the stability of the liquid front in the impregnation of soil layer in the model experiment 1988 *Pochvovedenie* (in Russian) **2** 163

[4] Bondarenko N 1975 *Physical Basis of Soil Reclamation* (Moscow: Kolos) p 260

[5] Guber A and Shein E 1997 *Pochvovedenie* (in Russian) **9** 1107

[6] Godeev P, Shmelina V and Shuyakov O 1990 *Hydrogeology* (Moscow: Nauka) (in Russian) p 448

[7] Verigin N 1953 *Proceedings of the Academy of Science* **10** 235

[8] Pachepski Y 1990 *Mathematical Models of Physical and Chemical Processes in Soils* (Moscow: Publ.House of the Academy of Science, AN SSSR) p 200

[9] Sysuev V 1986 *Modeling Processes in Landscape-geochemical Gystems*. (Moscow: Moscow State University Press) p300

[10] Verigin N 1962 *Fundamentals of the Theory of Dissolution and Leaching of Salts in Water Filtration in Rocks and Soils* (Moscow: Publ.House of the Academy of Science, AN SSSR) pp 59-70

[11] Mironenko V and Rumynin V 1986 *Experimental Work Migration in Aquifers* (Moscow: Nedra) p240

[12] Voronin A 1986 *Fundamentals of Soil Physics* (Moscow: Moscow State University Press) p244

[13] Istomin A, Korablyova S and Noskov M 2005 *Proceeding of Tomsk Polytekhnich University* **308** 74

[14] Kazachyonok N 2013 *Int. Conf. Radioactivity and Radioactive Elements in Enviroment* (Tomsk) 234-8

[15] Kundas S, Gishkelyuk I, Kovalenko I and Khil'ko O 2011 *Computer Simulation of the Migration of Contaminants in Natural Dispersive Media* (Minsk: Nauka) p 212

[16] Mikaiylov F. 2014 *Bull. Altai State Agricultural University* **7** 59

Ecotoxicology of heavy metals: Liquid-phase extraction by nanosorbents

A Burakov, I Romantsova, A Babkin*, E Neskoromnaya, A Kucherova and Z Kashevich

Department of equipment and technologies of nanoproduct manufactory, Tambov State Technical University, 106, Sovetskaya Street, Tambov, 392000, Russia

*E-mail: Flex_Trol@mail.ru

Abstract. The paper considers the problem of extreme toxicity heavy metal compounds dissolved in wastewater and liquid emissions of industrial enterprises to living organisms and environment as a whole. The possibility of increasing extraction efficiency of heavy metal ions by sorption materials was demonstrated. The porous space of the latter was modified by carbon nanotubes (CNTs) during process of the chemical vapour deposition (CVD) of carbon on metal oxide catalysts. The increasing of the sorption capacity (10-30%) and the sorption rate of nanomodified activated carbons in comparison with standard materials in the example of absorption of Co^{2+} and Ni^{2+} ions from aqueous solutions was proven.

1. Introduction

Among all the known to humanity chronic poisoning the most significant place belongs to intoxication by heavy metal salts. They cause abnormalities, mutations and intoxication, reaching a certain concentration in the human body. In addition to their toxic impact, their ability to settle on the walls of the finest systems of the body and clog the channels of the liver and kidneys is not less dangerous, thus reducing their ability to filter. This inevitably leads to the accumulation of toxins and products of vital activity of various cells of the body, i.e., its self-poisoning [1-9].

One of the major anthropogenic source of heavy metal ions entering into the environment and, as a consequence, into the human body, is industrial wastewater enterprises of ferrous and non-ferrous metallurgy, mechanical engineering, petrochemical industry, thermal power stations, etc. Water is the host environment of heavy metals migration in the earth's crust. At an excessive density heavy metals begin to behave as toxicants and ecotoxicants. This has a devastating impact not only on biocenosis individual organisms, but also on the whole ecosystem. For example, the accumulation of heavy metals in plants leads to their accumulation in the food chain and can cause serious diseases of man and animals. Increasing of heavy metals concentrations in water, soil and plants cause specific toxicosis and various mutagenic effects to human beings. The result of such cell deviation of the human body can become imbalance regulation of their division, which usually leads to malignant disease: mutation, degeneration of peripheral nerves, liver cirrhosis and blindness [9, 10].

Thus, research in the creation and modification of new materials that effectively adsorbed heavy metals, and the search for ways of their key characteristics optimization are very relevant for the direction of environmental protection and human health in general. This is confirmed, in particular, by positive dynamics of the volume growth on industrial enterprises wastewater stream for cleanup from

Content from this work may be used under the terms of the Creative Commons Attribution 3.0 licence. Any further distribution of this work must maintain attribution to the author(s) and the title of the work, journal citation and DOI.
Published under licence by IOP Publishing Ltd

heavy metal compounds, as well as the discrepancy between the existing materials with modern requirements [10-12].

Modern methods of aqueous solutions purification of heavy metals salts are not generalizing models and characteristic patterns. The ion exchange and complexation reactions due to the formation of hydroxides and physical adsorption, directly related to the specific surface of the adsorbents can be considered as a possible mechanism of ions sorption from solutions [2, 13].

One of the possible ways to intensify the liquid phase sorption processes of heavy metals is a modification of the sorbent using carbon nanostructures. The literary sources analysis shows that CNTs have unique absorption properties. Since the CNTs are surface structure, all of their mass is concentrated in the surface of layers. This defines an abnormally high specific surface of CNTs. This kind of characteristics allows us to consider the CNTs as a surface modifying material of the sorbent.

2. Experiment

The technology obtaining of modified materials experimental samples based on the CVD process [14, 15]. Nanomodified samples were synthesized in an industrial reactor of the synthesis CNTs (NanoTechCenter Ltd., Tambov, Russia). The porous material (a carrier (sorbent)) previously impregnated by a solution of a metal oxide catalyst obtained by the sol-gel method or by thermal decomposition of multicomponent systems (Fe, Co, Mo, Mg, Al). Impregnated sorbent was dried (80°-140°C) and annealed (500°-550°C). Technical grade propane-butane was used as a carbon gas mixture. The process of the CVD synthesis was conducted for 60 min at 650°C.

Complex studies of sorption inorganic substances (ions and hydroxide ions of toxic heavy metals) from aqueous solutions were made to determine the effectiveness of the obtained nanomodified sorbents.

The study of the adsorption of Co^{2+} and Ni^{2+} ions was performed on activated carbons AG-5, AG-5-N, NWC and NWC-N under static conditions. For carry out of investigation a seven tubes with 50 ml of the solutions and 1.75 g of the sorbent were taken. All samples were shaken during each 3 minutes, sorption time is 40 min (for the Co^{2+} ions) and 30 min (for ions Ni^{2+}). After that the solution was filtered, and the optical density was measured spectrophotometrically. Then the dependence of adsorption and the equilibrium concentration - adsorption isotherm, was constructed.

3. Results and discussion

The proposed technology allows to stably synthesize the CNTs with predetermined parameters (porosity and layer height, etc.) on the porous support surface (Figure 1). The structure of as-growing CNTs (Figure 2) was obtained by scanning electron microscopy (SEM) method (Carl Zeiss, Neon 40, Tambov, Russia).

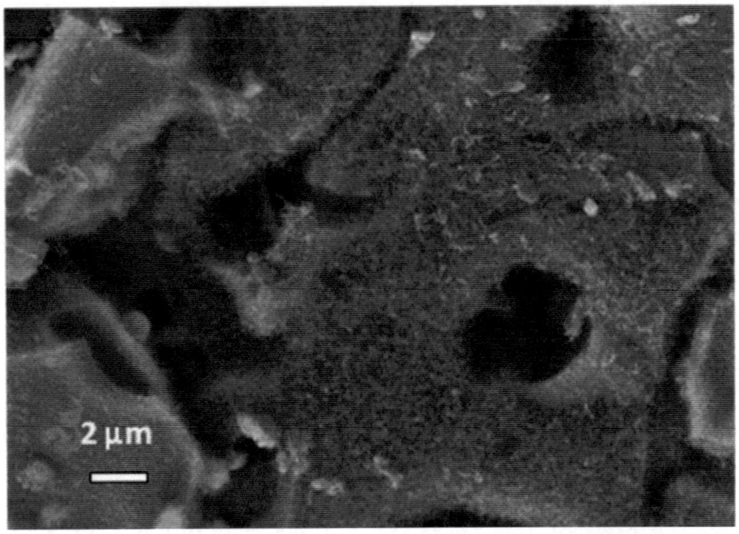

Figure 1. The granules surface of coconut activated carbon modified by the CNTs

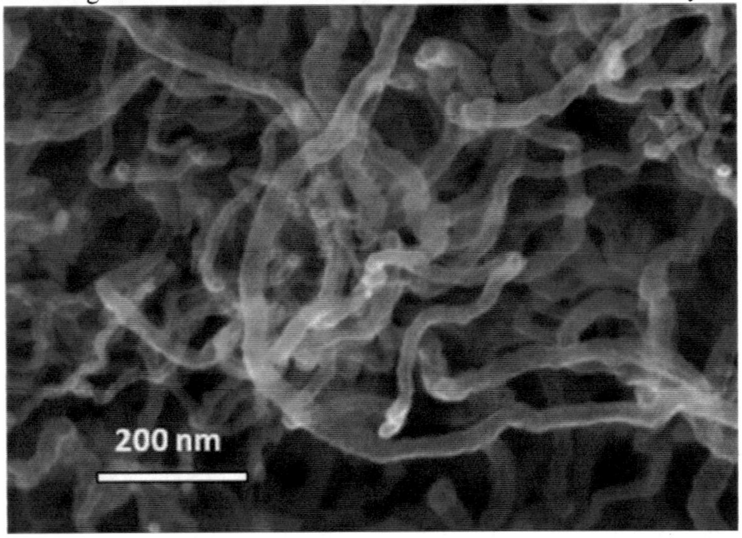

Figure 2. SEM-image of the CNTs structure

From the SEM images (Figures 1, 2), it can be seen that the products are nanotubes without visible admixture of other particles: CNTs diameter near 10-20 nm. There are small amount of catalyst nanopartical with size about 25 nm.

After separating the adsorbent from the aqueous solution, the amount of adsorbed heavy metal ions can be evaluated according to the following equation:

$$Q = \frac{(C_0 - C_e) \cdot V}{m}$$

where Q is the equilibrium metal uptake (mol g^{-1}), C_0 is the initial metal concentration in the solution (mol L^{-1}), C_e is the equilibrium metal concentration in the solution (mol L^{-1}), V is the volume of the solution (mL), and m is the mass of the adsorbent (g).

Obtained comparative dependence - the adsorption isotherm of Co^{2+} and Ni^{2+} ions are shown in Figures 2 and 3, respectively.

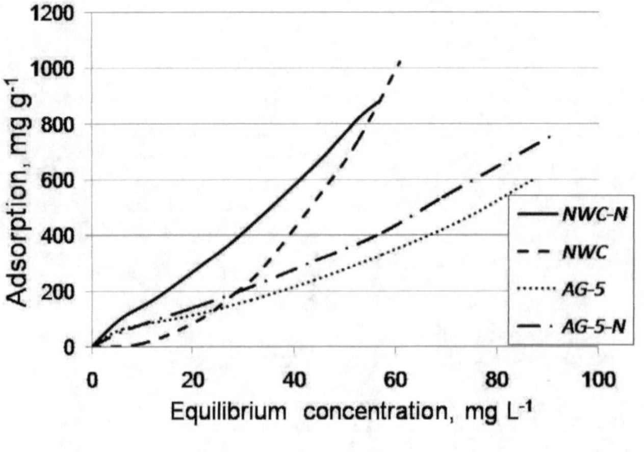

Figure 3. The adsorption isotherms of Co^{2+} ions at different types of adsorbents

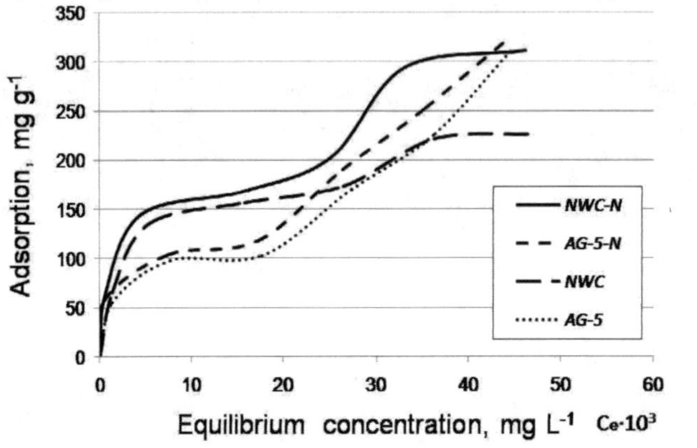

Figure 4. The adsorption isotherms of Ni^{2+} ions at different types of adsorbents

The effectiveness of the nanomodified materials in adsorption processes is confirmed, in particular, that nanomodified samples possess higher adsorption capacity at 30% of Co^{2+} and 10-15% for the Ni^{2+} in comparison with initial materials. These results also show that the nanomodified sorbents have higher adsorption rate.

The studies showed that there is a real possibility of applying the modification process on an industrial scale with an appropriate technical and economic effect due to a marked increase in indicators of quality cleaning.

4. Conclusion

The pollution of the biosphere by a group of contaminants, which has been generally termed "heavy metals", took the particular importance. Many heavy metals, when they come into a living organism, it leads to poisoning or death. These metals belong to the class of xenobiotics. Penetration toxicant in living organism occurs mainly through the water medium, so the waste water which is highly toxic, cannot be reset in open water without proper processing.

The experimental researches show that the complex sorbents based on activated carbons modified by carbon nanostructures are more effective in comparison with conventional counterparts and allow to deep treatment of aquatic environments.

The adsorption capacity of the nanomodified sorbent was shown that nanomodification of activated carbons increased the equilibrium adsorption of heavy metals (Ni^{2+}, Co^{2+}) by 10-30%. In addition, the developed technology of nanocarbon modification allows obtaining samples of sorbents based on carrier at different chemical nature with improved extraction of heavy metal ions from aqueous solutions.

Acknowledgments
This work was supported with Ministry of education and science of the Russian Federation, project # 16.711.2014/K.

References

[1] Abu-Eishah S I 2008 *Appl. Clay Sci.* **42** 201
[2] Aziz H A, Adlan M N and Ariffin K S 2008 *Bioresour. Technol.* **99** 1578
[3] Babel S and Kurniawan T A 2003 *J. Hazard. Mater.* **B97** 219
[4] Bose P, Bose M A and Kumar S 2002 *Adv. Environ. Res.* **7** 179
[5] Chen G H 2004 *Sep. Purif. Technol.* **38** 11
[6] EPA, EPA National Library Network. U.S. Environmental Protection Agency, URL:http://www.epa.gov/natlibra/ols.htm: 2014
[7] Agbozu I E and Emoruwa F O 2014 *African Journal of Environmental Science and Technology* **8** 239
[8] Keng Pei-Sin, Lee Siew-Ling, Ha Sie-Tiong, Hung Yung-Tse and Ong Siew-Teng 2014 *Environmental Chemistry Letters* **12** 15
[9] Dhananjay Chaturvedi and Omprakash Sahu 2014 *Global Journal of Environmental Science and Technology* **2(3)** 20
[10] Kurniawan T A, Chan G Y S, Lo W H and Babel S 2005 *Sci. Total Environ.* **366** 409
[11] Alekseenko V A, Suvorinov A V et al. 2002 *Metals in the Environment* (Moscow: Logos) pp. 220 (in Russian)
[12] Bingham F T, Costa M and Eichenberger E 1993 *Some questions toxicity of metal ions* (Moscow: Mir Publishing) pp. 158 (in Russian)
[13] Bailey S E, Olin T J, Bricka M and Adrian D.D. 1999 *Water Res* **33** 2469
[14] Burakov A E, Romantsova I V, Burakova E A, Tkachev A G and Tugolukov E N 2013 *Sorption and chromatographic processes* **13** 334 (in Russian)
[15] Romantsova I V, Mikhaleva Z A and Tkachev A G 2013 *Glass Physics and Chemistry* **39** 89 (in Russian)

IOP Conf. Series: Materials Science and Engineering **98** (2015) 012024 doi:10.1088/1757-899X/98/1/012024

Fabrication and characterization of polycaprolactone cross-linked and highly-aligned 3-D artificial scaffolds for bone tissue regeneration via electrospinning technology

SN Gorodzha[1], MA Surmeneva[1] and RA Surmenev[1, 2]

[1]Department of Theoretical and Experimental Physics, National Research Tomsk Polytechnic University, 634050, Tomsk, Russian Federation

[2]Fraunhofer Institute for Interfacial Engineering and Biotechnology IGB, Stuttgart, 70569, Germany

Email: rsurmenev@mail.ru

Abstract. Novel technologies allowed the scientific community to develop scaffolds for regeneration of bone tissue. A successful scaffold should possess specific macroscopic geometry and internal architecture to perform biological and biophysical functions. In this study the process of polycaprolactone microfibrous development with either cross-linked or highly-aligned three-dimensional artificial mats via electrospinning technology for potential application in tissue engineering is described. The morphology and size of electrospun fibers were assessed systematically by varying the rotation speed of grounded collector. It was found that the diameter of the fibers decreased by increasing the rotation speed of collector. The morphology of the fibers changed from cross-linked to highly-aligned at appr. 1000-1100 rpm.

1. Introduction

With the development of the modern bone regenerative medicine, more and more novel materials based on the concept of the nature are used [1]. These materials possess bioinspired, multiple bone-like structures with unique properties close to that of native human skeletal system.

Nowadays, the development of biomaterials for substitution of bone tissue defects with predetermined architectonics and individual functional optimization as alternative for available prostheses and implants is one of the most important challenges [2].

The three-dimensional (3-D) scaffolds define as porous solid 3-D structures to perform various functions, including promotion of the cells adhesion and formation of the extracellular matrix (ECM). Several characteristics, such as morphology, porosity, biodegradation rate, acceptable biocompatibility are required for the fabrication of 3-D scaffolds. The physical characteristics as pores geometry of the scaffold play a crucial role in the formation of the natural ECM by providing support for the cells and vascularization. Insufficient vascularization can lead to inaccurate cell integration and proliferation, resulting in cell death in the constructs [3].

Presently, there are three methods accessible for the synthesis of 3-D fiber structure, using different physical principles: phase separation [4], self-assembly [5] and electrospinning [6]. Electrospinning is the most broadly used technique and present the most promising results for tissue engineering applications, when compared with other methods. The reason of success is that this technique allows to fabricate fibers with the diameter from 3 nm to 5 μm or even more [7].

Content from this work may be used under the terms of the Creative Commons Attribution 3.0 licence. Any further distribution of this work must maintain attribution to the author(s) and the title of the work, journal citation and DOI.
Published under licence by IOP Publishing Ltd

After the development of the elecrospinning process by Zeleny [8] in 1914 it was patented in 1934 by Formhals [9]. During electrospinning an electric field is applied to convert polymer solution into the form of micro- or nano-fibers. Schematic diagram of the electrospinning process for manufacturing of fiber-structured mats is shown in Figure 1. When an external electric field is applied between a pendant droplet of a polymer solution and a metal collector which serves as a counter electrode, the solution gets charged and electrostatic repulsion affects the surface tension of the droplet and it is deformed into a conical shape (Taylor cone).

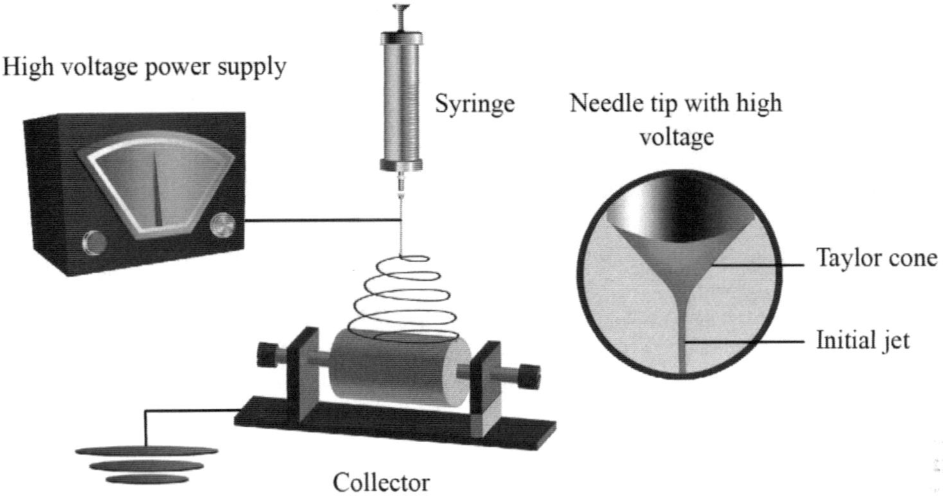

Figure 1. Schematic representation of the electrospinning system

When the electric field strength approaches a critical value, which is sufficient to overcome the surface tension of the droplet, a charged, continuous jet of a polymer solution is ejected from the needle towards a collector. The fluid jet of a solution accelerates towards the collector, undergoes significant bending instability and whipping.

As the polymer is moved from the needle towards the grounded collector, the solvent evaporates quickly. On the surface of the collector, a mat of solid nanofibers in nonwoven form is deposited. A variation of the rotation speed results in increase of the mechanical properties of scaffolds [10].

The method of electrospinning is well suited to produce natural and synthetic biocompatible polymers for diverse biomedical applications. These materials can be potentially used as wound dressings, artificial blood vessels and bone tissue engineering scaffolds [11].

Polymers are organic materials which consist of large macromolecules represented by covalently bonded chains of atoms. Unless they are cross-linked, the macromolecules interact with one another via weak secondary bonds (hydrogen and van der Waals bonds). The mechanical, thermal and other properties of the polymers are influenced by several factors, including the composition of backbone, side groups, structure of chains, and the molecular weight of the molecules [12]. Polycaprolactone (PCL) is biodegradable aliphatic polyester with desirable mechanical properties and slow degradation rate, which makes it possible to be used for hard and soft tissue repair. This study aimed to develop PCL electrospun scaffolds for bone tissue engineering with enhanced biocompatibility and vascularization. This polymer was used in the study due to higher resistivity to degradation rate in biofluids compared with other bioresorbable polymers [13].

In the present study, the feasibility of electrospinning process of polymers for bone tissue engineering applications was demonstrated. The effect of the rotation speed of the collector on the scaffolds fibers orientation and morphology was studied.

2. Materials and methods

2.1 Polymer

The PCL polymer with an average molecular weight of 80 kDa and density of 1.145 g/mL at 25 °C and chloroform ($CHCl_3$) were obtained from Sigma-Aldrich (St. Louis, MO, USA) Electrospinning solutions were prepared by dissolving PCL in chloroform with a stirrer and ultrasonic homogenizer to a concentration of 9% (w/v). Viscosity of the polymer is a crucial characteristic to prepare the uniform and homogeneous structure of the scaffolds [14]. It was measured with Rotational Rheometer: RheolabQC (Anton Paar GmbH, Austria) and the value of the viscosity for 9% (w/v) PCL solution was 1.47 kg/(s·m).

2.2 Electrospinning

A custom-made electrospinning apparatus with a high-voltage generator (Weil am Rhein, Germany) was designed. A syringe pump was used to feed the solutions through extension tube capped with blunted 24-gauge needles (inner diameter of 0.55 mm). The polymer solution was delivered at a constant flow rate of 2 ml/h. The cylindrical stainless steel collector rotating with the speed ranging from 200 to 1100 rotation per minute (rpm) was used. The fixation of the syringe was made in a horizontal plane of the infusion pump. Delivery of the polymer solution to the needle was done through a plastic tube, which was fixed vertically at a distance of 5 cm above the collector. A high voltage potential (7 kV) was applied between the tip of the needle and the collector. The polymer solution was forced going through the needle and collected as fibers on the rotating cylinder. The well-aligned and cross-linked fibrous membranes were separated from the collector and used for further analysis.

2.3 Surface characterization

A Fourier transform infrared (FT-IR) spectrometer Bruker Equinox 55 (Saarbrücken, Germany) was used for ATR–FT-IR measurements. Topographic images of the fibrous scaffolds were examined using a Keyence VHX Portable Digital Optical Microscope (Neu-Isenburg, Germany), which allows performing microscopic examination of the samples.

3. Results and discussion

To study the effect of the rotating collector speed on the morphology and structure of the microfiber mats, the rotation speed of the collector was varied. Optical micrographs of electrospun PCL scaffolds are shown in Figure 2. The thickness of the scaffolds is ~ 0.2 mm.

It was revealed that with the rotation speed of collector in the range from 200 rpm to 800 rpm randomly oriented structures and smooth morphology of the electrospun PCL microfibers were obtained. An average interconnected pores size of three-dimensional fibrous mesh is ~10-50 μm, which are composed of randomly oriented microfibers with diameter ranging from 4 μm to 6 μm (Figure 3: a-d). Was noted that the by increasing the collectors speed to 1000 and 1100 rpm, the fibers diameter ranged from 4 μm to 2 μm (Figure 3: e, f). The morphology of the scaffolds has become oriented along the longitudinal axis which resulted in an aligned structure of the scaffold. When the rotation speed of the collector increases, the number of pores decreases.

This controlled alignment in two- and three-dimensional architecture of the fibers can positively affect cell differentiation and proliferation, since ECM found in living tissues has regular and defined structure [15]. It is reported that the aligned structure of the scaffold can serve as a base for creation *in-vivo*-like morphology and mechanical deformations can be uniformly delivered to all cells in the cultures [16].

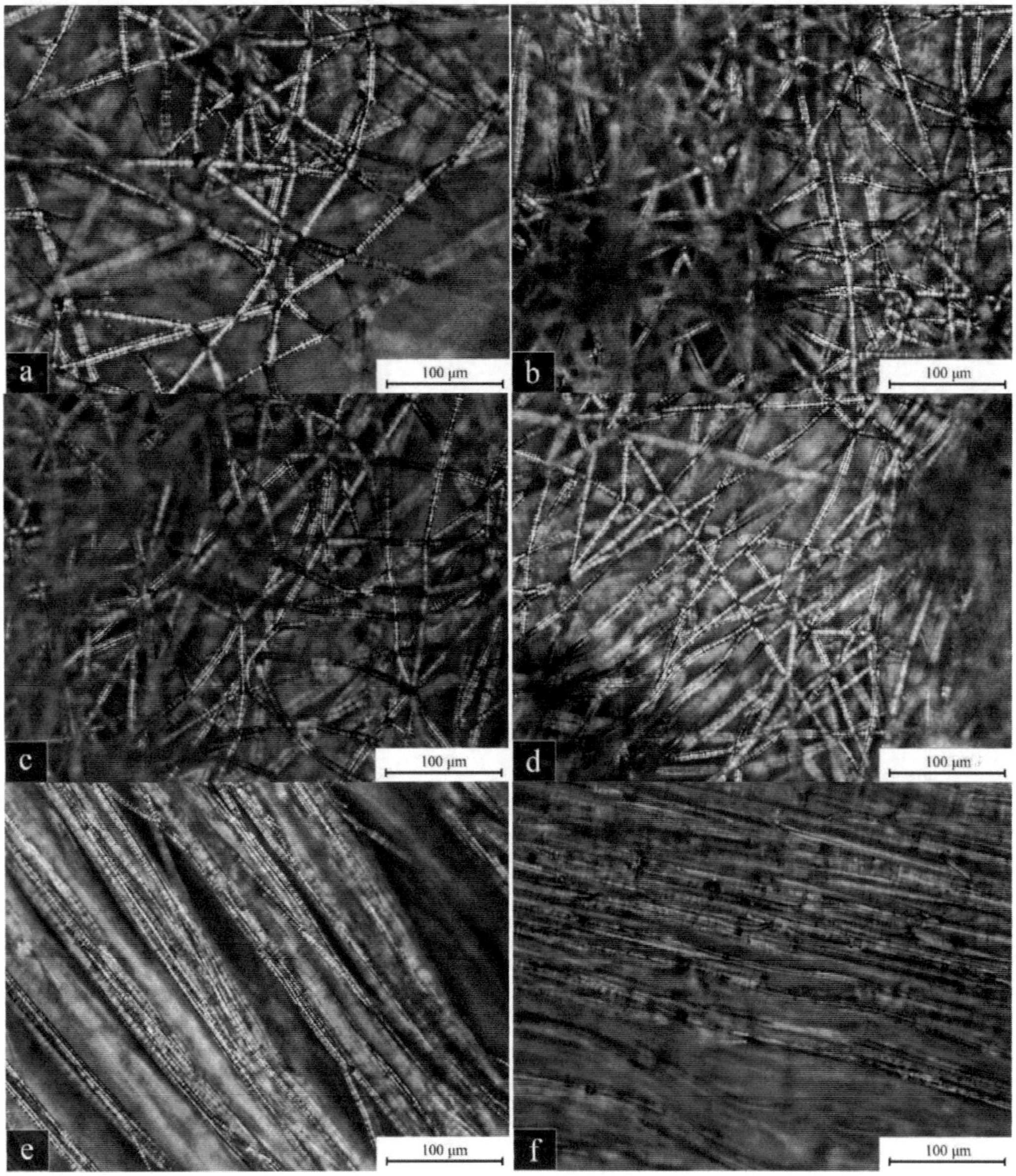

Figure 2. Optical micrographs of microfibrous PCL scaffolds: (a) 200 rpm, (b) 400 rpm, (c) 600 rpm, (d) 800 rpm, (e) 1000 rpm, (f) 1100 rpm.

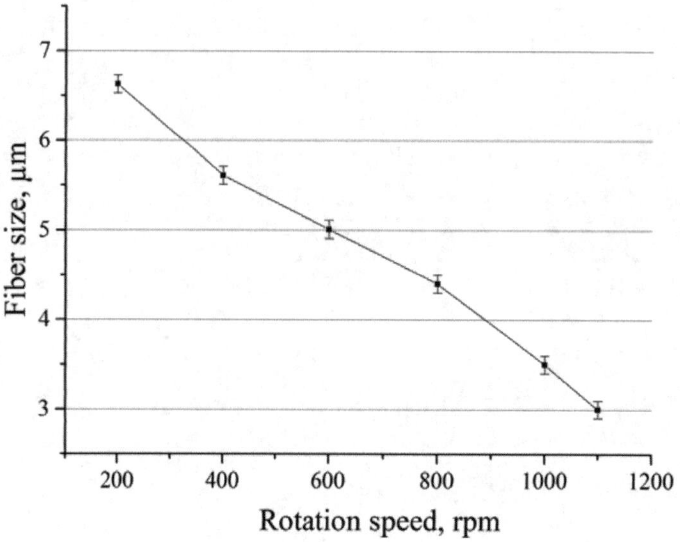

Figure 3. Dependence of the PCL fiber size from rotation speed of the collector

In Figure 4 a FT-IR spectrum of PCL in the wavelength range of 3000-600 cm^{-1} is shown. From these data carbonyl stretching mode around 1727 cm^{-1} can be observed. The most intensive bands which correspond to PCL are presented in Table 1. It is known that the band at 1293 cm^{-1} relates to the backbone C–C and C–O stretching modes of the crystalline phase in the PCL (see Table 1), which also characterize the structure of the scaffold [17]. So, after different manipulations with the polymer, the composition of the PCL was not changed.

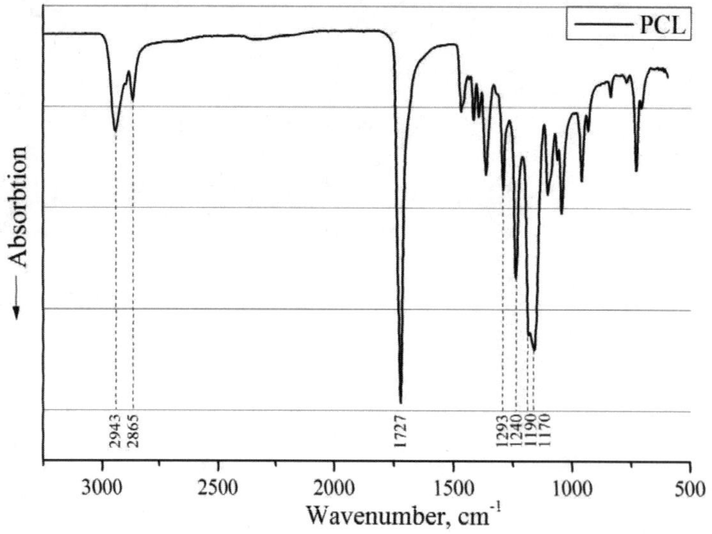

Figure 4. FT-IR spectrum of PCL mats in the region 3000-600 cm^{-1}.

Table 1. Characteristic infrared bands of PCL [17], [18]

Value [cm^{-1}]	Vibration	Abbreviation
2943	Asymmetric CH_2 stretching	ν_{as} (CH_2)
2865	Symmetric CH_2 stretching	ν_s (CH_2)
1727	Carbonyl stretching	ν (C=O)
1293	C–O and C–O and C–C stretching in the crystalline phase	ν_{cr}
1240	Asymmetric COC stretching	ν_{as} (COC)
1190	OC–O stretching	ν (OC–O)
1170	Symmetric COC stretching	ν_s (COC)

4. Conclusion

In this study, electrospun microfiber PCL scaffolds were developed and the influence of the rotation speed of the collector during electrospinning process on the structure of the polymer fibers was investigated. The results revealed that by increasing the speed rotation, the fibers diameter decreases, presenting a structure preferentially oriented in one particular direction, as confirmed by optical microscope analyses. This study serves as a basis for future development of electrospun 3-D polymer PCL scaffolds for repair of bone defects.

Acknowledgements

The authors would like to thank Dr. C. Oehr and Dr. T. Schiestel for support in performing experiments. This research was supported by *a FP7 Marie Curie grant* (327701), *the Russian Fund for Basic Research* (15-08-08652A), *the Russian President's Stipend* SP-6664.2013.4, grant MK-485.2014 and *the State order NAUKA* #11.1359.2014/K.

References

[1] Xia F and Jiang L 2008 *Adv. Mater.* **20** 2842
[2] Nerem RM and Sambanis A 1995 *Tissue Eng.* **1** 1
[3] Borenstein JT, Megley K, Wall K, Pritchard EM, Truong D, Kaplan DL, Tao SL and Herman IM 2010 *Materials* **3** 1833
[4] Liu S, He Z, Xu G and Xiao X 2014 *Mater. Sci. Eng. C: Mater. Biol. Appl.* **44** 201
[5] Zhao Y, Tanaka M, Kinoshita T, Higuchi M and Tan T 2010 *Control. Release* **142** 354
[6] Ma G, Fang D, Liu Y, Zhu X and Nie J 2012 *Carbohydr. Polym.* **87** 737
[7] Bhardwaj N and Kundu SC 2010 *Biotechnol. Adv.* **28** 325
[8] Zeleny J 1917 *Phys. Rev.* **10** 1
[9] Formhals A 1934 Germany Patent 500 283
[10] Thomas V, Jose MV, Chowdhury S, Sullivan JF, Dean DR and Vohra YK 2006 *J. Biomater. Sci. Polym.* **17** 969
[11] Sill TJ and von Recum HA 2008 *Biomaterials* **29** 1989
[12] Dee KC, Puleo DA and Bizios R 2002 *John Wiley & Sons*, Inc., New York, USA. doi: 10.1002/0471270598.ch7
[13] Williams JM, Adewunmi A, Schek RM, Flanagan CL, Krebsbach PH, Feinberg SE, Hollister SJ and Das S 2005 *Biomaterials* **26** 4817
[14] Huang ZM, Zhang YZ, Kotaki M and Ramakrishna S 2003 *Compos. Sci. Technol.* **63** 2223
[15] Singhvi R, Kumar A, Lopez G, Stephanopoulos G, Wang D, Whitesides G and Ingber D 1994 *Science* **264** 696
[16] Deutsch J, Motlagh D, Russell B and Desai TA 2000 *J. Biomed. Mater. Res.* **53** 267
[17] Coleman MM and Zarian J 1979 *J. Polym. Sci. Polym. Phys. Ed.* **17** 837
[18] Hummel DO, 1966 *New York: Interscience* 207

Investigation of the HA film deposited on the porous Ti6Al4V alloy prepared via additive manufacturing

M Surmeneva[1], E Chudinova[1], M Syrtanov[1], A Koptioug[2] and R Surmenev[1,3]

[1] Tomsk Polytechnic University, Institute of Physics and Technologies, 30, Lenina ave., Tomsk, 634050, Russia

[2] Mid Sweden University, Akademigatan 1, Östersund, 83125, Sweden,
[3] Fraunhofer Institute for Interfacial Engineering and Biotechnology IGB, Nobelstraße 12, Stuttgart, 70569, Germany

E-mail: surmenevamaria@tpu.ru

Abstract. This study is focused on the use of radio frequency magnetron sputtering to modify the surface of porous Ti6Al4V alloy fabricated via additive manufacturing technology. The hydroxyapatite (HA) coated porous Ti6Al4V alloy was studied in respect with its chemical and phase composition, surface morphology, water contact angle and hysteresis, and surface free energy. Thin nanocrystalline HA film was deposited while its structure with diamond-shaped cells remained unchanged. Hysteresis and water contact angle measurements revealed an effect of the deposited HA films, namely an increased water contact angle and contact angle hysteresis. The increase of the contact angle of the coating-substrate system compared to the uncoated substrate was attributed to the multiscale structure of the resulted surfaces.

1. Introduction

Additive manufacturing (AM) of plastics and metals is already well-established method in engineering and technology and is commonly used for rapid prototyping and small series production as well as manufacturing of the unique components with complex 3D shapes [1]. One of the rapidly growing application areas of the AM is medicine, especially orthopedics and prosthetics. These applications are utilizing strongest advantages of the AM technologies such as freedom of component shapes, possibilities of computer optimization of the manufactured-to-be component functionality and properties, and reasonable prices in case of manufacturing of small series of products. These benefits are already recognized by medical implant manufacturers and practical surgeons [1].

It is known that calcium-phosphate (CaP) coated titanium implants have been widely used in orthopedic and dental fields due to their excellent biocompatibility attributable to the surface properties [2]. Radio frequency (RF) magnetron deposited hydroxyapatite-based (HA) coatings have become a promising candidates for the promotion of a faster and more effective titanium implant osseointegration at early implantation times [3, 4]. In this study, the focus is based on the use of RF-magnetron sputtering to modify the surface of porous Ti6Al4V alloy fabricated via AM.

2. Materials and methods

All the samples used were fabricated using Ti6Al4V ELI powder (Arcam AB) in ARCAM A2 EBM® machine (Arcam AB, Mölndal, Sweden) [5]. All flat coin like sample had a dimension of 10 and 30 mm in diameter had flat surfaces normal to the melt layer surface and were made in a single

Content from this work may be used under the terms of the Creative Commons Attribution 3.0 licence. Any further distribution of this work must maintain attribution to the author(s) and the title of the work, journal citation and DOI.
Published under licence by IOP Publishing Ltd

batch. Porous 3D-lattice scaffolds were made in two different batches with identical parameter settings and same diamond-shaped basic lattice cells. All samples were carefully blasted in the ARCAM powder recovery system using the same precursor powder. Electrochemical etching was carried out at 30° C in the ethanol and 2-propanol solution of the aluminum chloride and zinc chloride with the stabilized DC current density 0.2 A/cm^2 [6].

A powder of HA ($Ca_{10}(PO_4)_6(OH)_2$) was prepared using mechanochemical activation and then used as a precursor-powder to prepare a target for sputtering (Prof. M. Chaikina, Institute of Solid State Chemistry and Mechanochemistry SB RAS). The RF-power (500 W), argon gas pressure (working pressure 0.4 Pa, base pressure 10^{-4} Pa) and the distance between the target and substrate (40 mm) were kept constant in all experiments. HA coatings were deposited for 8 hours which resulted in the coating thickness of 700 nm.

The phase composition and the structure of the coatings were identified by X-ray diffraction (XRD-7000, Shimadzu, Japan) with CuKα radiation (λ=0.154 nm) in the 2 theta range from 10° to 60° with a scan speed of 2.0°/min, sampling pitch of 0.03°, preset time of 5.0 sec at 30 kV and 30 mA. The average crystallite size was determined using the Scherrer's equation from the broadening of the diffraction peaks taking into account the instrumental broadening and with the PowderCell 2.4 software. An instrumental broadening of 0.14 in 2θ was determined by the full width at half maximum of a silicon powder. As references for the patterns of HA and Ti #9-0432 and #44-1294 (ICDD database) were used, respectively. The morphology of the HA-coated samples was observed with the field-emission scanning electron microscopy (FESEM, Quanta-200, FEI, USA) equipped with an Energy Dispersive X-ray (EDX) system. Contact angle analyses were performed with an optical contact angle apparatus (OCA 15 Plus Data Physics Instruments GmbH, Germany), using the SCA20 software (Data Physics Instruments GmbH, Germany). The contact angle (CA) of water in air was measured using a sessile drop method. A minimum of 10 droplets (2 μL, 5 μL·s^{-1}) of water, 3 droplets of diiodomethane or ethylene glycol were examined for each sample, and the resulting mean CA values were then used for the calculations. The surface free energy σ was calculated using Owens-Wendt-Rabel-Kaelble (ORWK) method.

3. Results and discussion

Figure 1 illustrates the morphology of the Ti6Al4V alloy cellular structure with diamond-shaped unit cells. Unit cells dimensions demanded by the design were 500 μm with start diameter of 150 μm, but the resulting ones are some larger. As clearly visible from these images metal surface contains a number of powder grains attached to the bulk metal (figure 1b) providing a micrometer scale roughness, a feature commonly met for the powder bed AM processes. Surfaces are also showing distinct roughness at the sub-micron scale (figure 1c).

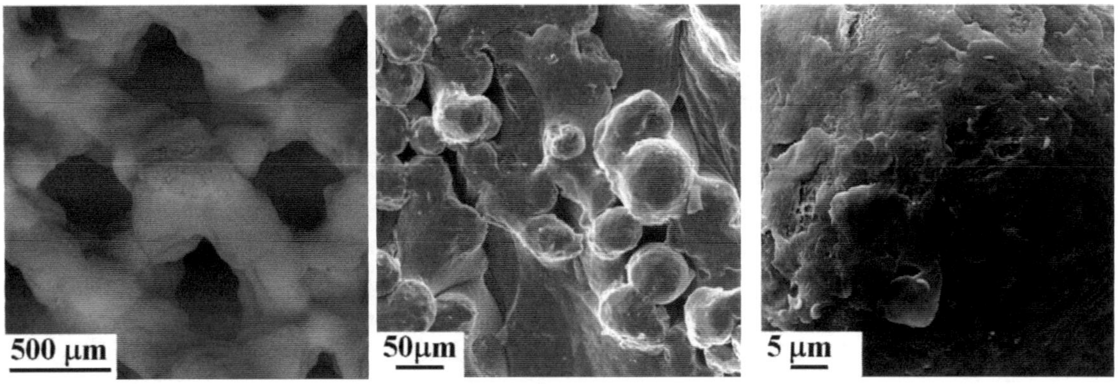

a b c

Figure 1. SEM images of the porous Ti6Al4V sample morphology prepared via AM. Images are acquired using both SE (b, c) and BSE (a) detectors

It has been extensively reported that the microscale morphology significantly influences cell behaviour such as adhesion, proliferation and differentiation *in vitro* and *in vivo* [7].

Figure 2 a. illustrates the morphology and composition of HA films prepared on the surface of Ti6Al4V substrates fabricated via AM technology. It can be seen that the surface of the HA-coated substrate is some smoother compared to that of uncoated Ti6Al4V. As already mentioned elsewhere [8], the initial island nucleation of the thin film was attributed to the preferred growth in the valleys. The authors observed that for a smooth and rough substrate, the film surface gets smoother with the increasing in the film thickness due to the preferred growth of the film in the valleys [8].

It was reported that the coating composition and thickness are strongly depended on the substrate that was used [9]. It has been also shown [10] that pretreatment of the titanium substrates can strongly influence the coated surface properties. A comparison of the HA films on titanium substrate surfaces modified by acid etching and pulse electron beam treatment revealed significant differences in the morphology of the films (roughness, grain size and shape) and the mechanical properties [10].

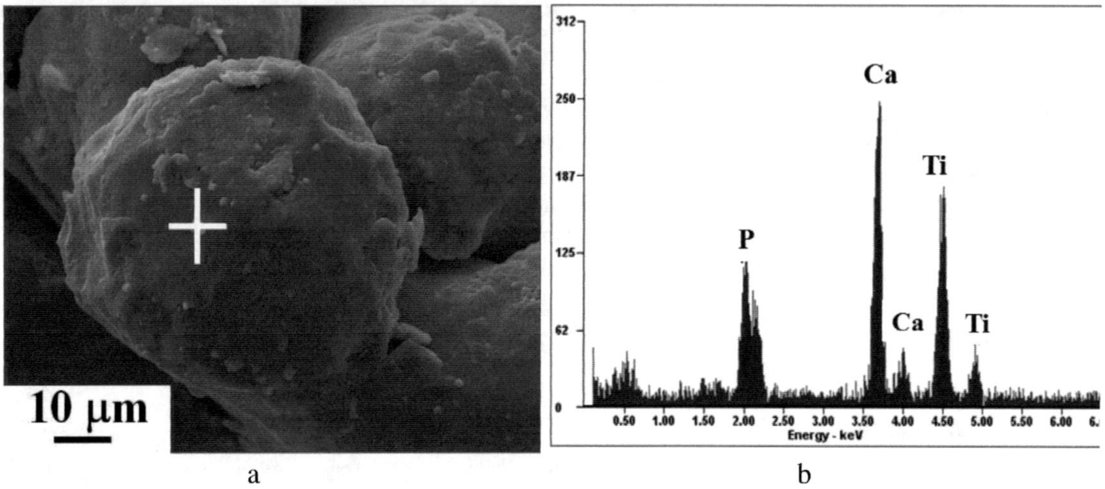

Figure 2. SEM image (a) and EDX spectrum (b) of the HA-coated Ti6Al4V prepared via AM

The Ca/P ratio of the CaP coatings deposited via RF-magnetron sputtering onto the Ti6Al4V substrate prepared via AM technology according to EDX-data was 1.20±0.03, which is lower than that of the 1.67 Ca/P ratio of stoichiometric HA (figure 2b).

The typical XRD-pattern of the HA coating fabricated via RF-magnetron sputtering on the surface of titanium prepared via AM is shown in figure 3.

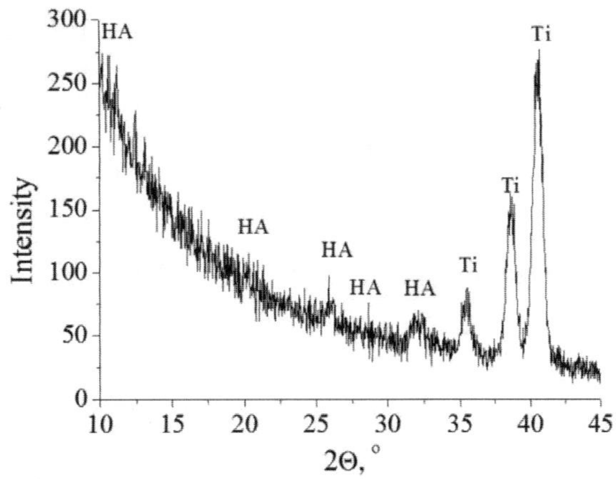

Figure 3. XRD pattern of the HA-coated titanium alloy prepared via AM

All diffraction peaks corresponded to the structure of HA, no diffraction peaks from other phases were detected. HA coating showed a nanocrystalline structure with the crystallites of an average size of 32±9 nm. Normally, the most intense peaks of HA are observed in the 30–35° 2θ range. The lattice parameters of the HA coating deposited on the surface of porous titanium alloy were determined to be a=9.4042 and c=6.8875 Å.

The data presented in Table 1 show that HA coating deposited on the surface of porous Ti6Al4V alloy resulted in increase of the CA. It should be noted that water CA did not reveal any significant effect from the coating. Surface energy of the Ti6Al4V porous alloy was decreased after depositition of the HA coating. Previously, we observed increase of the water CA of the coating on titanium in case of the silicate-substituted HA prepared at different bias [11].

Table 1. Surface wettability of Ti6Al4V porous alloy with and without HA coating

Sample	CA (degrees)				CA hysteresis (degrees)	Surface free energy ($mN{\cdot}m^{-1}$)		
	Water	Water in 5 min	diiodomethan	ethylene glycol		dispersive	polar	total
Bare Ti6Al4V	106±1	88±1	34±1	42±2	20±3	48.75	0.32	49.07
HA-coated Ti6Al4V	109±2	101±1	56±2	80±1	24±3	29.20	0.02	29.22

The increase of the CA of the coating-substrate system compared to the uncoated substrate was attributed to the hierarchical organization of the resulting surfaces. Thus, to study the process of surface wettability of a porous Ti6Al4V substrates a water droplet was placed and contact angle was monitored for a 5 minute time interval with a 1 minute step. It can be seen that in case of the coated substrates the water CA decreases slowler than for the uncoated substrates. This allows to suggest that HA coating prevents water spreading over the surface of the porous Ti6Al4V alloy. Most probably it is because the water is becoming trapped in the microscale cavities of the surface of HA-coated titanium Ti6Al4V alloy resulted in a strong adhesion between the surface and a liquid.

4. Conclusion

RF-magnetron sputtering provides the ability to deposit thin HA coatings onto the surface of porous Ti6Al4V scaffold with the diamond lattice structure fabricated via AM. The HA coating showed a nanocrystalline structure with the crystallites of an average size of 30 nm. The lattice parameters of the HA coating prepared on the surface of porous titanium alloy were determined to be a=9.4042 and c=6.8875 Å. Water contact angle and hysteresis increase was revealed as an effect of the deposited HA films. Surface energy of the Ti6Al4V porous alloy was decreased after deposition of the HA coating. We attributed the increase of the contact angle of the coating-substrate system compared to the uncoated substrate to the multiscale structure of the resulted surfaces. Water sealed within the microscale structure of HA-coated titanium alloy surface resulted in a strong adhesion between the surface and a liquid.

Acknowledgments

This research was supported by the *Russian Science Foundation* (15-13-00043).

References

[1] Koptyug A, Rännar L-E, Bäckström M, Franzén, Sanna F and Dérand P 2013 *Int. J. of Life Sci. and Med. Res.* **3** 15.

[2] Arcos D, Boccaccini AR, Bohner M, Díez-Pérez A, Epple M, Gómez-Barrena, Herrera A, Planell JA, Rodríguez-Mañas L, and Vallet-Regí M 2014 *Acta Biomater.* **10** 1793.

[3] Jansen JA, Wolke JGC, Swann S, Van Der Waerden JPCM and K De Groot 1993 *Clin. Oral Impl. Res.* **4** 28.

[4] Urquia Edreira ER, Wolke JGC, Aldosari AA, Al-Johany SS, Anil S, Jansen JA, and van den Beucken JJJP 2015 *J. Biomed. Mater. Res.* A **103** 300.

[5] Standard Specification for Additive Manufacturing Titanium-6 Aluminum-4 Vanadium ELI with Powder Bed Fusion; ASTM F2924–14; American Society for Testing Materials: West Conshohocken, P., USA, 2014.

[6] Tajima K, Hironaka M, Chen K-K, Nagamatsu Y, Kakigawa H and Kozono Y 2008 *Dent. Mater. J.* **27** 258.

[7] Li L and Xie T 2005 *Ann. Rev. Cell Dev. Biol.* **21** 605.

[8] Zhong M, Zhang C, and Luo J 2008 *Appl. Surf. Sci.* **254** 6742.

[9] Feddes B, Vredenberg AM, Wolke JGC and Jansen JA 2004 *Surf. Coat. Technol.* **185** 346.

[10] Surmeneva MA, Surmeneva RA, Tyurin AI, Mukhametkaliyev TM, Teresov AD, Koval NN, Pirozhkova TS, Shuvarin IA and Oehr C 2014 *Thin Solid Films* **571** 218.

[11] Surmeneva MA, Kovtun A, Peetsch A, Goroja S, Sharonova AA, Pichugin VF, Grubova IY, Ivanova AA, Teresov AD, Koval NN, Buck V, Wittmar A, Ulbricht M, Prymak O, Epple M and Surmenev RA 2013 *RSC Adv.* **3** 11240.

IOP Conf. Series: Materials Science and Engineering **98** (2015) 012026 doi:10.1088/1757-899X/98/1/012026

Correlation between surface properties and wettability of multi-scale structured biocompatible surfaces

SN Gorodzha[1], MA Surmeneva[1], O Prymak[2], A Wittmar[3], M Ulbricht[3], M Epple[2], A Teresov[4], N Koval[4], and RA Surmenev[1, 5]

[1]Department of Theoretical and Experimental Physics, National Research Tomsk Polytechnic University, 634050, Tomsk, Russian Federation

[2]Inorganic Chemistry and Center for Nanointegration Duisburg-Essen (CeNIDE), University of Duisburg-Essen, 45117, Essen, Germany

[3]Technical Chemistry and Center for Nanointegration Duisburg-Essen (CeNIDE), University of Duisburg-Essen, 45141, Essen, Germany

[4]Institute of High-Current Electronics, Siberian Branch, Russian Academy of Sciences, 634055, Tomsk, Russia

[5]Fraunhofer Institute for Interfacial Engineering and Biotechnology IGB, Stuttgart, 70569, Germany

Email: rsurmenev@gmail.com

Abstract. The influence of surface properties of radio-frequency (RF) magnetron deposited hydroxyapatite (HA) and Si-containing HA coatings on wettability was studied. The composition and morphology of the coatings fabricated on titanium (Ti) were characterized using atomic force microscopy (AFM) and X-ray diffraction (XRD). The surface wettability was studied using contact angle analysis. Different geometric parameters of acid-etched (AE) and pulse electron beam (PEB)-treated Ti substrates and silicate content in the HA films resulted in the different morphology of the coatings at micro- and nano- length scales. Water contact angles for the HA coated Ti samples were evaluated as a combined effect of micro roughness of the substrate and nano-roughness of the HA films resulting in higher water contact angles compared with acid-etched (AE) or pulse electron beam (PEB) treated Ti substrates.

1. Introduction

Understanding and controlling of the surface wetting has a significant impact on many fields of science. In particular, wettability is a fundamental property of implant surface, which influences the initial cell adhesion [1]. With the development of modification technologies of biomaterials surfaces, the field of the coating fabrication has been broadened, from the wet-chemical deposition to the physical deposition [2], [3].

Radio-frequency (RF) magnetron sputtering is a known method for producing pure hydroxyapatite (HA) [4, 5] as well as HA with different admixtures, such as Si [6] or Ag [7]. Although RF-magnetron sputtering was widely used for fabrication of HA-based coating, the number of reports related to the wetting of the surface of HA thin film and influence of the surface topography on wettability is limited. It is known, that predefined topography can affect surface wettability. Recently, hydrophobic surface of silicate containing HA (Si-HA) film with 0.5 mol of silicate has been fabricated [6]. The deposited film was hydrophobic as a result of the formation of the nanostructured surface. The field of

Content from this work may be used under the terms of the Creative Commons Attribution 3.0 licence. Any further distribution of this work must maintain attribution to the author(s) and the title of the work, journal citation and DOI.
Published under licence by IOP Publishing Ltd

new biomaterials development, particularly fabrication of multiscale surfaces, has progressed tremendously during the last years. Recent studies revealed that the combination of the multiscale (micro- and nanoscale) structures of implant surfaces with the predetermined chemical composition plays a crucial role in achieving the desired biocompatibility [8].

The aim of this study is to examine the role of the surface structure of pure HA and Si-HA-based coating fabricated via RF-magnetron sputtering on wetting characteristics.

2. Materials and methods

The HA coatings were deposited on the grounded Ti substrates via RF (13.56 MHz) magnetron sputtering. The deposition parameters were reported elsewhere [4]. The resulting coating thickness was determined by ellipsometry. A powder of Si-containing HA ($Ca_{10}(PO_4)_{6-x}(SiO_4)_x(OH)_{2-x}$, x=0, 0.5 and 1.72) was prepared by mechanochemical activation and then used as a precursor to prepare a target for sputtering. The mixture was slowly compacted in a press at the pressure of 70 MPa. The pressed powder was sintered in air at 1100°C for 1 h. The submicrometer structure of Ti was produced via acid etching (AE) and pulsed electron beam (PEB) treatment. The detailed information of treatment procedures was reported elsewhere [4].

Surface measurements at the nanoscale were evaluated using an atomic force microscope (AFM) Solver P47-PRO (NT-MDT, Moscow, Russia) in contact mode. An X-ray diffractometer (PANalytical, MA, USA) with Cu K_α radiation operated at 40 kV and 40 mA was used to characterize the coatings. The samples were scanned for 2θ from 5° to 100° with a step size of 0.05° using grazing incidence X-ray diffraction (GIXRD) with an incident beam angle of Φ = 1.0°. To calculate an average crystallite size and lattice parameters, Rietveld refinement (using the Le Bail method) with the program package TOPAS 4.2 from Bruker was performed. For each Rietveld refinement, the instrumental correction as determined with a LaB_6 standard powder sample from NIST (National Institute of Standards and Technology) as the standard reference material (SRM 660b). HA (#4-0432) and Ti (#44-1294) patterns from the ICDD database were used as references.

Contact angle analyses were performed with an optical contact angle apparatus (OCA 15 Plus Data Physics Instruments GmbH, Germany) using the SCA20 software (Data Physics Instruments GmbH, Germany). The static captive bubble method was chosen to measure contact angle. Five independent measurements have been realized on surface of each sample and the mean values were calculated.

3. Results and discussion

It was noted that Si addition, under the same deposition conditions, has influenced the coatings morphology. It is well-known that the surface morphology of the microstructures strongly depends on the content of dopants. Figure 1 shows AFM morphological images of the pure HA and Si-HA films.

In the case of HA-based film deposited on acid etched Ti, many grains ranging from 200 to 900 nm are observed on the surface. The grains can be either separated from each other or overlapped. 3D AFM images have revealed that Si-HA film deposited on acid etched Ti presents grains with dimensions ranging between 70-100 nm which are densely built into conglomerate formation (Figure 1). The surface roughness is the key factor to affect the hydrophobicity [9].

Figure 2 shows the typical XRD patterns of pure HA and Si-containing HA films. The diffraction peaks for pure HA and Si-HA (x=0.5 mol) can be indexed as hexagonal HA. No other diffraction peaks for any secondary crystalline phases were found. Moreover, the peaks of Si-containing HA coatings were slightly shifted to lower angles compared to that of the undoped HA coating. As shown in Figure 3, it can be an evidence of replacement of phosphate ions with silicate ions. However, for the samples doped with either 0.5 or 1.72 mol Si the intensity of HA peaks decreases.

IOP Conf. Series: Materials Science and Engineering **98** (2015) 012026 doi:10.1088/1757-899X/98/1/012026

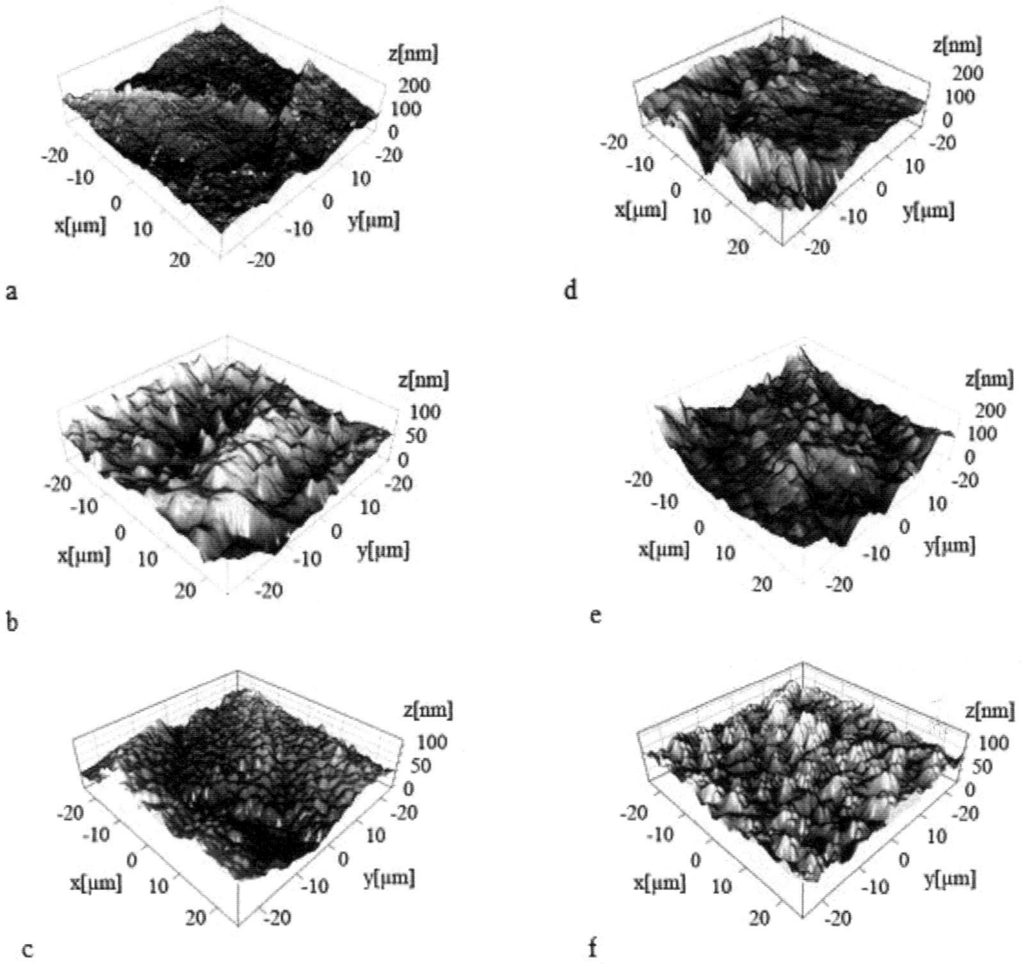

Figure 1. 3D AFM images of the pure and Si-containing HA coatings deposited on Ti substrate with different initial morphology obtained using AE and PEB: a) HA coating on Ti after PEB; b) Si-HA (x=0.5 mol) coating on Ti after PEB; c) Si-HA (x=1.72 mol) on Ti after PEB; d) HA coating on Ti after AE; e) Si-HA (x=0.5 mol) on Ti after AE; f) Si-HA (x=1.72 mol) on Ti after AE.

The effect of Si doping on the structural changes such as the lattice parameters, unit cell volume and average crystallite size was analysed. The lattice parameters are estimated from the XRD peaks positions and refined using Rietveld analysis. The lattice parameters (a, b, c), unit cell volume (V), and average crystallite size calculated for different Si content in HA are summarized in Table 1.

The lattice parameters of undoped HA film were a = 9.451(2) Å and c = 6.901(2) Å. For Si-containing HA films, the lattice parameters are slightly higher than that of undoped HA film. The lattice parameters and volume increases with increasing Si content in the film. The modification observed in the unit cell's volume could be attributed to the differences in the ionic radii. The calculated crystallite size was found to decrease with increase of Si content. The average crystallite size was found to be 27 nm for undoped HA film and 17-20 nm for the Si-containing HA films, which indicated their nanocrystalline structure.

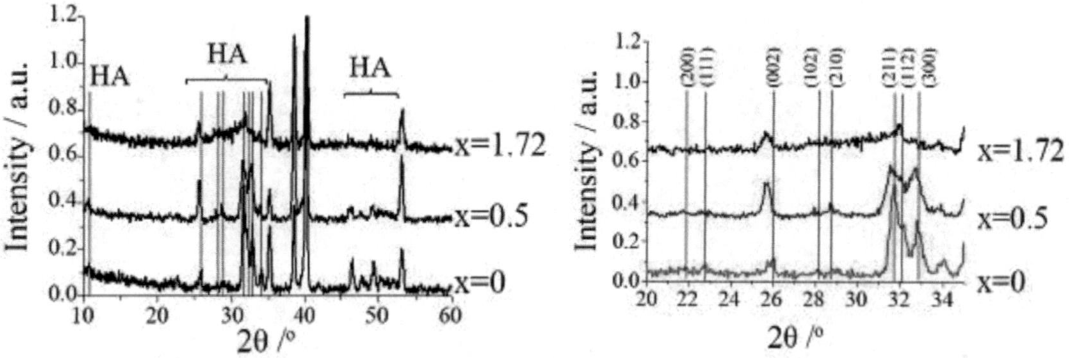

Figure 2. XRD patterns of the pure HA and Si-containing HA coating deposited on Ti after PEB treatment

Table 1. Lattice parameters, unit cell volume and crystallite size estimated for the coatings deposited via RF magnetron sputtering.

Film	Lattice parameters, [Å]	Crystallite size, [nm]	V, [Å^3]
HA	a = 9.451(2) c = 6.901(2)	27±1	533.8(3)
Si-HA, x=0.5 mol	a = 9.482(3) c = 6.927(3)	20±1	539.4(4)
Si-HA x=1.72 mol	a = 9.538(5) c = 6.947(5)	12±1	547.3(3)

The average water contact angles on Ti after AE and PEB treatment were determined as 138.9° and 91.8°, respectively, exhibiting hydrophobicity. The static contact angle measurements demonstrated that, among all other films studied, pure nanocrystalline HA coating revealed the lowest contact angle. In figure 3 the Si-HA-based (x=1.72 mol) film revealed a higher water contact angle of 143.2 and 150° for PEB and AE Ti surfaces, respectively, showing hydrophobic surface. These results are in good agreement with that of the recent study [6]. It was reported that the wettability of the Si-HA (x=0.5 mol) coating is time-dependent with a tendency to become more hydrophilic over time.

An interesting fact of the influence of the substrate topography on the growth mechanism and, therefore, morphology and mechanical features of the HA-based coating is reported [4]. Wetting of the structured surface can be described using two distinct modes [10]. In the first mode, a rough surface is completely wetted by a liquid, known as a Wenzel state [11]. The apparent contact angle is described in this case via the Wenzel equation:

$$\cos \Theta^* = r \times \cos \Theta \qquad [11]$$

where r is the roughness factor, defined as the ratio of the actual surface area and the projected surface area of the solid.

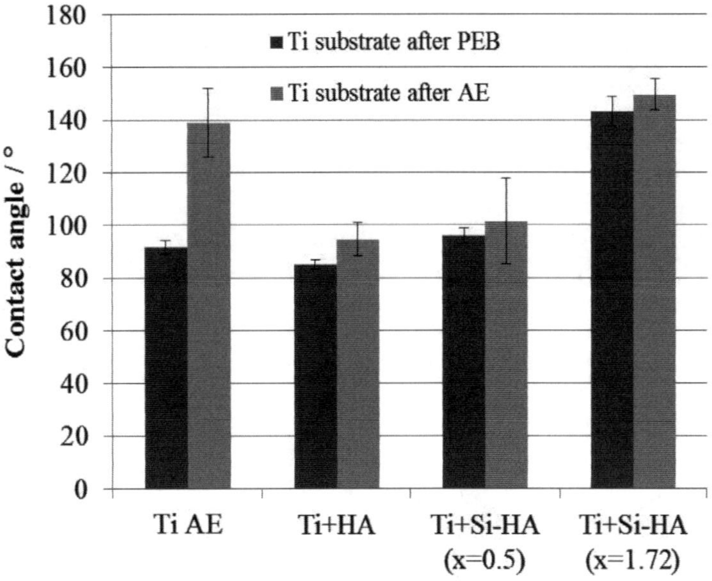

Figure 3. Water contact angle of the uncoated substrate and pure HA, Si-containing HA coating deposited on Ti with different initial morphology obtained via AE and PEB

The Wenzel equation indicates that the roughness can amplify the wettability of a surface. If the surface is intrinsically hydrophobic, roughness will further enhance the surface hydrophobicity. In the case of the coated surface, a water droplet on the deposited film is expected to penetrate into the microscale grooves. However, air gaps are present in the nanoscale structure, thus forming a partial wetting state. Water sealed in microscale structure of Ti surface, resulting in strong adhesion between the surface and a liquid.

4. Conclusions

The morphology and composition of RF magnetron sputter-deposited HA coating fabricated on Ti were characterized using AFM and XRD. AE and PEB-treatment of Ti substrates and silicate content in the deposited HA films resulted in the different morphology of the coatings at micro- and nano-length scales. The influence of micro- and nanoscale geometrical structures on wettability of the HA- and Si-HA-based was investigated. Water contact angles for the HA coated Ti samples were evaluated as a combined effect of micro roughness of the substrate and nano roughness of the HA films resulting in higher water contact angles compared with acid-etched (AE) or pulse electron beam (PEB) treated Ti substrates.

Acknowledgements

This research was supported by the *Russian Fund for Basic Research* (15-08-08652A), *the Russian President's Stipend* SP-6664.2013.4, grant MK-485.2014 and *the State order NAUKA* #11.1359.2014/K.

References

[1] Zhu X, Chen J, Scheideler L, Reichl R and Geis-Gerstorfer J 2004 *Biomater.* **25** 4087

[2] Bosco R, Edreira ERU, Wolke JGC, Leeuwenburgh SCG, van den Beucken JJJP and Jansen JA 2013 *Surf. Coatings Technol.* **233** 91

[3] Arcos D, Boccaccini AR, Bohner M, Díez-Pérez A, Epple M, Gómez-Barrena E, Herrera A, Planell JA, Rodríguez-Mañas L and Vallet-Regi M 2014 *Acta Biomater.* **10** 1793

[4] Surmeneva MA, Surmenev RA, Tyurin AI, Mukhametkaliyev TM, Teresov AD, Koval NN, Pirozhkova TS, Shuvarin IA and Oehr C 2014 *Thin Solid Films* **571** 218

[5] Alghamdi HS, Cuijpers VMJI, Wolke JGC, van den Beucken JJJP and Jansen JA 2013 *J. Dent. Res.* **92** 982

[6] Surmeneva MA, Kovtun A, Peetsch A, Goroja SN, Sharonova AA, Pichugin VF, Grubova I Y, Ivanova AA, Teresov AD, Koval NN, Buck V, Wittmar A, Ulbricht M, Prymak O, Epple M and Surmenev RA 2013 *RSC Adv.* **3** 11240

[7] Ivanova AA, Surmeneva MA, Grubova IY, Sharonova AA, Pichugin VF, Chaikina MV, Buck V, Prymak O, Epple M and Surmenev RA 2013 *Mat.-wiss. u.Werkstofftech.* **44** 218

[8] Tan J and Saltzman WM 2004 *Biomaterials* **25** 3593

[9] Liu M, Wang S and Jiang L 2013 *MRS Bull.* **38** 375

[10] Wong T-S, Sun T, Feng L and Aizenberg J 2013 *MRS Bull.* **38** 366

[11] Wenzel RN 1936 *Ind. Eng. Chem.* **28** 988

Thin hydroxyapatite coating on AZ91D magnesium alloy fabricated via RF-magnetron sputtering

T M Mukhametkaliyev[1], M A Surmeneva[1], M Mueller[2], O Prymak[3], M Epple[3], R A Surmenev[1, 2]
[1]National Research Tomsk Polytechnic University, 30 Lenina ave., 634050, Tomsk, Russia
[2]Fraunhofer Institute for Interfacial Engineering and Biotechnology IGB, 70569, Stuttgart, Germany
[3]Inorganic Chemistry and Center for Nanointegration Duisburg-Essen (CeNIDE), University of Duisburg-Essen, 45117, Essen, Germany

E-mail: rsurmenev@gmail.com

Abstract. In this study the morphology, composition, structure and wettability of radio-frequency (RF) magnetron sputter-deposited hydroxyapatite (HA) coating deposited on the surface of AZ91D magnesium alloy were investigated. The results revealed that the fabricated coating is uniform, homogeneous with the structure of the stoichiometric HA. The deposition of the HA coating did not change significantly the surface wettability of the bare alloy, however water contact angle dynamics in the case of the HA coated substrates revealed a lower rate of a droplet spreading over the surface.

1. Introduction

The development of biodegradable materials for bone injuries repair is an attractive scientific topic [1-5]. Biodegradable materials allow to avoid the problems associated with non-resorbable implants for example long-term complications like restenosis, thrombosis [6]. The concept is based on the idea to use implants inserted into the human body which are slowly degraded and gradually replaced by natural bone [6].

Nowadays attention of researchers is devoted to the development of magnesium alloy implants [7]. Magnesium alloys have been applied as implants since 1930s and possess good biocompatibility. Over the last years, magnesium alloys have been proposed for applications as biodegradable implants for bone repair along with cardiovascular stenting [1, 8-10]. Magnesium is degradable in physiological environment and has no toxic effect. Moreover, magnesium is an essential microelement for human metabolism Magnesium is also necessary for bone strength and can intensify bone tissue growth [5]. In addition Young's modulus and density of the alloy are closer to that of bone which allows to reduce stress-shielding effect and thus induce bone growth and improve implant stability [11].

However, magnesium reveals a high degradation rate in physiological environment [12, 13]. In recent years, there are several studies reported the investigation on the degradation behavior of the magnesium alloys in body fluids [7, 12, 14]. Due to the fact that magnesium alloys actively corrodes in the body there is a need to improve its corrosion resistance. Biocompatible coatings such as calcium phosphates (CaP), in particular HA, are prospective ways to enhance the corrosion resistance of

Content from this work may be used under the terms of the Creative Commons Attribution 3.0 licence. Any further distribution of this work must maintain attribution to the author(s) and the title of the work, journal citation and DOI.
Published under licence by IOP Publishing Ltd

magnesium alloys. Furthermore CaPs are widely used in orthopedic and dental surgery due to their biocompatibility and high osteoinductivity [15]. There are different techniques to deposit HA coatings [16]. Radio-frequency (RF) magnetron sputtering allows to deposit thin uniform, dense, high-purity films. It is possible to tailor the structure (amorphous or crystalline) and the Ca/P ratio of the coating. Moreover, the coating has a strong adhesion to the substrate [15, 18].
The corrosion resistance of magnesium alloys was improved [2, 4, 5, 7, 12, 13, 17, 18], though RF magnetron sputtering, to the best of our knowledge, was not applied for this purpose. Therefore, the aim of this study was to investigate the structure, morphology, chemical composition and wettability of the HA coating deposited via RF magnetron sputtering on AZ91D magnesium alloy.

2. Materials and methods
The AZ91D magnesium alloy (DSM, Israel Chemicals Ltd) with the composition of 91% Mg, 8% Al and 1% Zn was used. The substrates were ultrasonically cleaned in acetone for 30 min. Then the samples were rinsed in distilled water and ultrasonically cleaned in ethanol for a second time.

Synthesized via mechanochemical activation pure HA powder (Ca10(PO4)6(OH)2 was used as a precursor-powder to prepare a target for sputtering. The uniaxially pressed powder was sintered in air at 1100 °C for 1 h according to the study [19]. The thickness of the prepared coating was investigated using optical ellipsometry (Ellipse 1891-SAG, Institute of Semiconductor Physics, RAS, Siberian Branch). Surface morphology was estimated using scanning electron microscopy (SEM) via ESEM Quanta 400 FEG instrument combined with an energy-dispersive X-ray analysis (EDX analysis system Genesis 4000, SUTW-Si (Li) detector) operated under high vacuum. The EDX spectra were collected for 60 s with a dead time of 30 % and electron beam energy of 15 keV. The structure and phase composition of the HA films were determined using X-ray powder diffraction (D8 Advance, Bruker, Germany) with Cu-Kα radiation (0.154 nm). The surface of the samples was irradiated using grazing incidence X-ray diffraction (incident beam angle was 1° and step size of 0.01°/2θ at 40 kV and 40 mA). As reference for the pattern of HA ICDD database #9-0432 was used. The contact angle analyses and water hysteresis were investigated (OCA 15 Plus Data Physics Instruments GmbH, Germany) using an SCA20 software (Data Physics Instruments GmbH, Germany). The calculations of the water contact angle and hysteresis were performed using 10 droplets (5 μL) to reveal an average θ value.

3. Results and discussion
The thickness of the deposited coating was estimated to be of 800±50 nm. Representative SEM-micrograph of the surface of HA coating is shown in Fig. 1.a. The coating is shown to be uniform, and dense, no cracks were observed.

One of the most important characteristics of CaP compounds is molar Ca/P ratio [15, 20]. A lower Ca/P ratio leads to a higher solubility of the HA coating. The Ca/P ratio of the coatings deposited for 8 h as determined by EDX (Fig. 1.b) was 1.6±0.1, which is close to the Ca/P ratio typical for stoichiometric HA.

Based on the XRD results (Fig. 2) it was found that main diffraction peaks at 25.8° (002), 53.1° (004), 31.8° (211), 32.2° (112), and 32.9° (300) are attributed to crystalline HA with the lattice parameters a=b=9.46 Å, and c=6.88 Å.

The wettability of the biomaterial surfaces was assessed by contact angle measurements. Wettability allows us to estimate possible mechanisms of protein absorption and cell adhesion [21, 22]. The average values of water contact angle, water contact angle change in dynamics, and contact angle hysteresis for bare substrate and the HA coating are given in Table 1. All the measurements were carried out at the temperature of 23 ± 1 °C and 45 ± 5% humidity. Water contact angle of the coated substrates was lower compared with the bare alloy. Contact angle hysteresis slightly increased in the case of the HA coating [23, 24]

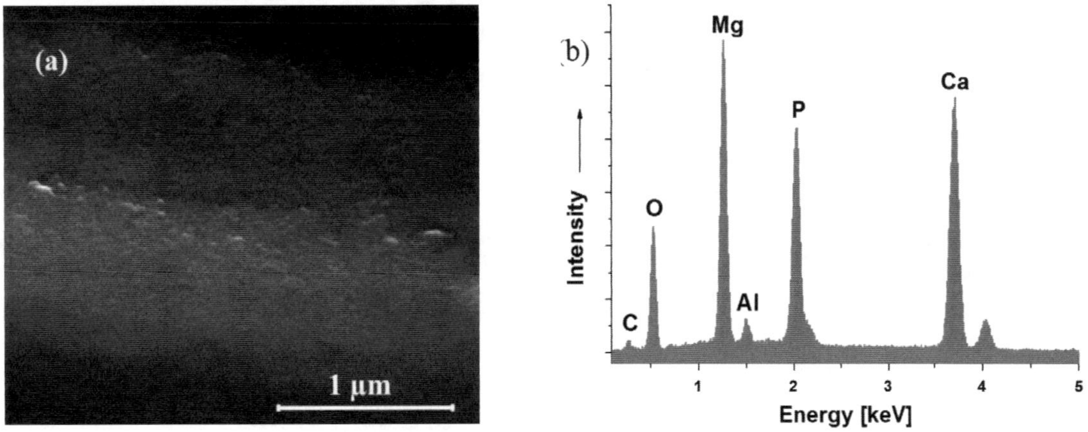

Figure 1. SEM image (a) and EDX spectrum (b) of the HA coating deposited on the surface of AZ91D magnesium alloy.

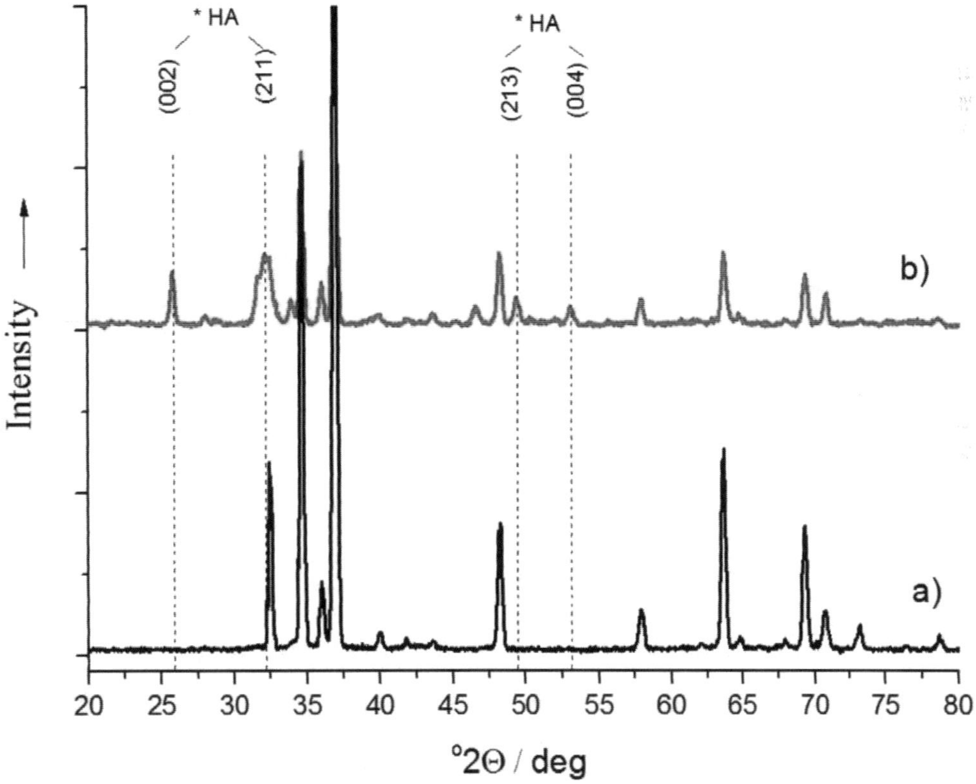

Figure 2. XRD patterns of the HA coating deposited at 500 W for 8 h on AZ91D magnesium alloy: a) bare substrate, b) HA coating on the substrate. "*" denotes the peaks attributed to HA. The peaks of HA (ICDD card number 09-432) are shown using vertical lines.

Table 1. Water contact angles and hysteresis for the HA coating on AZ91D magnesium alloy

Surface type	Water contact angle (°, n = 10)	Contact angle hysteresis (°)
Bare substrate	104±5	40
HA coating	100±3	44

In spite of the some decrease in the water contact angle it was observed that the surface of the HA coating still revealed hydrophobic behaviour [25]. It was reported that wettability of the surface can change over the time [23, 24]. For this reason dynamic contact angle measurements were performed and the results are presented in Fig. 3.

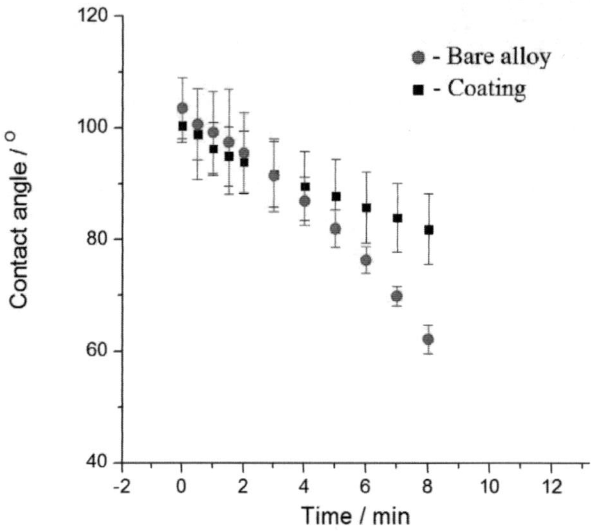

Figure 3. Water spreading behavior of a single droplet on the surface of the HA coating and bare alloy.

The wettability of the bare alloy and HA coating deposited on Mg alloy was monitored. It was revealed that after only 8 min of water droplet spreading over the surface a significant decrease in the water contact angle (from 100° to 66°) was observed. A significantly higher water spreading rate was observed in the case of bare alloy compared with that of the HA coated samples. The observed changes in the surface wettability over time indicated a strong time-dependent tendency to turn initially hydrophobic behavior to hydrophilic.

Conclusion

In this study, fabrication process and the properties of the HA coating deposited onto AZ91D magnesium alloys via RF-magnetron sputtering are described. Surface morphology and structure results suggested that the coating is crystalline HA with the uniform, homogeneous, and dense structure. Investigations did not reveal improvement of the surface wettability of the HA coated samples compared to the bare alloy, however water contact angle dynamics in the case of the HA coated substrates revealed a lower rate of a droplet spreading over the surface. The initially hydrophobic surface of the HA coating tend to be more hydrophilic with time.

Acknowledgements
We would like to thank Prof. M. Epple, Prof. M. Ulricht, Dr. A. Wittmar (Duisburg-Essen University) and Dr. M. Mueller and Dr. C. Oehr (Fraunhofer IGB) for the assistance with the experiments. This research was supported by the *Russian Science Foundation* (14-13-00274).

References
[1] Kirkland N T, Birbilis N and Staiger MP 2012 *Acta Biomater.* **8** 925
[2] Kirkland NT, Lespagnol J, Birbilis N and Staiger MP 2010 *Cor. Sci.* **52** 287
[3] Witte F 2010 *Acta Biomater.* **6** 1680
[4] Zeng R, Dietzel W, Witte F, Hort N and Blawert C 2008 *Adv. Eng. Mater.* **10** B3
[5] Staiger MP, Pietak AM, Huadmai J and Dias G 2006 *Biomaterials* **27** 1728
[6] Shaw BA, Sikora E and Virtanen S 2008 *Electrochem. Soc. Interf.* **17** 45
[7] Zheng YF, Gu XN and Witte F 2014 *Mater. Sci. Eng. R* **77** 1
[8] Erne P, Schier M and Resink T 2006 *Cardiovasc. Intervent. Radiol.* **29** 11
[9] Waksman R 2005 *ACC Cur. J. Rev.* **14** 36
[10] Heublein B, Rohde R, Kaese V, Niemeyer M, Hartung W and Haverich A 2003 *Heart* **89** 651
[11] Nagels J, Stokdijk M and Rozing PM 2003 *J. Shoulder Elbow Surg.* **12** 35
[12] Witte F, Hort N, Vogt C, Cohen S, Kainer KU, Willumeit R and Feyerabend F 2008 *Cur. Opinion in Solid State and Mater. Sci.* **12** 63
[13] Song G 2007 *Cor. Sci.* **49** 1696
[14] Vormann J 2003 *Molec. Aspects of Med.* **24** 27-37
[15] Surmenev RA, Surmeneva MA and Ivanova AA 2014 *Acta Biomater.* **10** 557
[16] Surmenev RA 2012 *Surf. Coat. Technol.* **206** 2035
[17] Dorozhkin S V 2014 *Acta Biomater.* **10** 2919
[18] Ren L, Lin X, Tan L and Yang K 2011 *Mater. Let.* **65** 3509
[19] Surmenev RA, Surmeneva MA, Evdokimov KE, Pichugin VF, Peitsch T and Epple M 2011 *Surf. Coat. Technol.* **205** 3600
[20] ASTM F1185–03(2009). Standard specification for composition of hydroxylapatite for surgical implants.
[21] Nishizawa K, Toriyama M, Suzuki T, Kawamoto Y, Yokogawa Y and Nagae H 1993 *J. Fermentation and Bioeng.* **75** 435
[22] Bodhak S, Bose S and Bandyopadhyay A 2009 *Acta Biomater.* **5** 2178
[23] Erbil H Y 2013 *Colloid Polym. Sci.* **291** 355
[24] Surmeneva MA, Kovtun A, Peetsch A, Goroja SN, Sharonova AA, Pichugin VF, Grubova IY, Ivanova AA, Teresov AD, Koval NN, Buck V, Wittmar A, Ulbricht M, Prymak O, Epple M and Surmenev RA 2013 *RSC Adv.* **3** 11240
[25] Mondon M and Ziegler C 2003 *Inter. J. Prosthodontics* **16** 49

IOP Conf. Series: Materials Science and Engineering **98** (2015) 012028 doi:10.1088/1757-899X/98/1/012028

The effect of Si content on structure and mechanical features of silicon-containing calcium-phosphate-based films deposited by RF-magnetron sputtering on titanium substrate treated by pulsed electron beam

M Surmeneva[1], A Tyurin[2], T Mukhametkaliyev[1], A Teresov[3], A Koval[3], T Pirozhkova[2], I Shuvarin[2], E Chudinova[1] and R Surmenev[1,4]

[1] Tomsk Polytechnic University, Institute of Physics and Technologies, 30, Lenina ave., Tomsk, 634050, Russia

[2] Derzhavin Tambov State University, Sovetskaya, 93, Tambov, 392000, Russia
[3] Institute of High-Current Electronics, Siberian Branch, Russian Academy of Sciences, Tomsk, 634040, Russia
[4] Fraunhofer Institute for Interfacial Engineering and Biotechnology IGB, Nobelstraße 12, Stuttgart, 70569, Germany

E-mail: surmenevamaria@tpu.ru

Abstract. Silicon-containing calcium phosphate (Si-CaP) coatings were fabricated by radio frequency (rf) magnetron sputtering using the targets prepared from hydroxyapatite (HA) powder with different silicon content. A powder of Si-HA ($Ca_{10}(PO_4)_{6-x}(SiO_4)_x(OH)_{2-x}$, x=0.5 and 1.72) was prepared by mechanochemical activation and then used as a precursor-powder to prepare a target for sputtering. The titanium substrate was acid etched and treated with pulsed electron beam with an energy density of 15 J/cm^2. The average crystallite size as determined by XRD was 28 nm for the coatings obtained using the target prepared from the Si-HA powder (x=0.5), whereas Si-CaP (Si-HA powder x=1.72) films showed an amorphous structure. The nanohardness and the Young's modulus of the Si-CaP coating (x=0.5) deposited on titanium treated by pulsed electron beam are enhanced to 4.5 and 113 GPa compared to titanium substrate. Increase of Si content resulted in a dramatic decrease of the nanohardness and Young's modulus of Si-CaP films. However, Si-CaP coatings with the highest Si content revealed significantly lower values of elastic modulus, but slightly higher values of H/E and H^3/E^2 than did the non-coated specimens. Rf-magnetron sputtering allowed us to produce Si-CaP coatings with higher nanohardness and lower elastic modulus compared to titanium substrate.

1. Introduction

Biomaterials used for producing load-bearing implants must be mechanically strong, possess high resistance to corrosion and wear to prevent weakening of the mechanical strength. Implant surface roughening using the mechanical, chemical and electrochemical methods would cause surface contamination. The electron beam treatment of metal surface gains more and more attention nowadays, because the electron beams can be used for roughening the implants surface, without direct

Content from this work may be used under the terms of the Creative Commons Attribution 3.0 licence. Any further distribution of this work must maintain attribution to the author(s) and the title of the work, journal citation and DOI.
Published under licence by IOP Publishing Ltd

contact and therefore without contamination. The coating on the surface of metallic implants has been successfully applied for the last decades to improve their biocompatibility [1]. Rf-magnetron sputter deposition allows close control over the processing conditions with the resultant possibility to develop nanostructured coatings with a high quality [2]. The most used calcium phosphate (CaP) for coating fabrication is hydroxyapatite (HA) because it is mineral component of the bones [3]. The silicate-containing HA-based (Si-HA) coatings show an improvement in biological performance over their undoped counterparts [4]. Due to this reason the study of silicon-doped CaP (Si-CaP) film deposited on modified surface of Ti substrates is an active area of research [2, 5-8]. Radio-frequency (rf) magnetron sputtering is a known method for producing pure hydroxyapatite (HA) [9, 10] as well as HA with different admixtures, such as Si [2] or Ag [11]. One of the technique most recently used to study the mechanical properties of thin CaP films is nanoindentation [9, 12]. In this study, the surface microstructure, phases (amorphous or nanocrystalline) composition, and mechanical features (nanohardess and Young's modulus) of the coatings with different silicate content are studied.

2. Materials and methods

Titanium substrates were acid etched and then treated with a pulsed electron beam (PEB) generator "SOLO" according to a proprietary process of Institute of High Current Electronics SB RAS (Tomsk, Russia). The conditions of the PEB treatment were as follows: duration of the electron beam pulse 50 μs, number of pulses 3, electron beam energy density 15 J cm^{-2}. Si-CaP coatings were fabricated by rf-magnetron sputtering using the targets prepared from HA powder with different silicon content. The Si-CaP coatings were deposited for 8 h, which resulted in the coating thickness of 680 and 700 nm for Si-CaP (x=0.5 and 1.72), respectively. A powder of Si-HA ($Ca_{10}(PO_4)_{6-x}(SiO_4)_x(OH)_{2-x}$, x=0.5 and 1.72) was prepared by mechanochemical activation and then used as a precursor-powder to prepare a target for sputtering (Prof. M. Chaikina, Institute of Solid State Chemistry and Mechanochemistry SB RAS). The rf-power (500 W), argon gas pressure (working pressure 0.4 Pa, base pressure 10–4 Pa) and the distance between the target and substrate (40 mm) were kept constant in all experiments.

The phase composition and the structure of the Si-CaP coatings were identified by X-ray diffraction (XRD-7000, Shimadzu, Japan) with CuKα radiation (λ=0.154 nm) in the 2Θ range from 10° to 60° with a scan speed of 2.0°/min, sampling pitch of 0.02°, preset time of 1.0 sec at 30 kV and 30 mA. The average crystallite size was determined by using the Scherrer equation from the broadening of the diffraction peaks taking into account the instrumental broadening and with the PowderCell 2.4 software. An instrumental broadening of 0.14 in 2θ was determined by the full width at half maximum of a silicon powder. As reference for the pattern of HA and Ti we used the ICDD database: #9-0432 and #44-1294, correspondingly.

Nanoindentation experiments were conducted with a Nanotriboindenter TI-950 (Hysitron Inc., USA). The software calculated the nanohardness (H) and the reduced modulus (E) of the coatings according to the analysis of Oliver and Pharr [13]. This method has been successfully used to analyze nano- and micromechanical properties of a number of materials and allowed to obtain interesting results. The Berkovich pyramid with an angle between opposite faces of 142.3° and tip radius of around 50 nm was 5 mN min−1 were used. Repeated indentations were performed on different coated and uncoated Ti samples, and the values of H and E were calculated as the average of 10 indentations. A maximum load of 0.62 mN was reached at 5 s, then 2 s of dwell and 5 s of unloading followed. The surface roughness of the Ti substrates was 0.4±0.1 μm according to the R_a parameter.

3. Results and discussion

The typical XRD-patterns of a Si-CaP coating with different Si content deposited by rf-magnetron sputtering are shown in figure 1. All diffraction peaks correspond to the structure of HA. No diffraction peaks from other phases were detected. We therefore tentatively assume that the PO43- groups were partially replaced by SiO_4^{4-}. As reported by Gibson et al. [14], the substitution of phosphate by silicate results only in a slight change in the unit cell dimensions. The Si-CaP coating (x=0.5) consisted of crystallites of an average size of 28 nm (table 1). Si-CaP coatings in case of Si-HA (x=1.72) powder showed a nanocrystalline structure with crystallites of an average size of 22 nm.

Mechanical properties are important for the long-term performance of Si-CaP coated implants and are related to the characteristics of the coating. For the load-displacement measurements of the nanoindentation tests, typical load-unloading curves were obtained. The results for the two types of coatings as well as for the substrate for the maximum load employed (0.62 mN) are presented in figure 2.

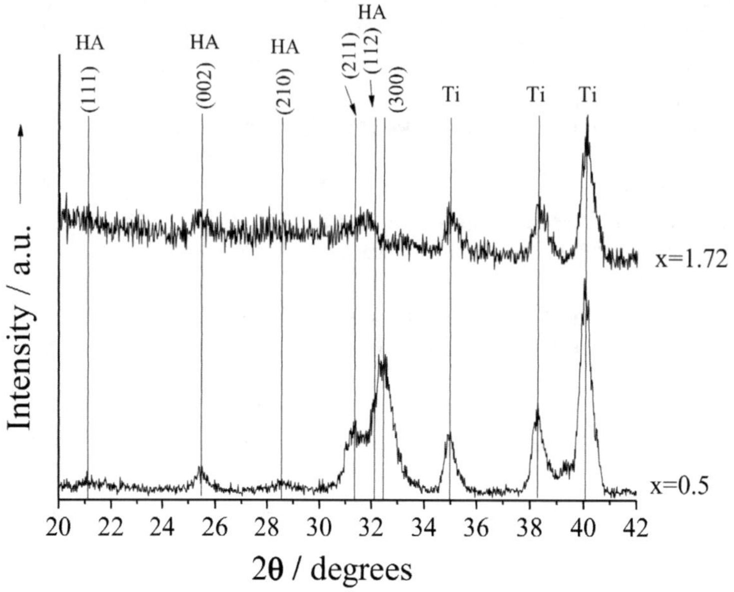

Figure 1. XRD pattern of Si-CaP films with different Si concentration deposited on titanium substrate via rf-magnetron sputtering.

When a low P_{max} (0.62 mN) was applied, the highest final penetration depth h_{max} (71 nm) was obtained for the Ti substrate. The minimal final penetration depth corresponded to the Si-CaP coating (55 nm).

Table 1. Lattice parameters and crystalline size estimated for Si-CaP coatings with different silicon content fabricated by rf-magnetron sputtering

Coating	Lattice parameters (Å)	Crystallite size (nm)
Si-CaP (x=0.5)	a = 9.4326 c = 6.7827	28
Si-CaP (x=1.72)	a = 9.4727 c = 6.9827	22

Table 2 shows the values of H and E, the elastic strain to failure ratio (H/E), the final displacement of the indenter into the sample h_{max}, and the percentage of elastic recovery of displacement on unloading ($\%R=[(h_{max}-h_f)/h_{max}]\times 100$) for the Si-CaP coatings with different Si concentration and for the uncoated Ti. The Ti substrate revealed a nanohardness of 3.0 GPa and an Young's modulus of 127 GPa. The Si-CaP coating in case of x=0.5 showed the highest nanohardness, which is 1.5 and 1.36 times higher than hardness of a pure titanium and Si-CaP coating (x=1.72), respectively (table 2). The coating with the highest amount of Si showed the lower Young's modulus and nanohardness. The Si-CaP (x=1.72) coatings had a hardness of 3.3 GPa and an elastic modulus of 96 GPa. This fact was attributed to the amorphous structure of this coating. It is generally accepted that wear resistance of a solid can be adjusted by tailoring its elastoplastic properties [15]. To predict the wear resistance of the Si-CaP coating, the ratio H/E was evaluated from the data obtained at the load of 0.62 mN. From the previous H and E results and from the H/E values, shown in Table 2, it is observed that silicon addition tends to decrease the wear resistance of the Si-CaP films. H^3/E^2 decreases from $7.1 \cdot 10^{-3}$ to $3.9 \cdot 10^{-3}$ MPa as Si is doped into HA powder.

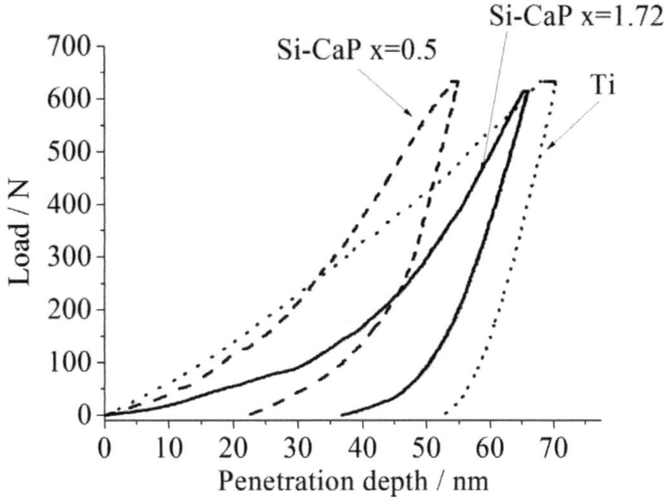

Figure 2. Representative load-displacement curves for the uncoated and CaP-coated Ti pretreataed via PEB at a maximum load of 0.62 mN

Table 2. The values of penetration depth hmax, nanohardness H and Young's modulus E, H/E ratio, H^3/E^2 factor and the percentage of elastic recovery of displacement at unloading %R for the CaP coatings and the uncoated Ti after PEB

Samples	h_{max} (nm)	*H* (GPa)	*E* (GPa)	*H/E*	H^3/E^2 (GPa)	%*R*
Ti after PEB	71	3.0	127	0.024	1.7×10^{-3}	27
Si-CaP coating (x=0.5)	55	4.5	113	0.040	7.1×10^{-3}	60
Si-CaP coating (x=1.72)	66	3.3	96	0.034	3.9×10^{-3}	45

According to the authors of the study [15] a material, which is highly resistant to plastic deformation, should have a high value of H and low value of E (consistently with the H/E parameter). The H^3/E^2 parameter proposed by Johnson [16] was associated with the film resistance to plastic deformation. Si-CaP film with the highest Si content showed significantly lower values of E but slightly higher values of H/E and H^3/E^2 than revealed the uncoated Ti substrate. Therefore, from the nanoindentation results discussed here it is possible to infer that even with x=1.72, the deposition of amorphous CaP coatings gained in nanohardness and elasticity. This association improved the elasticity index and the resistance to plastic deformation, suggesting a better mechanical performance of the surface for medical application. The percentage of elastic recovery of Si-CaP (x=1.72) coatings was 45%, which was higher than that of the uncoated Ti (27%). The Si-CaP (x=0.5) coatings showed the highest percentage of elastic recovery of 60% during unloading as well as nanomechanical properties, namely H/E; H^3/E^2; H and E of 0.04, 7.1×10^{-3} GPa, 4.5 GPa and 113 GPa, respectively.

4. Conclusion

Si-CaP coatings were fabricated via rf-magnetron sputtering using the targets prepared from HA powder with different Si content. The phase composition and structure, nanohardness and Young's modulus of the Si-CaP coating deposited on Ti substrate were investigated using XRD and nanoindentation test, respectively. The increase of Si content in Si-CaP coating resulted in the decrease of the film crystallinity. The structure of Si-CaP coatings has significant effect on their nanomechanical properties. Rf- magnetron sputtering allowed us to produce Si-CaP coatings with higher values of nanohardness and lower elastic modulus compared to uncoated Ti substrate.

Acknowledgments
The authors would like to thank Prof. M. Chaikina for the synthesis of HA-powder; Dr. A. Lider and Mr. M.S. Syrtanov for the assistance with XRD measurements. This research was supported by the *Russian Fund for Basic Research* (15-08-08652), *the Russian President's Stipend* SP-6664.2013.4, grant MK-485.2014, *the state - order NAUKA* #11.1359.2014/K and *Russian Science Fund* (nanoindentation studies, 14-13-00274).

References
[1] Bosco R, Edreira ERU, Wolke JG, Leeuwenburgh SC, van den Beucken JJ and Jansen JA 2013 *Surf. Coat. Technol.* **233** 91.
[2] Surmeneva MA, Kovtun A, Peetsch A, Goroja SN, Sharonova A, Pichugin VF, Teresov AD, Koval NN, Buck V, Wittmar A, Ulbricht M, Prymak O, Epple M and Surmenev RA 2013 *RSC Adv.* **3** 11240.
[3] Dorozhkin SV and Epple M 2002 *Ang. Chemie Int. Ed.* **41** 3130.
[4] Thian ES, Huang J, Best SM, Barber ZH and W Bonfield 2005 *Biomaterials* **26** 2947.
[5] Balas F, Perez-Pariente J and Vallet-Regi M 2003 *J. Biomed. Mater. Res. A* **66** 364.
[6] Bohner M 2009 *Biomaterials* **30** 6403.
[7] Solla EL, González P, Serra J, Chiussi S, León B and Garcıá Lopez J 2007 *J. Appl. Surf. Sci.* **254** 1189.
[8] Thian ES, Huang J, Best SM, Barber ZH and Bonfield W 2006 *J. Biomed. Mater. Res.* **76*B*** 326.
[9] Surmeneva MA, Surmenev RA, Tyurin AI, Mukhametkaliyev TM, Teresov AD, Koval NN, Pirozhkova TS, Shuvarin IA and Oehr C 2014 *Thin Solid Films* **571** 218.
[10] Alghamdi HS, Cuijpers VMJI, Wolke JGC, van den Beucken JJJP and Jansen JA 2013 *J. Dent. Res.* **92** 982.
[11] Ivanova AA, Surmeneva MA, Grubova IY, Sharonova AA, Pichugin VF, Chaikina MV, Buck V, Prymak O, Epple M and Surmenev RA 2013 *Mat.-wiss. u.Werkstofftech.* **44** 218.
[12] Snyders R, Bousser E, Music D, Jensen J, Hocquet S and Schneider JM 2008 *Plasma Proc. Polymers* **5** 168.
[13] Oliver WC and Pharr GM 1992 *J. Mater. Res.* **7** 1564.
[14] Gibson IR, Best SM and Bonfield W 1999 *J. Biomater. Mater. Res. Symp.* **44** 422.
[15] Leyland A and Matthews A 2000 *Wear* **246** 1.
[16] Johnson KL 1985 Cambridge University press, London, UK, ISBN: 0-521-34796-3 464p.

Formation of patina on a copper surface in polyacrylate gels with gold nanoparticles

E T Kim[1,a], E V Abdulmenova[1], Lyamina[1] and A Yu Godymchuk[1,2]

[1] Tomsk Polytechnic University, 30, Lenina ave., 634050, Tomsk, Russia
[2] National University of Science and Technology "MISiS", 4, Leninsky pr., Moscow, 119049, Russia

E-mail: [a]kim_elena11@mail.ru

Abstract. The experimental research demonstrates the effective use of patination of the copper and copper alloys in containing gold nanoparticles in order to increase their protective properties. The copper samples were examined by cyclic voltammetry method, optical microscopy, and X-ray analysis. It has been shown that patinas obtained in the polymer gel solution have a better corrosion resistance in comparison with the patinas obtained with chemical methods.

1. Introduction

There are many architectural and culturally valuable objects made of copper metal and copper alloys [1], which require the restoration. Such objects are covered by the layer of patina to make the decorative painting and keep the surface protection against external influence [2–5]. However, even if patination recipes are well-known it is highly difficult to achieve reproducible phase composition on the surface of pure copper and copper alloys. Therefore, the search for new methods of patina creation on the copper surface of culture and art objects is still an issue of concern. In this work, we propose to use the polymer a gel, the mixture of homopolymers of polymethyl methacrylate and polymethacrylic acid, polyethylene glycol containing gold nanoparticles.

2. Experimental part

The polymeric gels were based on copolymer of methyl methacrylate and methacrylic acid filled with polyethylene glycol. The reagents were immediately put into a polyetilen polymerization form and solidified by thermal polymerization at 60°C; benzoyl peroxide was used as an initiator. In experiment we obtained films with thickness of 0.5 mm by two approaches. The first method included the films formation by thermal pressing, and the second one – films formation from the solution of the polymeric gel in composition solvents [6, 7]. Methylcellosolve, butylacetate, toluene (in ratio 75:15:10) were used as the composite solvents [8]. The nanoparticles in gel solution were obtained by the method of laser ablation.
Chemical patination was used for the creation of some resistance layers as the reference standard (table 1).

Table 1. The solutions for patination

Solution	Composition

Content from this work may be used under the terms of the Creative Commons Attribution 3.0 licence. Any further distribution of this work must maintain attribution to the author(s) and the title of the work, journal citation and DOI.
Published under licence by IOP Publishing Ltd

Solution	Composition
1	$Cu(CH_3COO)_2$ – 5 g, NH_4OH- 7 ml, CH_3COOH (p) – 3 ml, distilled water – 85 ml
2	$NiSO_4$ – 20 g, $KClO_3$ – 3 g, distilled water – 77 ml
3	$Cu(NO_3)_2$ – 200 g., $AgNO_3$ – 8 g., HNO_3 – 6.5 ml
4	PMMA – 20 ml, MAA – 10 ml, PEG – 40 ml
5	GEMA – 20 ml, PMMA – 10 ml, PEG – 20 ml, CF_3COONa – 0.5g, $NaClO_4$ – 0.3 g
6	PMMA – 20 ml, MAA – 10 ml, PEG – 40 ml, Au nanoparticles – 44 mg/l
7	PMMA – 20 ml, MAA – 10 ml, PEG – 40 ml, Au nanoparticles – 32 mg/l

The solutions were placed on the surface of the 1x1 cm samples. The morphology analysis of the samples was carried out by means of an optical microscope "Metam PB-21-1" (Russia). Surface phase transformations after contact "metal – gel" were examined by the method of X-ray diffraction (XRD-7000, Shimadzu). Corrosion resistance of the coatings was evaluated using a cyclic voltammetry method with the help of a potentiostat-galvanostat IPC-Pro MF (Russia). The cyclic current-voltage curves registration (CV) was performed in the three-electrode cell: the reference electrode (chloride saturated silver electrode), the auxiliary electrode (graphite rod), and a test samples (copper plates, 5x5 mm). Patina obtained on the surface was used as the indicator electrode. Registration of the cyclic current-voltage curves was performed in potential change range at1000 … 700 mV with scan rate of 10 mV/c in three media: HCl, KCl, NaOH.

3. Results and discussion

The figure 1 presents the results of X-ray analysis after patination of copper. All X-ray diffractograms (XRD) show the signals of the elemental copper. Most of patinas contain cuprite. The patina obtained in polymer gel solution also contain phase of tenorite (Cu $(OH)_2H_2O_2$).

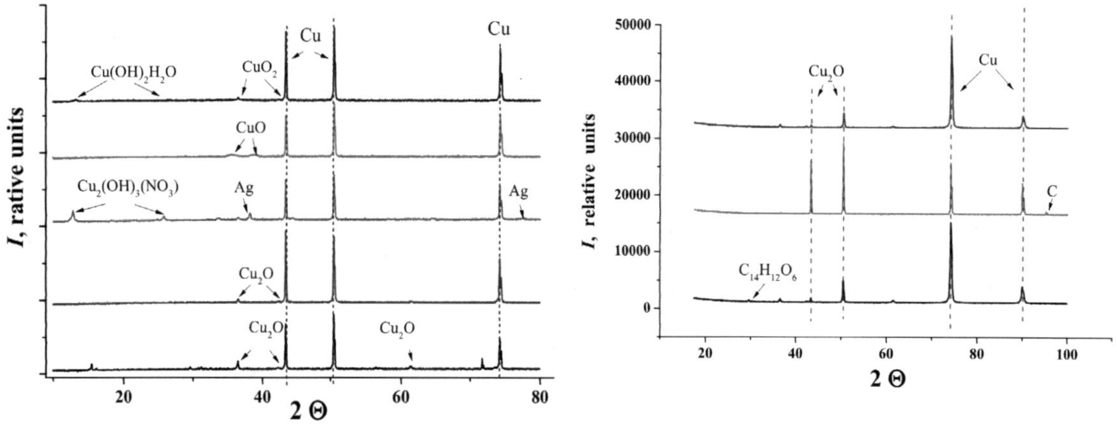

Figure 1. X-ray diffractograms of copper objects after patination.

The figure 1b shows the XRD of patinas formed in polymer solution with nanoparticles at different temperatures. It can be seen that the patina formed at 40 °C is the most stable because the copper peaks were weaker; it means that the patina is homogeneous, that proved optical images (figure 2). Patina formed at 80 °C has a higher copper XRD signal, which indicates a thinness of the film. The areas with green color point of copper carbonates on the surface, which are generally have less corrosion resistant than cuprite.

Values of the potential and the corrosion current are shown in the table 2. Among the chemically produced patinas only one has the best protective properties in all three aggressive mediums: the patina formed in the solution of copper and silver nitrate (solution #3). This patina has the lowest value of the current and the highest value of corrosion potential. The patina formed in gel solution

IOP Conf. Series: Materials Science and Engineering **98** (2015) 012029 doi:10.1088/1757-899X/98/1/012029

(MMA, MAA, PEG) has stronger protective properties. When the polymeric gels include the gold nanoparticles, the reproducibility of patina increases as shown by the error value.

Most informative data are received after overnight exposure in the electrolyte solution. Patina's amperage has increased by an order, but patinas prepared by the polymer gel electrolyte have smaller corrosion current than chemically prepared patinas. The results demonstrate that the addition of gold nanoparticles into the gel solution increases the corrosion resistance of patina. However, increasing the number of particles in the volume of the gel has low effect on the properties of patina.

Table 2. Corrosion potential and current of some patinas

Solution		$I_{corr.}$ $A \cdot 10^{-9}$	$E_{corr.}$ mV	$I_{corr.}$ $A \cdot 10^{-9}$	$E_{corr.}$ mV
		KCl		after 24 hour in 0.1M KCl	
Solution #4		31±11	206±113	113±59	138±22
Solution #5		62±31	192±72	130±98	133±44
Solution #6		70±38	174±50	106±59	149±31
Solution #7		53±23	193±78	107±34	174±27
Solution #6	40 °C	63±38	253±23	75±63	121±121
	80 °C	92±59	185±15	88±25	96±22
Solution #1		23±6	206±24	132±86	198±23
Solution #2		73±61	215±86	183±51	129±12
Solution #3		28±29	291±10	113±38	247±64
		HCl		after 24 hour in 0.1M HCl	
Solution #4		45±18	304±5	42±13	238±73
Solution #5		47±12	270±76	77±29	256±19
Solution #6		21±2	219±61	27±5	202±59
Solution #7		14±2	220±15	89±24	182±48
Solution #6	40 °C	18±7	322±15	105±92	231±49
	80 °C	69±16	280±9	156±71	238±53
Solution #1		32±7	351±12	32±28	230±15
Solution #2		38±23	243±24	24±17	258±45
Solution #3		18±4	257±12	62±7	265±15
		NaOH		after 24 hour in 0.1M NaOH	
Solution #4		34±6	97±7	39±8	80±16
Solution #5		25±2	117±11	105±56	155±152
Solution #6		29±8	47±5	44±21	111±9
Solution #7		38±5	83±6	17±3	91±10
Solution #6	40 °C	43±11	112±80	66±30	109±87
	80 °C	21±20	118±10	87±13	178±102
Solution #1		39±10	88±18	114±3	118±17
Solution #2		32±3	53±44	183±51	129±12
Solution #3		40±36	128±7	160±9	191±21

IOP Conf. Series: Materials Science and Engineering **98** (2015) 012029 doi:10.1088/1757-899X/98/1/012029

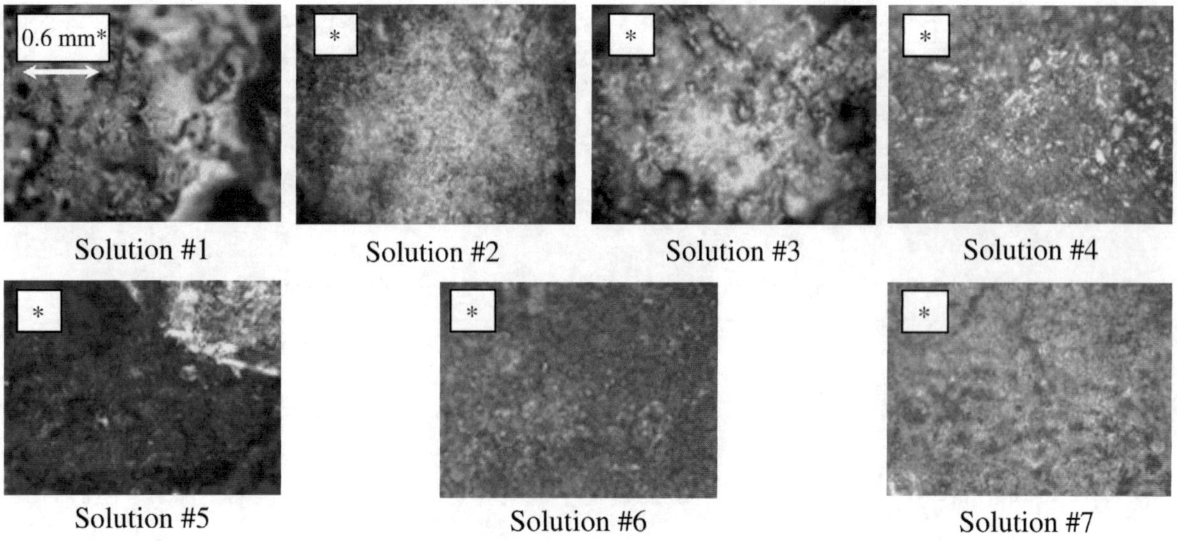

Figure 2. Optical images of copper after patination.
* – Scale is the same as on the images oft he solution #1

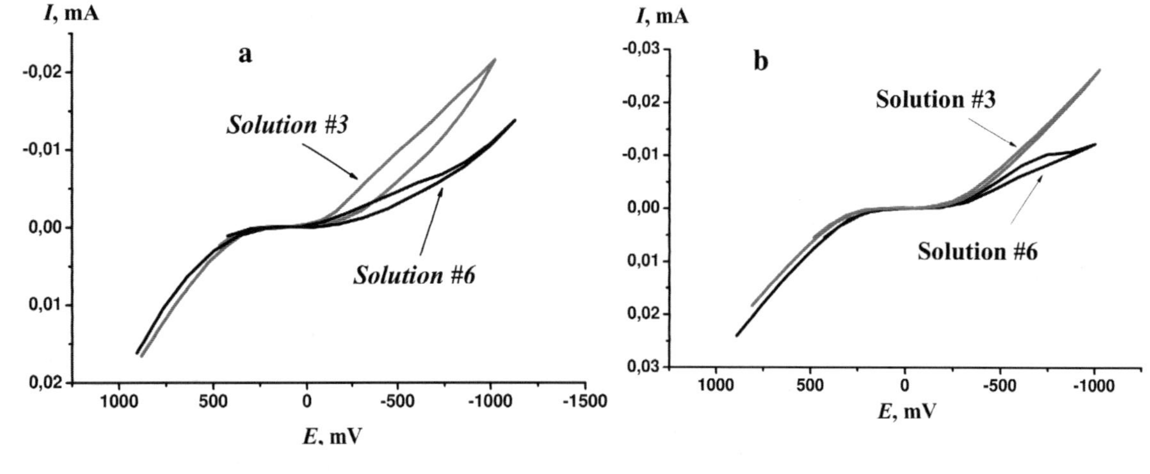

Figure 3. Cyclic voltammograms of the patinas in 0.1M KCl and 0.1M NaOH (b)

The figure 3 shows the cyclic voltammetry curves of the patinas obtained by the chemical method in solution of silver and copper nitrate. Also there is a patina formed by the polymer gel electrolyte on the figure 3. According to the data, the oxidant reduction passes more actively on the surface of patina formed in copper and silver nitrate. It testifies the presence of juvenile-free areas of copper, that may be seen on the optical images on the figure 2. However, the oxidant reduction passes not actively on the surface of patina formed in the solutions of MMA, MAA, PEG polymer gels with gold nanoparticles (figure 3).

Thus, patinas formed by polymer gel electrolyte solutions with the addition of gold nanoparticles at the temperature of 40 °C showed the greatest stability.

4. Conclusion

Cathode reduction of oxidants on the surface of patina, formed by chemical method, extends at a higher rate. Patinas obtained by polymer gel electrolytes have stronger protective properties than the patinas formed by the chemical methods. The adding of gold nanoparticles in polymeric gel increases the stability of the patina and its reproducibility. The most stable patina is formed in the MMA-MAA-PEG gel with gold nanoparticles at 40 °C.

References

[1] Nicitin MK 1990*Chemistry in the restoration: Reference guide " Chemistry"* (Leningrad, Chemistry)
[2] Alberghina MF and Barraco R 2011 Integrated analytical methodologies for the study of corrosion processes in archaeological bronzes *Spectrochimica Acta* **66** (Part B) 129
[3] Ling H, Junyan L, Xiang Z and Baolian Jiang 2011 Corrosion behavior and morphological features of archeological bronze coins from ancient China *Microchem. J.* 99 203
[4] D'Ars de Figueiredo Junior JC, Lins VFC and Vito Modesto De Bellis 2007 Surface characterization of a corroded bronze-leaded alloy in a salt spray cabinet *Appl. Surf. Sci.*. **253** 7104
[5] Valério P, António M, Monge Soares and Araújo M F 2014 Arsenical copper and bronze in Middle Bronze Age burial sites of southern Portugal: the first bronzes in Southwestern Iberia *J. Arch. Sci.* **42** 68
[6] Izaak TI, Lyamina GV, Mokrousov GM and Firhova EB 2013 A method for cleaning surfaces of metal materials *RF Patent # 2495156*. Published 10.10.2013
[7] Dubinina OV, Lyamina GV, Mokrousov GM 2014 Application of polymer gel-electrolytes for cleaning and restoration of steel objects (*Adv. Mater. Res.* vol **1040** 8
[8] Lyamina GV, Kamchatnaya OV, Plotnikova AY, Izaak TI, Anishenko EV and Mokrousov GM 2012 Application of a voltammetric method for investing the formation mechanisms of silver nanoparticles in multicomponent solutions *J. Solid State Electrochem.* **16** 2473

Combined effect of pulse electron beam treatment and thin hydroxyapatite film on mechanical features of biodegradable AZ31 magnesium alloy

MA Surmeneva[1], AI Tyurin[2], AD Teresov[3], NN Koval[3], TS Pirozhkova[2], IA Shuvarin[2] and RA Surmenev[1, 4]
[1]Department of Theoretical and Experimental Physics, Tomsk Polytechnic University, 30, Lenina ave., Tomsk, 634050, Russia
[2]NanoCenter "Nanotechnology and Nanomaterials", G.R. DerzhavinTambov State University, 33, Internationalnaya st., 392000, Tambov, Russia
[3]Institute of High-Current Electronics, Siberian Branch, Russian Academy of Sciences, 2/3, Akademichesky ave., Tomsk, 634050, Russia
[4]Fraunhofer Institute for Interfacial Engineering and Biotechnology IGB, 12, Nobelstraße, Stuttgart, Germany

E-mail: tyurin@tsu.tmb.ru

Abstract. The morphology, elemental, phase composition, nanohardness, and Young's modulus of the hydroxyapatite (HA) coating deposited via radio frequency (RF) magnetron sputtering onto the AZ31 surface were investigated by atomic force microscopy (AFM), scanning electron microscopy (SEM), energy dispersive X-ray spectroscopy (EDX), X-ray diffraction (XRD), and nanoindentationtechniques. The calcium phosphate (Ca/P) molar ratio of the HA coating deposited via RF-magnetron sputtering onto AZ31 substrates according to EDX was 1.57±0.03. The SEM experiments revealed significant differences in the morphology of the HA film deposited on untreated and treated with the pulsed electron beam (PEB) AZ31 substrate. Nanoindentation studies demonstrated significant differences in the mechanical responses of the HA film deposited on the initial and PEB-modified AZ31 substrates. The nanoindentation hardness and the Young's modulus of the HA film on the magnesium alloy modified using the PEB treatment were higher than that of the HA layer on the untreated substrate. Moreover, the HA film fabricated onto the PEB-treated surface was more resistant to plastic deformation than the same film on the untreated AZ31 surface.

Key words: *hydroxyapatite coating, magnesium alloy, RF-magnetron sputtering, pulsed electron beam treatment, nanohardness, Young's modulus*

1. Introduction

Magnesium and its alloys may potentially be applied as degradable metallic materials in surgical orthopedics and traumatology due to their degradability and resemblance to human cortical bone [1-4]. The control of biodegradation of the magnesium alloys has been the subject of research and development for decades [5]. In an attempt to improve the quality of the bone–implant interface and corrosion resistance, numerous implant surface treatments have been used [6]. The demand for new

Content from this work may be used under the terms of the Creative Commons Attribution 3.0 licence. Any further distribution of this work must maintain attribution to the author(s) and the title of the work, journal citation and DOI.
Published under licence by IOP Publishing Ltd

modifications of biodegradable magnesium alloys to control their degradation rate will continue to grow.The fabrication of hydroxyapatite (HA) coatings is a very popular approach to control the degradation rate of the magnesium alloys.

Electron beam treatment is a novel way for the modification of implant surfaces to produce a high degree of purity with enough roughness for good osseointegration [7]. Electron beam treatment can result in unique microstructures with increased hardness, corrosion resistance, or other useful surface properties of metallic materials. Moreover, recent study showed that electron beam irradiation is a new method of treating implant surfaces to produce a high degree of purity with improved corrosion resistance [7].

Combination of the electron beam treatment and the deposition of ceramic coatings onto the surface of magnesium alloys can open up new opportunities for the biodegradable composite developmentwith the improved mechanical features and corrosion resistance.

Since a nanostructure and high adhesion of the coating are necessary to maintain a healthy connection with the soft tissues, radio frequency (RF) magnetron sputter deposition of HA thin films was investigated [8]. Other important properties of the implant surface are its strength and elastic characteristics (hardness and the Young's modulus).

Therefore, the aim of the present study was to obtain results relating to mechanical features of the magnesium alloy AZ31 after the modifications, such as the pulsed electron beam (PEB) treatment and RF-magnetron sputter deposition of HA thin film.

2. Materials and methods

The AZ31 magnesium alloy substrates were purchased from GoodFellow (Germany). The size of the samples was set to 10 × 10 × 1 mm^3 (width × length × thickness). The surface of the magnesium alloys was treated by a high-energy PEB irradiation with a PEB generator "SOLO" (Institute of High Current Electronics SB RAS, Tomsk, Russia). The conditions of the PEB treatment were as follows: electron beam energy density: 14 J cm^{-2}, electron beam pulse duration: 50 μs, and number of pulses: 3. An average electron energy in the beam was 15 keV.

A pure HA target was prepared according to the previously described procedures [9]. A commercially available apparatus with an RF (13.56 MHz, COMDEL) magnetron source was used to deposit HA coatings. HA coatings with the thickness of 700±60 nm were deposited at an RF-power level of 500 W in pure Ar atmosphere onto the grounded substrate holder. Optical ellipsometry (Ellipse 1891-S AG, Institute of Semiconductor Physics, RAS, Siberian Branch) was used to evaluate the thickness of the thin HA film. The coating thickness was derived from the changes in ellipsometric parameters between the bare and the coated substrates using a three-phase model (substrate-layer-air).
Surface roughness examined by Atomic force microscope (AFM) Solver P47-PRO (NT-MDT, Moscow, Russia), operating in a tapping mode. The roughness parameters of the surface were calculated three times at different spots for each specimen from AFM scans over the surface areas of 5 × 5 μm^2 using Nova SPM software (NT-MDT).

The surface morphology and surface composition of the deposited coatings were investigated using MERLIN field emission scanning electron microscope (FE-SEM) equipped with energy dispersive X-ray spectroscopy (EDX, Carl Zeiss).

The phase composition and structure of the surface were investigated using an X-ray diffractometer (XRD) (XRD-7000, Shimadzu, Japan) in Bragg-Brentano mode with a monochromatic CuK_α radiation source (a wavelength of 1.5406 Å) operated at 30 mA and 40 kV. The HA, magnesium XRD patterns (#9-0432 and #04-0770) from the ICDD database were used as references.

In order to fully investigate the indentation behaviour of HA films an investigation of the films was carried out by nanoindentation. Nanoindentation has been established as an important tool for measuring material hardness and elastic properties of thin films on the submicron scale [9-14].
Nanoindentation tests were performed using a Nanotriboindenter TI-950 (Hysitron Inc., USA) equipped with a Berkovich indenter with an angle between the opposite faces of 142.3° and tip radius of around 50 nm. A special software to calculate the nanohardness (*H*) and the reduced modulus (*E*) of

the coatings according to the analysis of Oliver and Pharr method was used [10]. Load–displacement curves with the load ranging from 0.1 mN to 10 mN were obtained in order to determine penetration depth (*h*), elastic modulus (*E*) and hardness (*H*) of the composites as a function of the applied load. Repeated indentations were performed on initial and modified AZ31 magnesium alloys and the values of *H* and *E* were calculated as an average of 10 indentations.

3. Results and discussion

Table 1 summarises the roughness parameters estimated by AFM for the uncoated initial substrates and AZ31 surface treated via PEB. All roughness parameters were measured for the scan areas of 5 × 5 μm^2. The initial substrate exhibited values higher than those measured for the AZ31 surface modified by PEB treatment.

Table 1. The surface roughness parameters of initial and PEB-treated AZ31 magnesium alloy substrate

Surface of AZ31	Roughness parameters			
	Sa, nm	*Sq*, nm	*Sz*, nm	*Smax*, nm
Initial surface	28.8	37.6	304	375
PEB treated surface	10.2	16.2	189	288

The typical SEM micrographs of the surface morphology of the initial and the HA-coated AZ31 substrates are shown in the figure 1.

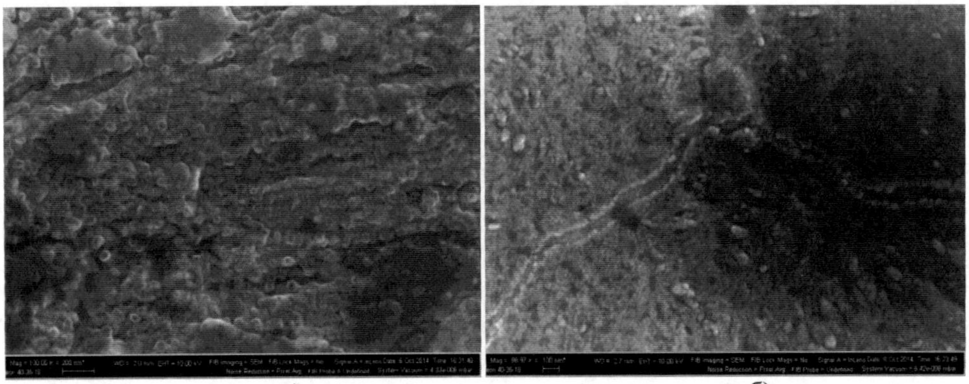

a) б)

Figure 1. SEM images of AZ31 magnesium alloy substrates before (a) and after (b) PEB treatment. The initial surface of AZ31 magnesium alloy was homogenous and revealed well-refined granular features

The typical XRD patterns of the untreated and treated via PEB AZ31magnesium alloy are shown in Fig. 2. XRD patterns showed that PEB treatment affected the preferential orientation of magnesium alloy substrates. The change in preferential orientation of the AZ31 was observed after the irradiation. Peaks located at 2θ=32.2°, 34.4°, 36.6°, 47.8°, 57.4° and 63.1° are assigned to the (100), (002), (101), (102), (110), and (103) planes of hexagonal magnesium (04-0770), respectively. It was interesting that the most intense peak (002), occurring at 2θ=34.4°, was found for the initial sample. When the substrate surface was modified by PEB the peaks at 2θ=32.2° and 36.6°, were more intense than the XRD reflexes for the initial Mg sample.

The evolution of the mechanical properties of the initial AZ31 sample as a function of the penetration depth revealed that the elastic modulus and hardness remain unchanged. The obtained increase in hardness and modulus at very low indentation depths is generally related to various effects like tip surface roughness and morphology.

The average values of *E* and *H* obtained for the penetration depths of 50 and 100 nm for the treated and untreated samples are presented in Table 2. The values of the elastic strain to failure ratio (*H/E*), and the parameter H^3/E^2 for the initial and modified substrates are also summarized in table 2. The initial magnesium alloy substrate revealed a nanohardness of 1.19 ± 0.32 GPa and a Young's modulus of 44.24 ± 3.40 GPa for a penetration depth of 100 nm.

The term *H/E* can be considered to be a useful indicator of a good wear resistance of the material [15, 16]. The material with a high plastic resistance ratio H^3/E^2 are more likely to resist plastic deformation during low load contact events and exhibit higher yield strength [17, 18]. Moreover, enhancement of the *H/E* ratio (and thus the resistance of the material to plastic deformation H^3/E^2) of the implant surface may offer advantages, such as less potential for surface damage and increased durability.

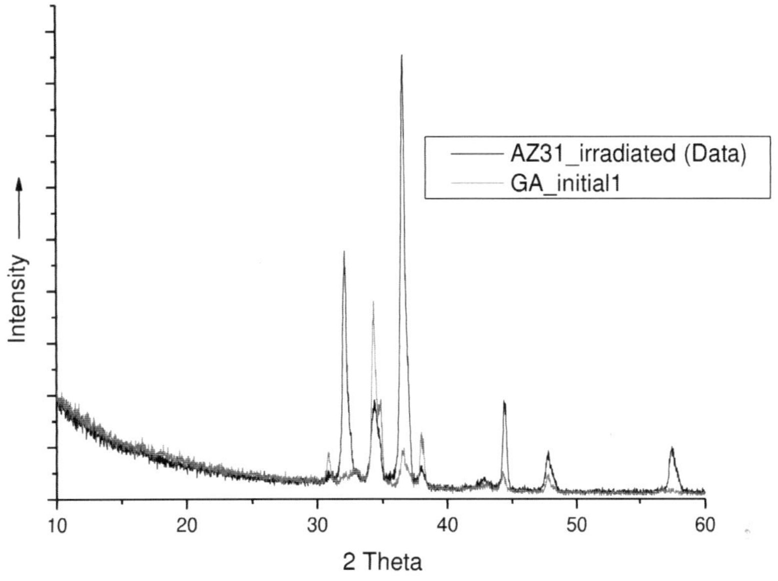

Figure 2. The typical XRD patterns of the AZ31 samples after (a) and before (b) PEB treatment

It could be seen that the magnesium alloy AZ31 treated with PEB demonstrated the heterogeneous elastic mechanical properties. However, the variation of values in that case was reducedsignificantly compared to the initial substrate. When the penetration depth was reduced to around 300 nm, the hardness decreased to 0.8 GPa, and the modulus increased to 45 GPa. These values of the hardness and Young's modulus were found to be slightly reduced compared to that of the initial substrate.
Therefore, according to XRD analysis the PEB treatment affected the preferential orientation of magnesium alloy substrates. The nanoindentation test revealed the decrease of the average values of the hardness and Young's modulus on the nano- and microscale for the magnesium alloy treated with PEB. However, the parameters *H/Eи*H^3/E^2 of the AZ31 surface remained unchangedafter the irradiation.

Examination of the coating microstructure by SEM revealed that the modification of the surface via the PEB irradiation has an influence on surface morphology. Fig. 4 showed the SEM images of the HA films fabricated on untreated and treated AZ31 magnesium alloy. All of the HA coatings deposited on untreated alloy revealed a regular grain-like morphology. However, the surface topography of HA film deposited onto the AZ31 substrate treated with PEB was smoother on the nano- and microscale level. Moreover, it can be observed that the coating on the treated substrate exhibited well-refined granular features. Therefore, the surface morphology of thin HA film is sensitive to the initial substrate topography.

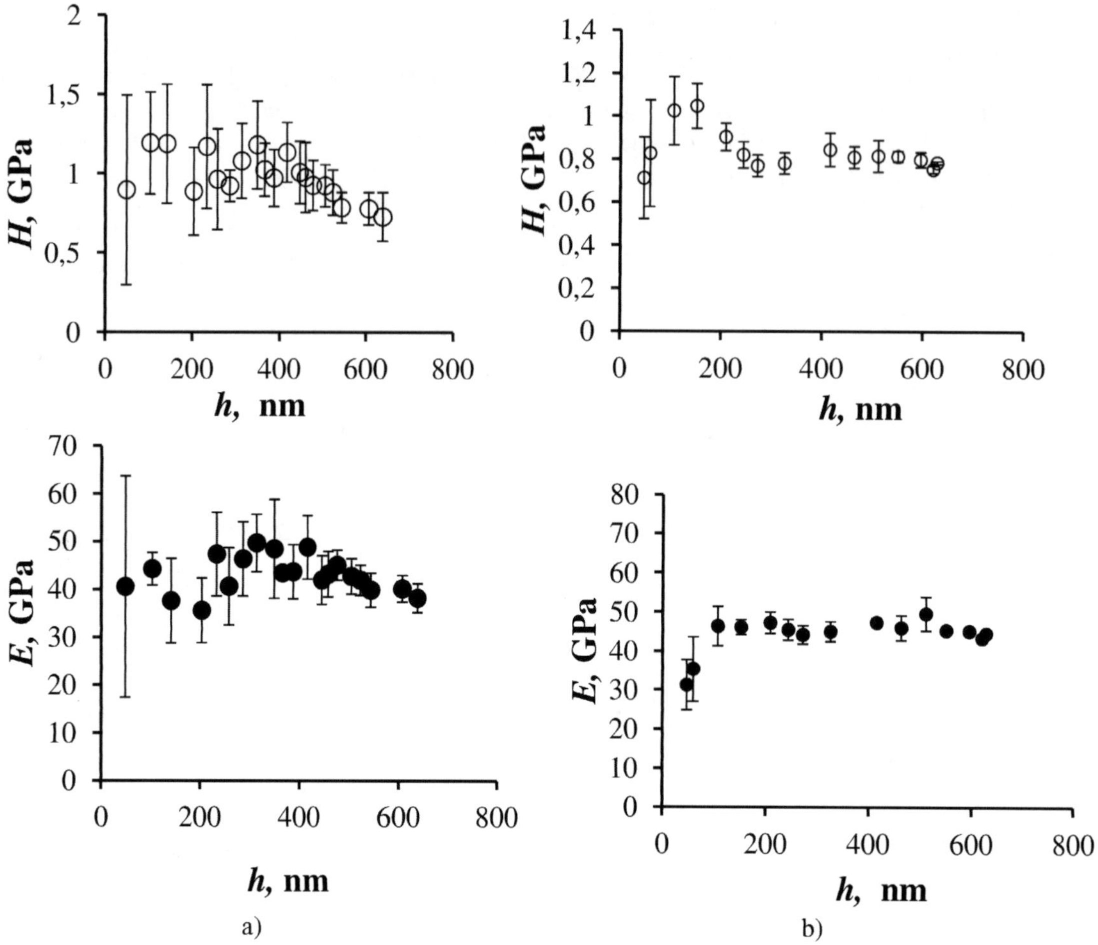

Figure 3. The hardness *H* and Young's modulus *E* values plotted as a function of the indentation depth for AZ31 magnesium alloy before (a) and after PEB treatment (b)

RF-magnetron sputter deposited HA coatings on the magnesium alloy have been described as dense, uniform, non-porous films [19] and improved corrosion resistance compared to uncoated substrate.

The most important parameters are the molar n(Ca)/n(P) ratio and the solubility [20]. The lower the Ca/P molar ratio is, the more soluble the calcium phosphate. The chemical formula of HA is $Ca_{10}(PO_4)_6(OH)_2$, and the stoichiometric Ca/P atomic ratio of 1.67 is close to that of natural bone. The Ca/P ratios of the HA films fabricated via RF magnetron sputtering onto the AZ31 substrate according to EDX were 1.57±0.03, which are slightly lower than the 1.67 Ca/P ratio of stoichiometric HA.

The evolution of *H* and *E* as a function of the penetration depth, determined from nanoindentation tests for the HA coating deposited on the untreated and treated AZ31 substrate, are shown in Fig. 5. The dispersion of *H* and *E* values in the case of the HA film on the AZ31 treated via the PEB were much smaller compared to that of *H* and *E* values obtained for the coating deposited on the untreated substrate. These results clearly indicated that the surface of the modified substrate with HA film exhibited more homogeneous properties. The evolution of the elastic modulus and hardness wasvery similar. The hardness as well as Young's modulus values decrease with increasing indentation depth independently of the HA film thickness. Saha and Nix observed a reduced hardness with increasing penetration depth in the case of a hard film on a soft substrate [21].

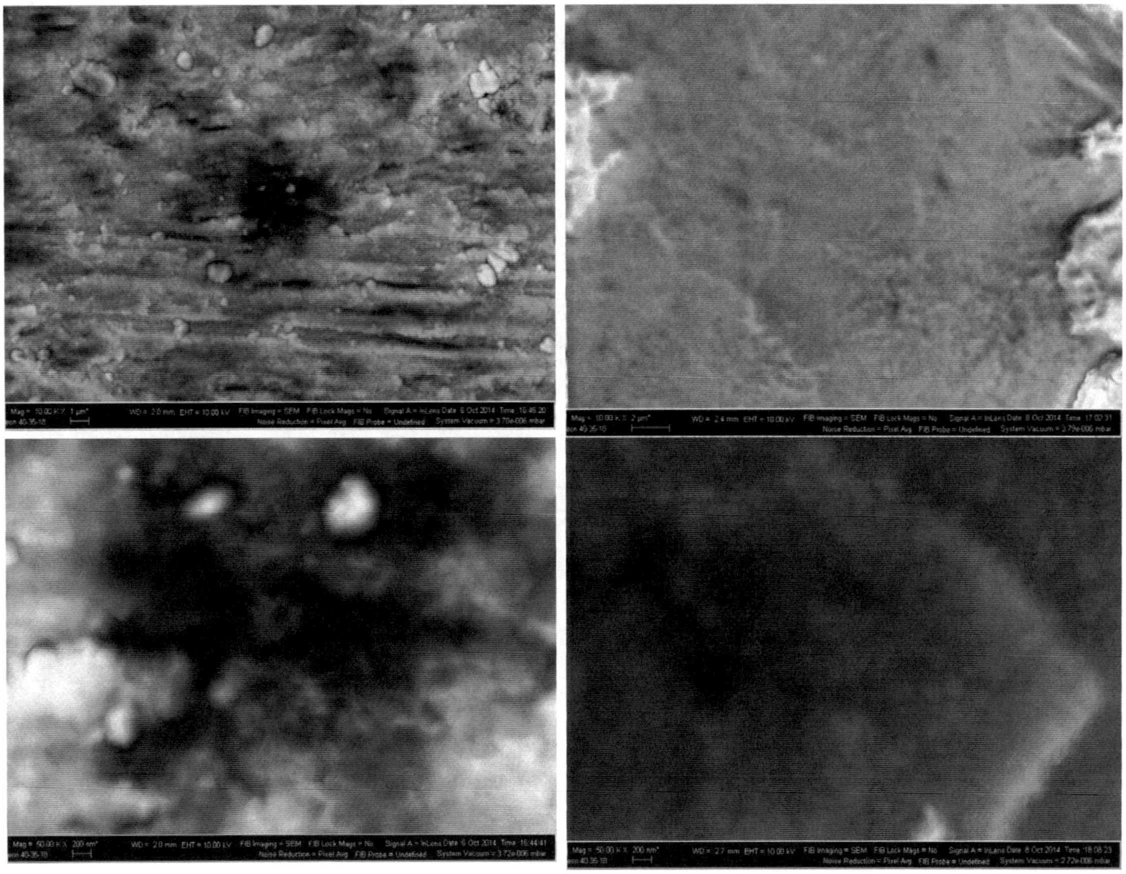

a) b)

Figure 4. SEM micrographs of HA coatings on the surface of AZ31 substrate before (a) and after PEB treatment (b)

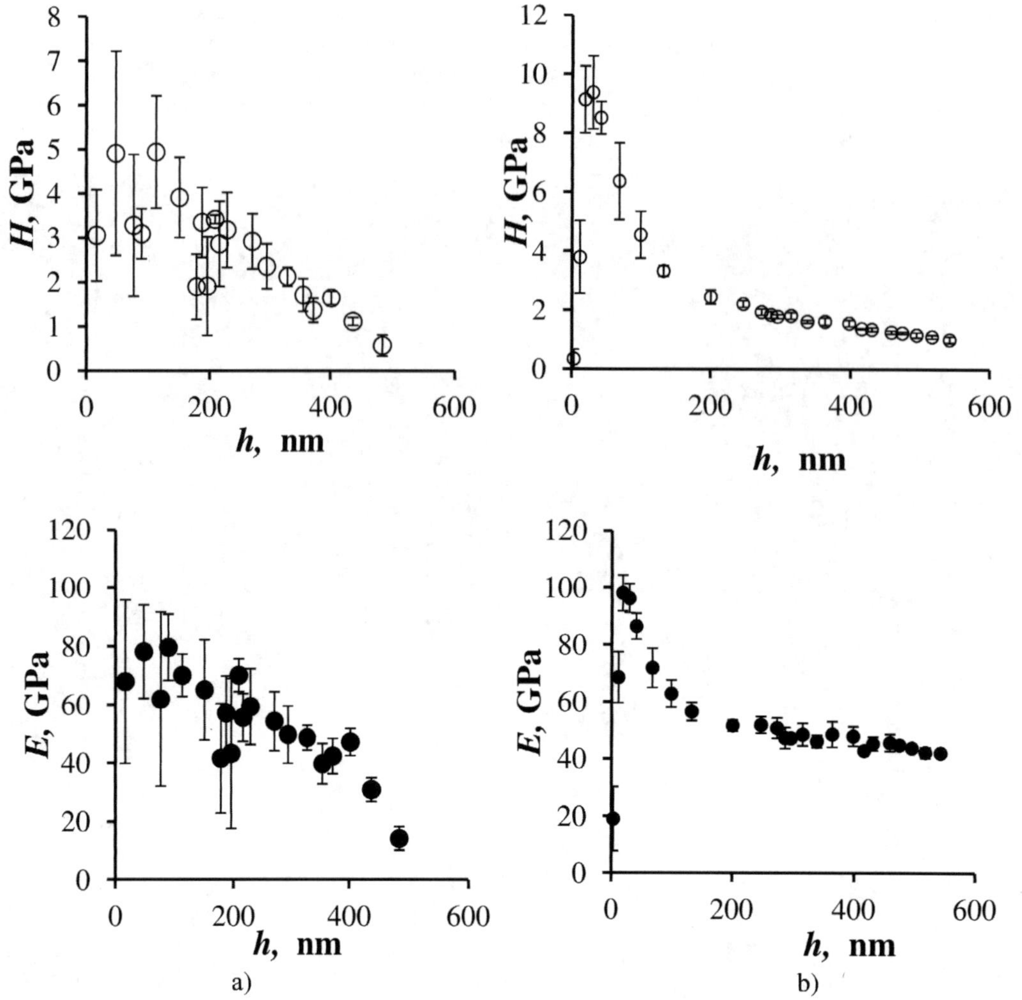

Figure 5. The hardness H and Young's modulus E values plotted as a function of the indentation depth for HA-coated AZ31 magnesium alloy before (a) and after PEB treatment (b)

Table 2. The values of the penetration depth h, nanohardness$H$ and Young's modulus E, H/E ratio, parameter H^3/E^2 for the HA-coated AZ31 magnesium alloy before and after PEB treatment

Sample	h, nm	H, GPa	E, GPa	H/E	H^3/E^2,GPa
Initial AZ31 surface	100	1.19 ± 0.32	44.24 ± 3.40	0.027	0.0009
	50	0.90 ± 0.80	37.01 ± 22.30	0.024	0.0005
HA coated AZ31 surface	100	3.08 ± 2.00	79.03 ± 11.00	0.039	0.0047
	50	4.90 ± 2.50	78.04 ± 16.00	0.063	0.0193
AZ31 surface after PEB treatment	100	1.02 ± 0.16	46.35 ± 5.00	0.022	0.0005
	50	0.71 ± 0.24	31.25 ± 6.41	0.023	0.0004
PEB treated and HA coated surface	100	4.55 ± 0.79	62.8 ± 4.7	0.072	0.0239
	50	8.52 ± 1.4	86.38 ± 4.6	0.099	0.0829

In the table 2 the results of nanoindentation of the HA coating deposited on the treated and precursor samples are summarized. Large differences are observed, which can be attributed to the surface structure of magnesium alloy treated differently. The HA coating on the AZ31 substrate processed by the PEB treatment was much harder (H = 8.52 ± 1.4 GPa, table 2) than the HA film on the rough initial substrate (H = 4.90 ± 2.50GPa, Table 2). Moreover, the Young's modulus of HA coating on the treated substrate was observed to be higher at the depth of 50 nm than HA film prepared by sputtering onto the initial AZ31 substrate (Table 2). The pre-treatment of the AZ31 substrates resulted in an increase in the H/E ratio and the plastic resistance ratio H^3/E^2. The same effect of the pre-treatment was observed for the HA coating deposited on the acid etched and pulse electron beam treated surfaces [9].

4. Conclusion

The morphology, elemental, phase composition, nanohardness, and Young's modulus of the HA coating deposited viaRF-magnetron sputtering were investigated by SEM, EDX, XRD, and nanoindentationtechniques. The Ca/P ratio of the HA coating deposited via RF-magnetron sputtering onto AZ31 substrates according to EDX was 1.57±0.03. The SEM experiments revealed significant differences in the morphology of the HA films deposited onto the initial AZ31 substrate and the PEB-treated substrate.

Nanoindentation studies demonstrated significant differences in the mechanical responses of the HA films deposited onto the initial and PEB-modified AZ31 substrates. The nanoindentation hardness and the Young's modulus of the HA films on the PEB-treated magnesium alloy are higher than that of the HA layer on the initial substrate. Moreover, the HA filmfabricated onto the treated surface is more resistant to plastic deformation than the same film deposited on the untreated AZ31 surface.

Acknowledgments

The authors thank T. Mukhametkaliyev for the deposition of the HA coating. This research was supported by the Russian Science Foundation (project number 14-13-00274).

References

[1] Staiger MP, Pietak AM, Huadmai Jand Dias G. 2006 *Biomater.* **27** 1728
[2] Surmeneva MA, Tyurin AI, Mukhametkaliyev TM, et. al. 2015 *J. of the mechanical behavior of biomedical materials* **46** 127
[3] Song BJ, Song WJ, Han JH, et. al. 2014 *J. of Nanosci. and Nanotechn.* **14**(12) 9124
[4] Li L, Bai L, Zhang QQ, et. al. 2014 *Mater. Res. Innov.* **18**(S4) 695
[5] Zheng YF, Gu XNand Witte F 2014 *Mater. Sci.* and *Engin.: Reports* **77**1
[6] Hornberger H, Virtanen Sand Boccaccini AR 2012 *Acta Biomater.* **8** 2442
[7] Gao B, Hao S, Zou J, et. al. 2005 *J. of Vacuum Sci. &Techn. A* **23** 1548
[8] Huang J, Jayasinghe SN, Best SM, et. al. 2005 *J. Mater. Sci.: Materials in Medicine* **16** 1137
[9] Surmeneva MA, Surmenev RA, Tyurin AI, et. al. 2014 *Thin Solid Films* **571** 218
[10] Oliver WCand Pharr GM 1992 *J. Mater. Res.* **7** 1564
[11] Fischer-Cripps AC 2011 Nanoindentation. NewYork: Springer 280
[12] Voyiadjis GZ, Yaghoobi M 2015 *Mater.Sci. and Engin.: A.***634** 20
[13] Akchurin MSh, Gainutdinov RV, Garibin EA, et. al. 2011 *Inorganic Mater.: Appl. Res.* **2**(2) 97
[14] Golovin YI, Iunin YL and Tyurin AI 2003 *Doklady Phys.* MAIK Nauka / Interperiodica **48**(9) 505
[15] Leyland A and Matthews A 2000 Wear **246** 1
[16] Ievlev VM, Kostyuchenko AV, Darinskii BMand Barinov SM 2014 *Physics of the Solid State* **56**(2) 321
[17] Musil J 2000 *Surface and coating technology* **125** 322
[18] Roy ME, Whiteside LA, Xu Jand Katerberg BJ 2010 *Acta Biomater.* **6** 1619
[19] Surmeneva MA and Surmenev RA 2015 *Vacuum* **117** 60
[20] Dorozhkin Sand Epple M 2002 *Angewandte Chemie Internat. Ed.* **114** 3260
[21] Saha R and Nix WD 2002 *Acta Mater.* **50** 23

IOP Conf. Series: Materials Science and Engineering **98** (2015) 012031 doi:10.1088/1757-899X/98/1/012031

Workers' cytokines profiling upon exposure to MWCNT aerosol in occupational settings

L M Fatkhutdinova[1], T O Khaliullin[1], R RZalyalov[1], O L Vasilyeva[1], I Kh Valeeva[2]and I G Mustafin[3]

[1]Department of Hygiene, Occupational Medicine, Kazan State Medical University, 49, Butlerova str., Kazan, Russia
[2]Central Laboratory, Kazan State Medical University, 49, Butlerova str., Kazan, Russia
[3]Department of Biochemistry, Kazan State Medical University, 49, Butlerova str., Kazan, Russia

[1]E-mail: liliya.fatkhutdinova@gmail.com

Abstract. Recent studies have found that upon pulmonary exposure to multi-walled carbon nanotubes (MWCNTs) animals develop primarily fibrosis and granulomas in lungs. *In vitro* and *in vivo* studies also give reason to assume that local exposure could be related to remote effects, including immune system and the endothelium. To investigate the remote effect hypothesis, we have analyzed blood, nasal lavage and induced sputum samples taken from workers in the frame of the Russian epidemiological study on Carbon Nanotubes Exposure and Risk Assessment (CNT-ERA).In serum and nasal lavage no significant differences between exposure and control groups were observed with a high variability to the cytokines content. In the samples of induced sputum from exposed workers the content of IL-1b, IL-6, IL-8, TNF-a, IL-4, IL-5, IFN-g exceeded the control group values, but after the regression models construction and bootstrap analysis, significant differences were found only for IL-1b. This study could not provide evidences of blood cytokines changes following local cytokine production in airways in workers exposed to MWCNTs. Cytokines variability in serum and nasal lavage may indicate the absence of severe systemic inflammatory response upon the existing occupational exposure to MWCNTs. Other systemic responses (including allergy-like or autoimmune reactions) should be regarded as well.

1. Introduction

Unique properties of multi-walled carbon nanotubes are in demand by different industries – in construction, engineering, energy-producing industry, medicine [1-3]. However, recently accumulated data suggest risk to human health upon the occupational contact with carbon nanotubes [4].Nanomaterials, depending on their characteristics and compositions, could interact with the immune system in several ways and either enhance or suppress immune system function [5].Measuring the levels of cytokines can be useful tools in evaluating nanoparticle immunotoxicity. Moreover, screening of a single reporter is inadequate; combinations of cytokine levels should be measured, due to the cytokine network influence on the immune system [5].

Th1 (Type-1) and Th2 (Type-2) cytokines are the two groups of cytokines produced by two different subpopulations of CD4+ T-cells [5,6]. They determine whether the immune response to the

Content from this work may be used under the terms of the Creative Commons Attribution 3.0 licence. Any further distribution of this work must maintain attribution to the author(s) and the title of the work, journal citation and DOI.
Published under licence by IOP Publishing Ltd

antigen would be humoral or cell-mediated. At the same time, a plurality of cells other than CD4+ may also express these cytokines – CD8+ T-cells, phagocytic cells (including monocytes, B-cells, macrophages), NK-cells, eosinophils, mast cells, basophils and others. Th1 cells secrete large quantities of interferon IFN-g, interleukin IL-2, IL-3, IL-12, granulocyte macrophage colony-stimulating factor (GM-CSF) and small quantities of tumor necrosis factor (TNF). Th2 cells secrete large quantities of IL-3, IL-4, IL-5, IL-6, IL-10, and small quantities of GM-CSF and TNF. The cytokines produced by specific T-cells (e.g. IFN-g from Th1 cells and IL-10 from Th2 cells) usually inhibit the action of other type of T-cells and thereby potentiate a particular pattern of immune response. IFN-gand IL-4 are the main markers of Th1 and Th2 cells, respectively. Various forms of cytokine secretion correspond to different functions of effector cells. Th1-cells promote cell-mediated immune response. Th2-cells are primarily helper cells that affect B-cells and induce humoral responses, such as secretion of antibodies by B-cells, mostly IgE. Both T-cell types affect each other, including the secretion of cytokines. Thus, these subpopulations are mutually antagonistic, and predominant subpopulation may determine the outcome of inflammatory or infectious process. In case of chronic inflammatory process Th1 cytokines inhibit tissue remodeling and collagen synthesis, but may lead to tissue damage, whereas Th2 response promotes fibrotic process, promoting healing of damaged areas [5] that may be related to pro-fibrotic activity of MWCNTs [7-9].

IL-1b is the most studied member of the IL-1 family [7]. This cytokine is produced by activated macrophages as a proprotein, which is proteolytically processed to its active form. This cytokine is an important mediator of the inflammatory response, and is involved in a variety of cellular activities, including cell proliferation, differentiation, and apoptosis. Increased production of IL-1B causes a number of different autoinflammatory syndromes.

Recent studies have found that upon pulmonary exposure to MWCNTs animals develop primarily fibrosis and granulomas in the pulmonary system [8-11]. However, *in vitro* and *in vivo* studies also give reason to assume that local exposure could be related to remote effects, including immune system [12] and the endothelium [13-15].

To investigate the remote effect hypothesis, we have analyzed blood, as well as nasal lavage and induced sputum samples taken from workers in the frame of the Russian epidemiological study on carbon nanotubes exposure and risk assessment (CNT-ERA).

2. Materials and Methods

2.1. MWCNT production facility

The study was conducted at two MWCNT-producing enterprises with the same reactor type. MWCNTs are produced using the catalytic vapor deposition method [16]. The technological process includes synthesis in reactor, raw product harvesting, optional grinding/disintegration and functionalization or acid treatment. MWCNT–characteristics are shown in the Table 1.

Table 1. MWCNT characteristics, provided by producer

Parameters	MWCNT
External diameter(nm)	8-15
Internal diameter(nm)	4-8
Length(um)	2 and more
Total amount of impurities– catalysts leftovers(%)	Up to 5
Bulk density(g/cm^3)	0.03-0.05
Specific surface area(m^2/g)	300-320
Thermal stability in air (°C)	Up to 600

IOP Conf. Series: Materials Science and Engineering **98** (2015) 012031 doi:10.1088/1757-899X/98/1/012031

Elemental carbon (EC) was evaluated in air samples taken in occupational settings, and the CNT presence was confirmed by transmission electron microscopy [17]. TEM microphotographs have shown the presence of MWCNT-containing agglomerates at all technological process stages. Agglomerates size ranged from 1 to 10 um, and no individual MWCNTs were found. Cobalt and nickel used as catalysts were found in agglomerates using EDXA analysis

The highest MWCNT aerosol short-term concentration was registered during the reactor cleanout process, reaching 157 ug/m^3; time-weighted average (TWA) concentration was 29.6 ug/m^{3}for the 8-hours shift respectively. Respirable fraction TWA concentrations (obtained using the cyclone) for EC at different workstations ranged from 0.53 to 6.11 ug/m^3.

2.2. Study group

25 workers of both genders aged 18-65 years with no history of diagnosed chronic respiratory diseases before they started working with CNTs were taken. 11 workers who had more than 1 year contact with MWCNT aerosol composed the exposure group. The control group (14 people) included persons who were not exposed to the MWCNT aerosol, but worked at the same enterprise - engineers, secretaries, research assistants. Average age (M±SD) of exposure group was 35.5±15.8 years, controls – 30,2±10,4. There were 4 smokers among exposed personnel and 3 in controls.

2.3. Biomaterial sampling

The blood sampling was carried out aseptically by a qualified staff using a BD Vacutainer® Safety-lok™ device («Beckton Dickenson», USA) with disposable vacuum tubes containing coagulation activator (Vacutainer®, USA). After clot retraction all tubes were centrifuged at 2500 rpm, serum was removed and frozen at -80 ° C.

Inhalation is a leading a potential nanoparticles exposure route in humans, and nasal cavities are the first part of respiratory tract, where direct contact with MWCNTs happens. Similar epithelial lining properties imply similar cellular responses in nasal cavities and lower parts of the respiratory system. Thus, a method of nasal lavage was selected to evaluate pro-inflammatory properties of inhalatory exposure to MWCNT [18].The induced sputum method allowed obtaining sputum samples from deeper parts of the respiratory tract, while being non-invasive and minimally uncomfortable for the subjects.

For the nasal lavage sampling, examinee tilted his head back at an angle of 45 degrees. 4.5 ml of physiological saline, heated to 37 ° C were introduced successively into each nostril using a soft catheter and 10ml syringe; after 10 seconds examinee bent to drain solution from the nasal cavity into the funnel (inner opening closed with polyamide mesh) and further into the vial. The resulting samples were centrifuged for 10 minutes at 9000 rpm and supernatant frozen at -80 ° C.

Induced sputum is gathered in the absence of its spontaneous production (in normal and healthy individuals) by inhalation of aerozolized hypertonic saline, which causes an increase in bronchial secretions [19]. This method allows obtaining sputum samples from the deeper parts of the respiratory tract, while being non-invasive and minimally uncomfortable for subjects. After centrifugation of the samples and separation of the cell pellet necessary indicators were determined in the supernatant.

2.4. Cytokine profiling

Evaluation of the cytokines content in blood, nasal lavage and induced sputum samples was conducted using a flowcytometry kit Th1/2 Multiplex (eBioscienceinc., USA) on the BD Canto flow cytometer (Beckton Dickenson, USA).

2.5. Statistical analysis of the results

For statistical analysis we used R statistical package. Initial statistical analysis of cytokine profiles obtained from the employees was carried out using t-test. Besides, generalized linear models including both main effects (exposure, age, smoking) and pairwise interactions were designed and developed.

Models fitting was performed using successive elimination and Akaike information criterion (AIC) evaluation with exhaustive search and bootstrap analysis [20]. Allcreated models were verified by the likelihood ratio test and analysis of variance. The confidence intervals of regression coefficients derived from a small sample were refined by bootstrap analysis. Bootstrap analysis included repeated sampling with replacement from the original sample, which represents the sampling distribution of the statistic. On the basis of the analyzed indicators values deviation obtained in the simulation, confidence intervals were built. Bootstrap confidence intervals (ICA) were constructed based on which we could make conclusions about the statistical significance of the relationship.

3. Results and discussion

The results of cytokines content analysis in serum, nasal lavage and induced sputum samples are shown in figures 1, 2 and 3.

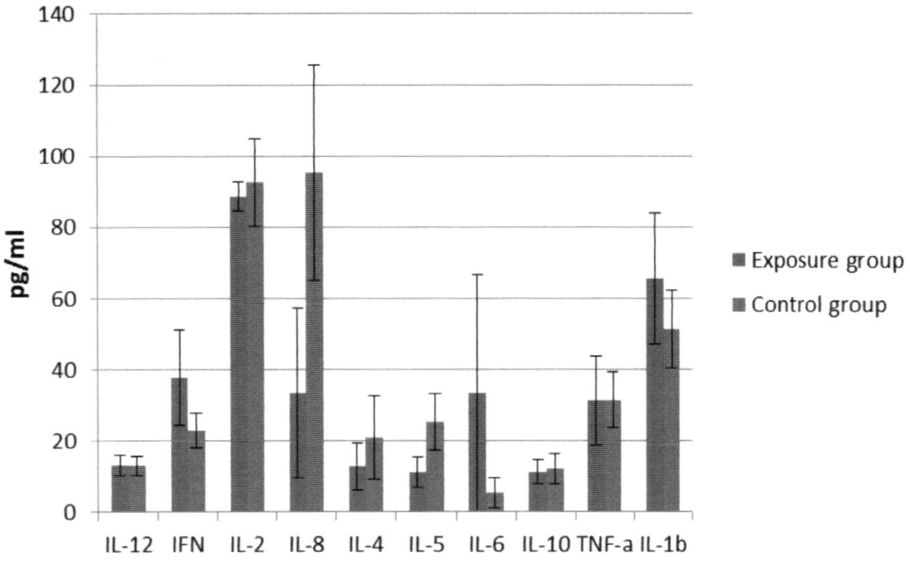

Figure1. Cytokines content in serum, pg/ml, M±m. TNF-b levels were below the detection limit in both groups.

In serum no significant differences between exposure and control groups was observed with a high variability to the cytokines content (figure 1).

In nasal lavage samples no statistically significant differences between the exposed group and unexposed individuals were detected as well (figure 2).

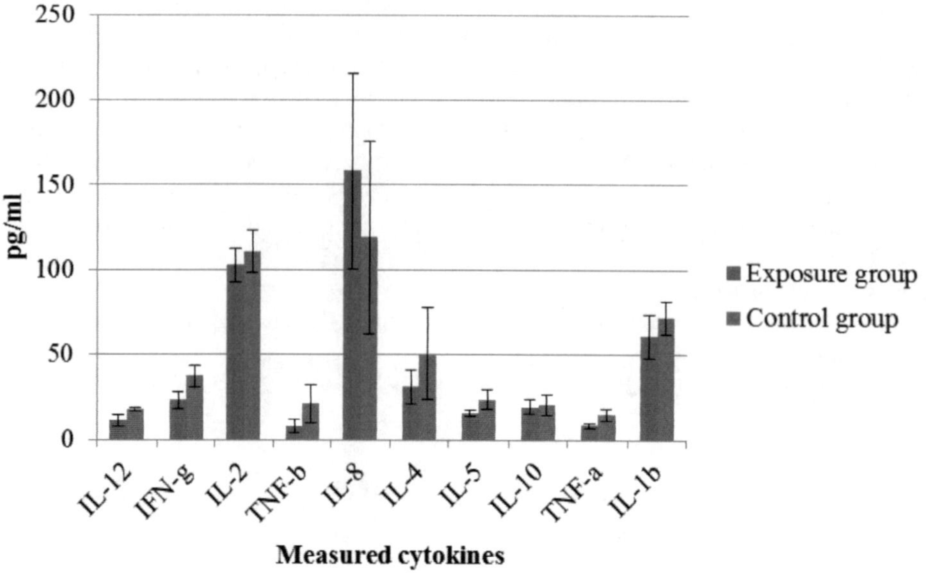

Figure 2. Cytokines content in nasal lavage, pg/ml, M±m. IL-6 levels were below the detection limit in both groups.

In the samples of induced sputum from exposed workers a definite trend has been traced – the content of IFN-g, IL-1b, IL-6, IL-8, TNF-a, IL-5, IL-4 exceeding the control group values (Fig. 3), but after the regression models construction and bootstrap analysis, significant differences were found only for IL-1b (b = 1030.9, CI=1.64, 2060.24).

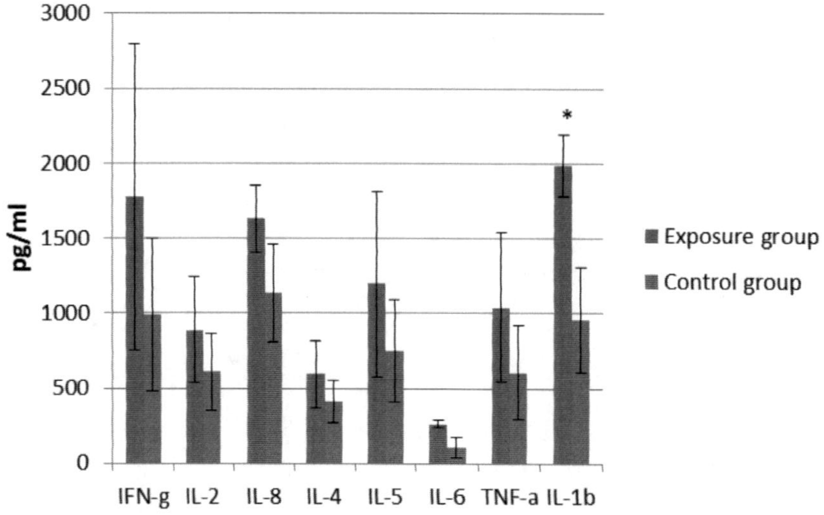

Figure 3. Cytokines content in induced sputum, pg/ml, * - p<0.1, M±m. IL-10, IL-12 and TNF-b levels were below the detection limit in both groups.

The content of IL-1b, which is involved in various processes including inflammation, cell differentiation, proliferation and apoptosis, was higher in induces sputum of exposed workers compared with controls.

In siRNA experiments on human primary macrophages Palomaki et al. [21] demonstrated that the NLRP3 inflammasome was essential for IL-1β secretion induced by long, needle-like MWCNT and asbestos. NLRP3 inflammasomes are multiprotein complexes that act as a major mediator for inflammatory responses. In addition, it was shown that double-walled carbon nanotubes enhanced the release of the pro-inflammatory cytokine IL-1b from human monocytes via NLRP3 inflammasome activation pathway [22]. Arnoldussen Y. et al [23] have recently presented results indicating that IL-1 contributes to the cellular and molecular effects of CNT exposure and that the type of CNT has an important effect on the cellular response. The author used cells from IL-1a/b wild type (IL1-WT) mice and cells with reduced inflammatory potential from IL-1a/b deficient (IL1-KO).

The IL-1b cytokine is of particular interest because its production has been shown to be associated with autoimmune disorders and fibrosis [21-23]. In report made by Ilves et al. on first results of the inhalation study of rigid MWCNT biological effects in mice, a significant increase in expression of Th2-cytokine in the lung tissues and asthma-like allergic response was shown [24].

4. Conclusion

Our preliminary study could not provide evidences of blood cytokines changes following local cytokine production in airways in workers exposed to MWCNTs. Cytokines variability in serum and nasal lavage may indicate the absence of severe systemic inflammatory response upon the existing occupational exposure to MWCNTs. At the same time, other systemic responses (including allergy-like or autoimmune reactions) should be regarded.

Application of transcriptomics methods, including the identification and determination of mRNA expression patterns of certain cytokines encoding genes, may be suggested instead of direct assessment of the cytokines content in blood, as the latter is often subjected to fluctuations related to changes in environmental parameters.

References

[1] Yakovlev G, Pervushin G, Maeva I, Keriene J, Pudov I, Shaybadullina A, Buryanov A, Korzhenko K and Senkov S 2013 *Procedia Eng* **57** 407

[2] Behabtu N, Young CC, Tsentalovich DE, Kleinerman O, Wang X, Ma AW, Bengio EA, ter Waarbeek RF, de Jong JJ, Hoogerwerf RE et al. 2013 *Science* **339**(6116) 182

[3] Saito N, Haniu H, Usui Yu, Aoki K, Hara K, Takanashi S, Shimizu M, Narita N, Okamoto M, Kobayashi Set al. 2014 *Chem Rev* **114**(11) 6040

[4] Fatkhutdinova LM, Khaliullin TO and Shvedova AA 2015 *Nanotechnol. Russ* **10**(5-6) 501

[5] Elsabahy M and Wooley KL 2013 *Chem Soc Rev* **42** 5552

[6] Wynn TA 2004 *Nat Rev Immunol* **4**(8) 583

[7] Dinarello Ch.A. 2009 *Annual Review of Immunology* **27** 519-550

[8] Mercer RR, Scabilloni JF, Hubbs AF, Battelli L, McKinney W, Friend S, Wolfarth MG, Andrew M, Castranova V and Porter DW 2013 *Part Fibre Toxicol* **10** 33

[9] Porter DW, Hubbs AF, Mercer RR, Wu N, Wolfarth MG, Sriram K, Leonard S, Battelli L, Schwegler-Berry D, Friend S et al. 2010 *Toxicology* **269**(2-3) 136

[10] Khaliullin TO, Shvedova AA, Kisin ER, Zalyalov RR and Fatkhutdinova LM 2015 *Bull Exp Biol Med* **158**(5) 684

[11] Shvedova AA, Yanamala N, Kisin ER, Tkach AV, Murray AR, Hubbs A, Chirila MM, Keohavong P, Sycheva LP, Kagan VE and Castranova V 2014 *Am J Physiol Lung Cell Mol Physiol* **306**(2) 170

[12] Mitchell L A, Lauer F T, Burchiel SW and McDonald JD 2009 *Nat Nanotechnol* **4**(7) 451

[13] Erdely A, Hulderman T, Salmen R, Liston A, Zeidler-Erdely PC, Schwegler-Berry D, Castranova V, Koyama S, Kim YA, Endo M and Simeonova PP 2009 *Nano Lett* **9**(1) 36

[14] Nurkiewicz TR, Porter DW, Hubbs AF, Stone S, Moseley AM, Cumpston JL, Goodwill AG, Frisbee SJ, Perrotta PL, Brock RW et al. 2011 *Res Rep Health Eff Inst* **164** 3

[15] Stapleton PA, Minarchick VC, Cumpston AM, McKinney W, Chen BT, Sager TM, Frazer DG, Mercer RR, Scabilloni J, Andrew ME et al. 2012 *Int J Mol Sci* **13**(11) 13781
[16] Cassell A M, Raymakers J A, Kong J and Dai HJ 1999 *J Phys Chem B* **103** 6484
[17] Khaliullin TO, Zalyalov RR, Shvedova AA, Birch EM and Fatkhutdinova LM, 2013 *Book of abstracts, II International School-Conference Applied Nanotechnology & Nanotoxicology* ed A Vedyagin 23
[18] Siegel PD, Short S, Jones WG 1995 *Proc. of the Sixth FIOH-NIOSH Joint Symposium on Occupational Health and Safety* (8-10 August 1995, Espoo, Finland) (People and Work - Research Reports: Finnish Institute of Occupational Health) 81
[19] Paggiaro PL, Chanez P, Holz O, Ind PW, Djukanović R, Maestrelli P and Sterk PJ 2002 *Eur Respir J* **20**(Suppl. 37) 3
[20] Moore D, McCabe G and Craig B2010 *Introduction to the Practice of Statistics 7^{th}ed* (W. H. Freeman)
[21] Palomäki J, Välimäki E, Sund, J Vippola M, Clausen PA, Jensen KA, Savolainen K, Matikainen S and Alenius H 2011 *ACS Nano* **5**(9) 6861
[22] Meunier E, Coste A, Olagnier D, Authier H, Lefèvre L, Dardenne C, Bernad J, Béraud M, Flahaut E and Pipy B 2012 *Nanomedicine* **8** 987
[23] Arnoldussen YJ, Skogstad A, Skaug V, Kasem M, Haugen A, Benker N, Weinbruch S, Apte RN and Zienoldiny S2015 *Cytokine* **73**(1) 128
[24] Ilves M, Rydman E, Savolainen K and Alenius H 2013 *Book of abstracts, II International School-Conference Applied Nanotechnology & Nanotoxicology* ed A Vedyagin 192

Aluminium oxide-hydroxides obtained by hydrothermal synthesis: influence of thermal treatment on phase composition and textural characteristics

A Zykova[1,2] A Livanova[1], N Kosova[1], A Godymchuk[2,3] and G Mamontov[2]
[1] Laboratory of catalytic research, Tomsk State University, 36, Lenina ave., Tomsk, 634050, Russia
[2] Division of nanomaterials and nanotechnologies, Tomsk Polytechnic University, 30, Lenina ave., Tomsk, 634050, Russia
[3] National University of Science and Technology "MISIS", 4, Leninsky ave., Moscow, 119991, Russia

E-mail: zykovaap@mail.ru

Abstract. Aluminium oxide-hydroxides obtained by means of the hydrothermal synthesis of aluminium nanopowder are of great interest in terms of the potential supports for heterogeneous catalysts due its high specific surface area (200…300 m^2/g) and pore size of 4…17 nm. In this work the influence of thermal treatment (150…1300 °C) on structural and phase composition, specific surface area and porosity of aluminium oxide-hydroxides has been investigated. Nanostructured γ-Al_2O_3 (T=400 °C) was found to have the specific surface area of 328 m^2/g and average pore size of ~9 nm. The thermal treatment of aluminium oxide-hydroxides at the temperature of higher and lower than 400 °C has caused the reduction of specific surface area and overall pore volume.

1. Introduction

Aluminium nanopowders obtained by means of electrical explosion of wires in various atmospheres are extensively investigated and widely used for composite and functional materials production. There is a particular interest in aluminium oxide-hydroxides produced through hydrothermal synthesis of aluminium nanopowder as advanced materials for catalyst supports [1, 2], sorbents [3], ceramics [4, 5] and filters [6]. The main advantages of such aluminium oxide-hydroxides are high specific surface area, high porosity, narrow pore size distribution, and capability to vary phase composition and basic-acid properties of surface. These properties enable using aluminium oxide-hydroxides as supports for heterogeneous catalysts for hydrocarbons reprocessing. That will to a large extent reduce the number of stages required for such catalysts production and help to develop supports and catalysts with desired structural and functional characteristics.

In literature there are more and more papers dedicated to the investigation of hydrothermal synthesis processes from ultradispersed and nanosized aluminium powder [3, 7-11]. However, the information about structure of aluminium oxide-hydroxide formed during hydrothermal synthesis and subsequent thermal treatment, presented in literature, is still diverse and often contradictory. This is because morphological and structural characteristics of aluminium oxide-hydroxides depend on the original nature of material, properties, method and conditions of production, the reactivity of ultra- and

Content from this work may be used under the terms of the Creative Commons Attribution 3.0 licence. Any further distribution of this work must maintain attribution to the author(s) and the title of the work, journal citation and DOI.
Published under licence by IOP Publishing Ltd

nanodispersed aluminium powders, conditions of thermal treatment, etc. This work is aimed at the estimation of the thermal treatment impact on the structural and textural characteristics of aluminium oxide-hydroxides produced through hydrothermal synthesis of aluminium nanopowder for further use of them as supports for heterogeneous catalysts.

2. Materials and methods

To obtain aluminium oxide-hydroxide (AlOH) the industrial metallic aluminium nanopowder (AlNP) produced by electrical explosion of wires (at LLC Novosibirsk Nanomaterials, Novosibirsk) was used. According to the manufacturer , the average size of AlNP particles was 115 ± 10 nm, specific surface area – 19.4 ± 3 m^2/g, loading of active aluminium – 80.8 ± 0.6 %.

The AlOH was produced using technique described in [12]. That was hydrothermal synthesis of an AlNP implying preparation of a suspension that consisted of the AlNP and preheated (60…65 °C) distilled water with mass ratio H_2O:Al=25. The prepared suspension was continuously stirred. After the induction period (3…5 min) the watering oxidation of AlNP was accompanied with self-heating of suspension with temperature growth up to 92…97 °C and intensive hydrogen release. Once the reaction was over the excess of water removed by decantation and solid products of reaction were dried at 21 ± 2 °C. The thermal treatment of the AlOH was conducted in a muffle furnace in the temperature range from 150 to 1300 °C with the heating rate of 10 °C/min and exposure at maximal temperature of 1 h. The size of the AlNP particles was defined using the particle analyzer SALD-7101 in circulation mode with ultrasound (Nano-Center Tomsk Polytechnic University, RF). The shape and size of AlNP particles were investigated by means of high resolution transmission electronic microscopy (HR TEM) using microscope JEM-2100F (Nano-Center Tomsk Polytechnic University, RF). The morphological characteristics of the AlOH were studied by means of scanning electron microscopy (SEM) using system Quanta 200 with focused electronic beam (Center for collective use, Tomsk State University, RF). The phase composition was determined by means of X-ray diffraction (XRD) using the X-ray diffractometer Rigaku Miniflex 600 with CuK_α-radiation at the scan rate of 2 °/min. The phase composition analysis was carried out in accordance with PDF-2 database. The AlOH specific surface area and porosity were determined by means of nitrogen low-temperature adsorption using the analyzer Tristar 3020 (Center for collective use, Tomsk State University, RF). The specific surface area (S_{BET}) was measured by multipoint BET method with rectifying adsorption isotherm in the relative pressure range of 0.05…0.30 p/p^o. The specific surface area of micropores (S_{micro}) was determined by t-plot method.

3. Characteristics of the original AlNP and products of hydrothermal synthesis

Hydrothermal synthesis of AlNP is accompanied by the huge heat release. Therefore thermal conductivity and heat transfer capability of oxide-hydroxide film of AlNP particles play an important role when AlNP interacts with water [6]. Meanwhile, these processes are accompanied by mass transfer and the formation of solids with totally different morphology and composition when a certain temperature in reaction zone is achieved occur [6]. The heating of aluminium hydroxide results in the following reaction [9, 13]:

$$Al(OH)_3 \xrightarrow{150\,^\circ C} \beta\text{-}Al_2O_3\cdot 3H_2O \xrightarrow{250^\circ C} AlOOH \xrightarrow{400\,^\circ C} \gamma\text{-}Al_2O_3 \xrightarrow{700\,^\circ C}$$
$$\xrightarrow{700\,^\circ C} \gamma\text{-}Al_2O_3 \xrightarrow{900\,^\circ C} \delta\text{-}Al_2O_3 \xrightarrow{1300\,^\circ C} \alpha\text{-}Al_2O_3$$

According to TEM data, AlNP particles had spherical shape and represented metallic aluminium with FCC lattice coated with oxide cover with thickness of 7…15 nm (Figure 1a). AlNP particle size distribution is monomodal and lies within the range of 10…150 nm. The average particle size was 54 ± 10 nm (Figure 1b). AlNP particles agglomerate and agglomerations had the size of 200…500 nm as seen from the TEM image (Figure 1a).

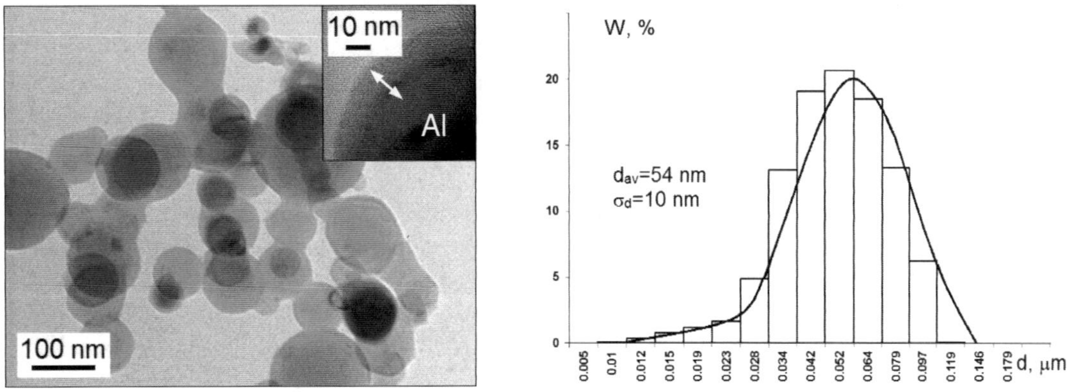

a) b)

Figure 1. Aluminum nanopowder: *a* – TEM image; *b* – particle size distribution

Hydrothermal synthesis of AlNP leaded to formation of a white friable powder. According to the X-Ray analysis, there were following products of hydrothermal synthesis: AlOOH (boehmite), $Al(OH)_3$ (gibbsite) and residual metallic Al (Table 1). The γ-AlOOH phase was mostly amorphous and have volume percent of 59 % (Figure 2b). The size of coherent scattering region (CSR) of the corresponding γ-AlOOH {020} plane D_{CSR} defined using diffraction line was equal to 1.6 nm. The gibbsite ($Al(OH)_3$) had good crystallinity (D_{CSR}=16.4 nm) with hexagonal lattice but its volume percent was negligibly small: ~1 % (Figure 2b). The residual aluminium had a crystalline phase (D_{CSR}=14 nm) with FCC lattice and its volume was 40 %. In [6, 10] is the aythors noted that the maximal suspension temperature obtained as the result of aluminium nanoparticles self-heating under hydrothermal synthesis achieved values of 92…94 °C. However, in the current work the maximal temperature of suspension during self-heating achieved 97 °C. The formation of phases γ-AlOOH and $Al(OH)_3$ as the result of the hydrothermal synthesis showed that in reaction zone there were possibly local overheatings with the temperature of at least 250 °C, that leaded to the formation of boehmite phase instead of gibbsite one even in the presence of multifold water excess. According to TEM data, the AlOH represented aggregates with the size of ~7.7 μm that consist of smaller particles (d_{av}=1 μm, σ_d=0.5 μm) of various shapes (Figure 2). The microstructure of synthesized AlOH was characterized by formed porous mesh structure with open pores (Figure 2b). In accordance with nitrogen low-temperature adsorption method data, the specific surface area of AlOH particles was 245 m^2/g and pore size – 4…17 nm.

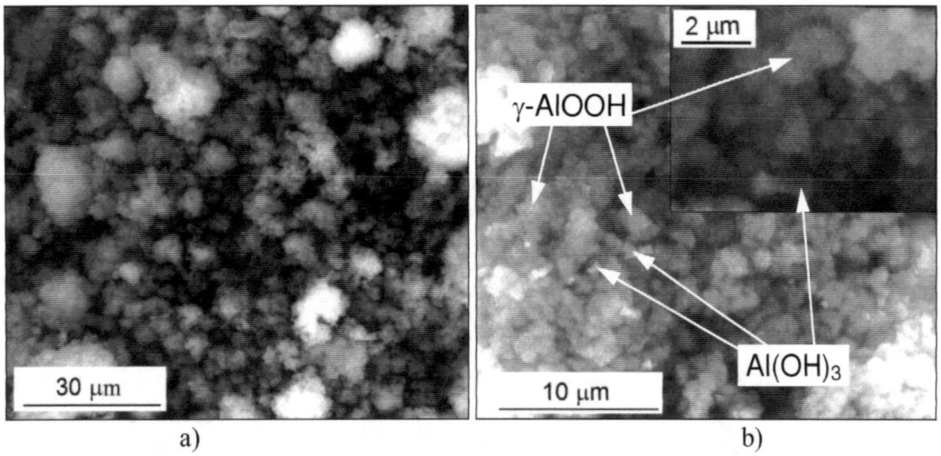

a) b)

Figure 2. The microstructure of AlOH particles after the hydrothermal synthesis of AlNP

4. The influence of thermal treatment on structural phase composition of AlOH

After the AlNP hydrothermal synthesis there were both adsorbed and structural waters in AlOH. The thermal treatment of the synthesized AlOH within the temperature range from 150 to 1300 °C was accompanied by the change of structural water content and phase composition. When heated up to 150 °C the synthesized AlOH loosed adsorbed water. These losses comprised ~3 wt. %. The phase composition of the AlOH heated up to 150 °C was equal to the one of the original AlOH (Figure 3). However, the phase volume percent and CSR size were different (Table 1). In spite of a slight rise of the γ-AlOOH phase CSR (D_{CSR}=1.9 nm), the phase stayed mostly amorphous and its volume comprised 49 % (*Sample Al-2* in *the* Table 1). After the thermal treatment the gibbsite ($Al(OH)_3$) and residual Al also had good crystallinity and their CSR rised in 3 and 1.5 times, respectively (Table 1).

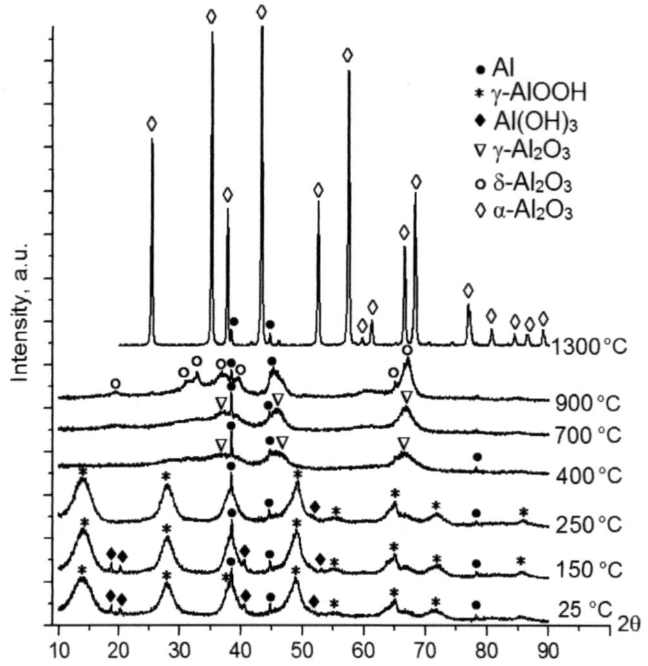

Figure 3. X-Ray diagrams for the products of hydrothermal synthesis after thermal treatment within the temperature range of 25…1300 °C.

Within the temperature range of 150…250 °C the structural water volume decreased to the value of ~ 4 wt. %. The basic phases of the *Al-3* sample were γ-AlOOH (boehmite) and residual metallic Al volume percent of which were 85 % and 15 %, respectively. Type, crystalline structure parameters and CSR of the γ-AlOOH phase stayed constant. As the temperature of thermal treatment raised up to the values of 400 °C and 700 °C, the water rating decreased by 9 and 3 wt.% respectively and the formation of γ-Al_2O_3 with FCC lattice was observed (*Samples Al-4, Al-5* in Figure 3 and Table 1). Alongside with the γ-Al_2O_3 phase in *Al-4* and *Al-5* samples there was residual Al with significantly fewer volume percent of 5 to 7 %. This was possibly due to the partial oxidation of the sample under the heat treatment in the air atmosphere. The γ-Al_2O_3 phase of the *Al-4* sample was mostly amorphous. The size of the CSR D_{CSR} defined by means of diffraction line corresponding to the {400} plane of the γ-Al_2O_3 structure had the value of 2.6 nm. As the temperature of thermal treatment rised up to 700 °C the CSR size of the γ-Al_2O_3 phase increased to the value of 10.5 nm. According to X-Ray data, the further rise of the treatment temperature up to the value of 900 °C resulted in the formation of δ-Al_2O_3

with tetragonal lattice (*Sample Al-6* in Table 1). The size of the CSR D_{CSR} defined by means of diffraction line corresponding to the {114} plane had the value of 4.6 nm that means that the δ-Al_2O_3 phase had microcrystalline structure (Figure 3 and Table 1).

Table 1. Phase composition of the products of hydrothermal synthesis after thermal treatment

Sample	*T*, °C	Phase	Type of crystal lattice, space group	Volume percent, %	CSR (D_{CSR}), nm
Al-1	25	Al	FCC, *Fm3m*	40	14
		γ-AlOOH	Orthorhombic, *Amam*	59	1.6
		$Al(OH)_3$	Hexagonal, *P-3*	~1	16.4
Al-2	150	Al	FCC, *Fm3m*	50	23.2
		γ-AlOOH	Orthorhombic, *Amam*	49	2
		$Al(OH)_3$	Hexagonal, *P-3*	~1	49.9
Al-3	250	Al	FCC, *Fm3m*	15	23
		γ-AlOOH	Orthorhombic, *Amam*	85	2
Al-4	400	Al	FCC, *Fm3m*	7	22.3
		γ-Al_2O_3	FCC, *Fd3m*	93	2.6
Al-5	700	Al	FCC, *Fm3m*	5	22.3
		γ-Al_2O_3	FCC, *Fd3m*	95	10.5
Al-6	900	Al	FCC, *Fm3m*	4	22
		δ-Al_2O_3	Tetragonal, *I*	96	4.6
Al-7	1300	Al	FCC, *Fm3m*	3	21.7
		α-Al_2O_3	Rhombohedral, $R\bar{3}c$	97	118

In the process of heating AlOH up to 1300 °C the α-Al_2O_3 crystalline phase with rhombohedral lattice was formed. According to X-Ray data, all samples contained residual metallic Al within the temperature range from 150 to 1300 °C. Herewith, as the temperature raised from 250 °C to 1300 °C the volume percent of Al dramatically drops from 15 down to 3 % (Table 1). This means that Al oxidizes when sample is heated in the air atmosphere. The presence of metallic aluminium in the sample heated up to 1300 °C indicated the formation of stable structures of core-shell type in which aluminium cores were covered by dense oxide-aluminium shell that prevented its further oxidation.

5. The influence of thermal treatment on specific surface area and pore size of AlOH

The porous AlOH structure characteristics before and after thermal treatment were investigated by means of low-temperature nitrogen adsorption (Table 2 and Figure 4). For the *Al-1*, *Al-2* and *Al-3* samples the presence of hysteresis loop at isotherms of nitrogen adsorption-desorption reveals their mesoporous structure (Figure 4a). The pore size distribution for the *Al-1*, *Al-2* and *Al-3* samples was monomodal (Figure 4b) with the average pore size of ~9 nm, S_{BET}=243...297 m^2/g, S_{micro}=33...41 m^2/g (Table 2).

Table 2. The textural characteristics of AlOH before and after thermal treatment

Sample	*T*, °C	Specific surface area, m^2/g		Pore volume, cm^3/g		Average pore size, nm
		S_{BET}	S_{micro}	V_{pore}	V_{micro}	
Al-1	25	243	33	0.53	0.016	8.7
Al-2	150	251	44	0.56	0.022	8.8
Al-3	250	297	41	0.65	0.019	8.6
Al-4	400	328	23	0.78	0.009	9.4
Al-5	700	256	13	0.76	0.005	11.8
Al-6	900	152	13	0.67	0.005	17.7

Al-7	1300	6.5	2	0.23	0.001	14.7

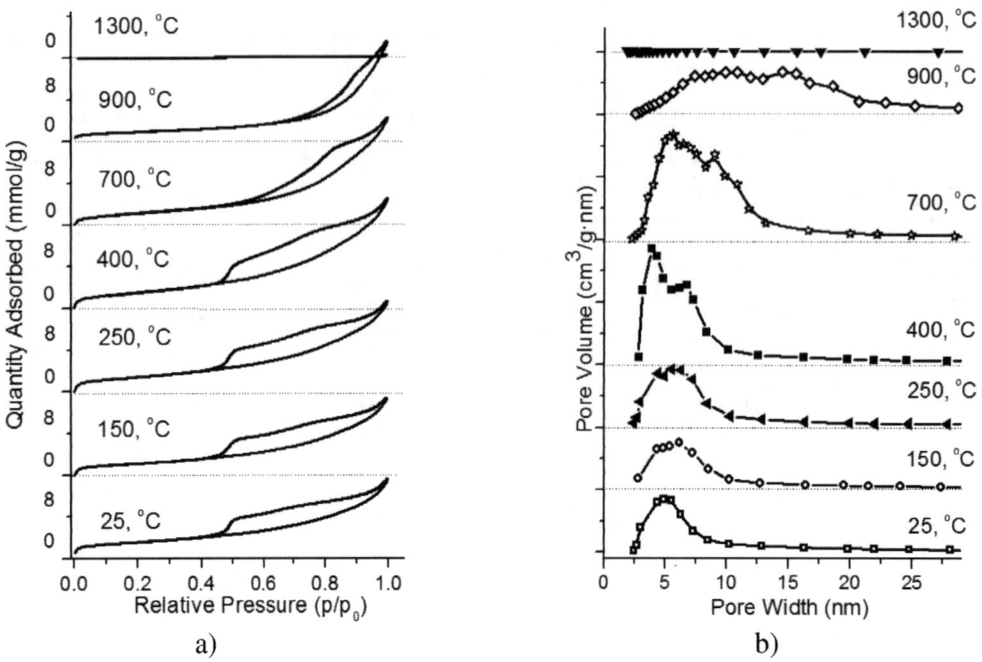

Figure 4. The textural characteristics of AlOH before and after thermal treatment: *a* – isotherms of N_2 adsorption-desorption; *b* – pore size distributions.

The rise of the AlOH thermal treatment temperature up to 400 °C for the *Al-4* sample resulted in changing of hysteresis loop shape marked by dual loop occurring in different domains of relative pressure (Figure 4a). For this sample we observed bimodal pore size distribution with the average pore size of ~9 nm (Figure 4b and Table 2). Although the values of S_{micro} and V_{micro} for the *Al-4* sample decreased by half comparing to the *Al-1*, *Al-2* and *Al-3* samples, the overall specific surface area ($S_{specific}$) and overall pore volume (V_{pores}) raised to the values of 328 m^2/g and 0.78 cm^3/g, respectively. The further treatment temperature increased up to 700 and 900 °C (*Al-5* and *Al-6* samples) also results in bimodal pore size distribution. However, the peak points of distribution curve shifted to the domain of pores with large diameter that leads to average pore size growth up to 12 nm for the *Al-5* sample and 18 nm for the *Al-6* sample. The overall specific surface area (S_{BET}) for the *Al-5* and *Al-6* samples went significantly down to the values of 256 m^2/g and 152 m^2/g, respectively. The AlOH heating up to the temperature of 1300 °C leaded to the porous structure disrupting confirmed by lack of hysteresis loop at the isotherms of nitrogen adsorption-desorption and the corresponding pore size distribution (Figure 4). The temperature growth up to 1300 °C caused the degradation of all textural characteristics due to domination of the α-Al_2O_3 phase in the sample structure (Table 2). Therefore the *Al-4* sample (*T*=400 °C) consisting of amorphous γ-AlOOH and residual Al had the most optimal textural characteristics for producing aluminium oxide supports.

6. Conclusion

This work investigates the influence of thermal treatment within the temperature range of 150-1300 °C on phase composition, textural characteristics and acid-base properties of aluminium oxide-hydroxides obtained through hydrothermal synthesis of aluminium nanopowder. It was shown that within the temperature range of 150…250 °C the $Al(OH)_3$ passes into γ-AlOOH, adsorbed and structural waters are eliminated and the porous structure is formed with the average pore size of ~9 nm and specific surface area of 243…297 m^2/g. It was established that under the AlOH heating up to 400 °C the amorphous γ-Al_2O_3 is formed that has the overall specific surface area and pore volume of 328 m^2/g and 0.78 cm^3/g respectively. Such structure has low values of S_{micro} и V_{micro} (23 m^2/g и 0.009 cm^3/g respectively) that is very important for developing catalyst supports. It was shown that further rise of treatment temperature (up to 900…1300 °C) results in increasing of pore volume and size and the formation of high-temperature aluminium oxides (δ-Al_2O_3, α-Al_2O_3) that have small specific surface area. It should be noted that in all samples treated within the temperature range of 150…1300 °C the size of particles stays small and residual metallic Al is still present. Therefore the most interesting substance for developing catalyst supports is γ-Al_2O_3 obtained at the temperature of 250…400 °C that has optimal structural and textural characteristics.

Acknowledgments
This work was supported by the *Russian Federal Target Program* "Research and development in the priority fields of scientific and technological complex of Russia in 2014–2020" (State Contract No. 14.578.21.0028).

References

[1] Galanov S, Sidorova O, Gryaznova E and Shiyan L 2014 *The Bulletin of the Tomsk Polytechnic University* **324(3)** 88
[2] Pahomov N, Tihov S and Bespalko Yu Russ. 2005 *J. of the Critical Technol. Membr.* **4(28)** 80
[3] Lerner M, Svarovskaya N, Glazkova E, Lozhkomoev A, Kirilova N 2006 *Rus. J. of the Phys. Mesomech.* **9** 201-204
[4] Amelkovich Yu and Ilyin A 2008 *The Bulletin of the Tomsk Polytechnic University* **313**(3) 14
[5] Bousquet C, Cansell F, Elissalde C, Maglione M and Aymonier C 2008 Available at: http://www.isasf.net/fileadmin/files/Docs/Barcelona/ISASF%202008/PDF/Key%20notes/KN_M_5.pdf
[6] Alemi A, Hosseinpour Z, Dolatyari M and Bakhtiari A 2012 *J. Solid State Phys.* **6** 1264
[7] Godymchuk A, Ilyin A and Astankova A 2007 *The Bulletin of the Tomsk Polytechnic University* **310**(1) 102
[8] Volkova G, Ivanov V and Kyharenko O 2005 *Russian Jour Chem. for Sustainable Development* **13** 427
[9] Al'myasheva O, Korytkova E, Maslov A and Gusarov V 2005 *J. of Inorg. Mater.* **41**(5) 460
[10] Zhu H, Gao X, Song D, Ringer S and Xi Y 2005 *J. Micropor. Mesopor. Mater.* **85** 226
[11] Hayashi H and Hakuta Y 2010 *Jour Materials* **3** 3794-3817
[12] Astankova A, Godymchuk A and Ilyin A 2005 *Russian Federation Patent* 2338684
[13] Ilyin A and Gromov A 1999 Oxidation of ultrafine powders of aluminum and boron, Tomsk Polytechnic university Publishing, 150

Antibacterial Properties of Copper Nanoparticle Dispersions: Influence of Synthesis Conditions and Physicochemical Characteristics

A Godymchuk[1,2,a], G Frolov[2], A Gusev[2,3], O Zakharova[3], E Yunda[1], D Kuznetsov[2] and E Kolesnikov[2]

[1] Tomsk Polytechnic University, 30, Lenina ave., Tomsk, 634050, Russia

[2] National Research Technological University "MISIS", 2, Leninsky ave., Moscow, 119049, Russia

[3]Tambov State University named after G.R. Derzhavin, Internatsionalnaya str., 33, Tambov, 392000, Russia

[a]E-mail: godymchuk@tpu.ru

Abstract. The production of bactericidal plasters, bandages and medicines with the inclusion of copper nanoparticles and copper ions may have a great potential in terms of their biomedical application. The work considers the influence of the synthesis conditions, size, aggregation status, and charge of nanoparticles in aqueous solutions as well as the type of microorganisms to the antibacterial properties of water suspensions of electroexplosive copper nanoparticles in the conditions *in vitro* in relation to strains *Escherichia coli*, *Pseudomonas aeruginosa*, *Staphylococcus aureus*, and *Bacillus cereus*. Water dispersions of copper nanoparticles were shown to inhibit the growth of test cells for both G+ and G- microbacteria but the degree of such an influence strongly depended on the type of a test strain. The authors have demonstrated that use of deeply purified water and alcohol-containing stabilizers at the synthesis of nanoparticles via metals electric erosion in the liquid prevents the copper nanoparticles coagulation and significantly influences on their physicochemical characteristics and, consequently, antibacterial properties.

1. Introduction

The treatment and cleaning of medical instruments and devices, water purification, food industry are closely connected with the development of effective, environmentally friendly, available and low cost bactericide compositions for pathogenic microorganism control. Due to a wide presence in nature, implementation of different functions within the majority of living organisms, relatively low cost and environmental safety, copper compounds (Cu) have a high potential for their application as antibacterial agents being capable to replace silver and composites of different precious metals [1].

Thus, it has been demonstrated that Cu nanosized particles are able to show antimicrobial properties in relation to a wide range of microorganisms including pathogenic bacteria [2]. On the one hand, according to [3], Cu nanoparticles are hypotoxic. On the other hand, they show a high antibacterial effect in relation to the cells of test cultures of G+ and G- microbacteria, which makes reasonable their application as part of wound-healing preparations. At the same time, Cu nanoparticles are demonstrated to have high efficiency for application in bactericide plasters or bandages due to high antimicrobial activity to pathogenic microorganisms and the illegible sensibility of human tissues to copper compounds [4,5]. Other authors have shown that Cu nanoparticles may significantly increase the antibacterial activity of sorbents exposed to *Escherichia coli* cultures owing to the copper capability to disrupt metabolism of a microbial cell while interacting with microorganisms [6]. In other

Content from this work may be used under the terms of the Creative Commons Attribution 3.0 licence. Any further distribution of this work must maintain attribution to the author(s) and the title of the work, journal citation and DOI.
Published under licence by IOP Publishing Ltd

words, the production of bactericidal plasters, bandages and medicines with the inclusion of copper nanoparticles and copper ions may have a great potential in terms of their biomedical application. However, the lack of experimental data in the literature on antibacterial properties linked to physicochemical parameters of nanoparticles in water suspensions prevents forecasting their antibacterial activity and, respectively, preparing widely used antibacterial suspensions.

The purpose of this work was to evaluate the influence of synthesis conditions and physicochemical characteristics on antimicrobial properties of water suspensions of copper nanoparticles in relation with gram-positive and gram-negative bacteria.

2. Materials and methods

Nanoparticle dispersions. Dispersions of Cu nanoparticles were produced via electro-erosion synthesis (Table 1). The process of electric erosion of copper cathodes was implemented according to approach described in [7]. The arc chamber was filled with liquid, and electric impulses of the capacity of 6.5 kW were fed to the copper electrodes. As a result of appearance of electric sparkles in the places of granule contact destruction, melting and evaporation of the surface took place which accompanied by water saturation with the products of electric erosion and formation of colloidal solution of copper in the liquid. The given method allowed receiving stable dispersions of nanoparticles, stabilization of which occurred due to the formation of hydrate (solvent) multilayer shells around the particle of a solid phase. Moreover, this method is defined with a high environmental friendliness, expressivity and possibility to control the size of particles due to the control of the explosion parameters.

Examination of physicochemical properties of nanoparticles. Information about dislocation structure of nanoparticles was obtained by means of transmission electronic microscopy (TEM) using a microscope JEM 2100 (Jeol, Japan). In this experiment studied suspensions were dropped on TEM copper nets with amorphous carbon substrate. All the measurements were conducted with a speeding voltage of 200 kV; cathode LaB6 was used as a source of electrons.

Table 1. Characteristics of Cu nanoparticle dispersions

Dispersion	WD-Cu	WDF-Cu	WAD-Cu
Name of the dispersion	Water dispersion of copper nanoparticles	Water dispersion of copper nanoparticles additionally purified with filtration	Water-alcoholic dispersion of copper nanoparticles
Characteristics (TEM-analysis)	Particles sizes: 1…40 nm Sizes of aggregates: 10…100 nm Particle composition: Cu_2O	Particle sizes: 1…40 nm Aggregates sizes: 100…1000 nm Particles composition: Cu_2O	Particle sizes: 3…10 nm Aggregates sizes: 100…1000 nm Particles composition: Cu
Notes	Synthesis of nanoparticles was implemented in distilled water ($pH = 6.11 \pm 0.2$, conductivity – 0.2 μS)	Synthesis of nanoparticles was implemented in distilled water ($pH = 6.11 \pm 0.2$, conductivity – 0.2 μS), additionally purified with reverse-osmosis filtration	Synthesis of nanoparticles was implemented in water solution of ethanol [8], synthesized suspension of copper nanoparticles was diluted in 100 times with distilled water ($pH = 6.11 \pm 0.2$, Conductivity – 0.2 μS)

Sedimentation stability of particle suspensions was evaluated with the spectrophotometry method measuring the alteration of a light transmission coefficient (%). Synthesized dispersions were treated with ultrasonication for 5 min and poured into quartz cuvettes (width – 1 cm, volume – 4 ml). The spectrum of distilled water ($pH = 6.11 \pm 0.2$, conductivity – 0.2 μS, electric distillatory DE-4, Russia) was used as a standard to compare the obtained spectra. Coefficient of light transmission was measured using a spectrophotometer Helios Alpha (Thermo Scientific, USA) with the wavelength of 540 nm. The measurements were conducted every 60 minutes during 24 hours.

To control the aggregation stability particle size distribution and zeta-potential in the dispersions were determined using dynamic light scattering technique (Nano Zetasizer, Malvern, USA). To run measurements a general-purpose capillary U-shape polystyrene cuvette was filled with 1 ml of dispersion avoiding the formation of bubbles. Triplicate samples of each nanopowder were measured at 25 °C (He-Ne laser, laser capacity – 4 mW, wavelength – 633 nm).

The change in the composition of the particle surface in water dispersions was defined with the IR-Fourier spectroscopy. The particle dispersions were centrifuged for 60 minutes with the speed of 12,000 r/min and dried up in the air within 72 hours. The received films were examined on the Fourier Spectrometer Nicolet 5700. Spectral resolution of the device was 0.09 cm^{-1}. The received spectra of reflection were compared with the known spectra of the chemical substances [9] and water.

Nutritional media preparation for microorganisms cultivation. Strains of the following bacteria were used to evaluate antibacterial activity: coliform bacterium *E. coli*, *Pseudomonas aeruginosa P. aeruginosa*, Stap. aureus *S. aureus*, soil bacterium *B. cereus*. According to various references, e.g. [10,11], the aforementioned bacteria are the most frequently used test-objects for studying nanoparticle ecotoxicity. Once entered the body, some of the listed bacteria propagate quickly and become the cause of infectious inflammatory diseases as well as contribute to the development of purulent complications in the post-surgical period.

Solid nutritional medium (agar) was made on the base of Hottinger broth (protein base used for nutritional media preparation for microorganisms. PanEco, Moscow.). Combination of the medium components provides the necessary nutritional elements for the growth of cultures in the form of certain colonies on the solid medium surface. Hottinger agar was melted on a boiling water bath and cooled to 45…50 °C. Then, 25 ml of agar was dispensed into sterile 100 mm Petrie dishes (RUSHIMSET, Tambov) with the resulting layer of 4.0 ± 0.5 mm. After setting the Hottinger agar was dried at 37 ± 1 °C for 45-60 min. The prepared agar was kept for not more than 10 days at 2…8 °C (thermostat TC1/80, Russia). pH of the agar was 7.3 ± 0.2 [12].

Preparation of the test strain suspensions (inoculation). Two days prior to carrying out the experiment the test culture was reinoculated from the storage medium onto slant nutritional agar in test tubes. The next day the obtained agar culture was used to produce suspension of the test strain employing turbidity standard. Sterile dilutant with the concentration of 10^9 kl/ml was used, and tenfold serial dilutions (till 8th dilution inclusive) were prepared.

To control dilution 0.1 ml of suspension from 6th and 7th dilutions of culture was plated to the dishes with nutritional agar by method of direct surface inoculation. Three such inoculations were made out of each culture. After inoculation test tubes with culture were immediately placed in a refrigerator. Dishes with cultures were incubated in the thermostat at 37 ±1 °C within 18…24 hours.

On the day of the experiment the average number of colonies grown in three dishes was calculated for each series of inoculations.

Before the measurement the inoculate was mixed thoroughly and, then, 2 ± 0.2 ml was applied with a pipette to the surface of a Petrie dish with nutritional medium and evenly distributed over the surface with roll motion, the excess was removed with the pipette. Petrie dishes filled with nutritional medium and test strain were installed onto a horizontal surface. Before the inoculation the absence of condensate on the internal surfaces of the caps was controlled.

The laboratory glassware was treated with 3-%water solution of phenol or 2% chloramine solution for disinfection [13].

Examination of antibacterial properties of nanoparticles. Disk diffusion test was used to assess antibacterial properties of the studied dispersions. It represents the measurement of the inhibition zone around the paper carrier of antibacterial preparation. The inhibition zone forms as a result of the preparation diffusion from the carrier into the media. Within certain limits the inhibition zone is reversely proportional to the minimum suppressing concentration (MSC). The result of the measurement is the definition of the microorganism as one from the sensitivity categories (sensitive, intermediate or resistant).

For the experiment small batches of disks (wafers) were made from filtering paper with the following properties: white band, filtering capability of not more than 45 seconds, neutral aqueous extract pH, ash content of the filter – 0.00042…0.00062 at the temperature not higher than 18 °C [14]. Wafers of 5 x 5 mm were made before the experiment under sterile conditions. They were soaked 10 min in the tested aqueous dispersion and then dried up in a suspended state to remove the excess liquid. Laboratory tools were thoroughly disinfected. Laboratory glassware was washed twice in 2% HNO_3 solution and rinsed 7 times with distilled water. All the surfaces including inner surfaces of air hoods and tables were treated with disinfecting compositions of sodium bicarbonate. After the experiment the wafers were disposed.

Incubation of nanoparticle dispersions in test strains. Application of the discs was implemented using sterile tweezers. The distance from the disc to the edge of the dish and between disks was 15...20 mm. Thus, one dish with a diameter of 100 mm had not more than 6 discs saturated with particle suspensions. The discs were thoroughly pressed to the agar by tweezers so that the discs evenly contacted the surface of agar. In the center of the plate a paper disc with gentamicin (control) was located. After application of the discs the Petrie dishes were placed in the thermostat (thermostat TC 1/80-SPU, Russia) upside down and incubated at the temperature of 35 ± 0.5 °C during 18 hours. As soon as the process of incubation was finished the dishes were placed upside down to the dark mat surface with the light falling down onto them under the angle of 45 °. The diameter of the zones of growth inhibition (DZGI) of bacteria around the discs was measured with a ruler. The accuracy up to 1 mm was checked with trammel.

While measuring the DZGI parameter the zone of a full suppression of the visible growth was taken into consideration without taking into account very little colonies revealed within the zone of growth inhibition at specific conditions and hardly noticed deposit at the edge of the zone.

3. Results and discussion

Characteristics of nanoparticles in dispersions. The Table 2 shows the average size and size distribution both for particles and aggregates of Cu dispersions based on the TEM-images analysis (figure 1a, 1b, and 1c). It can be seen that particles connect to each other mostly due to colloidal interparticle interaction and rarely with phase contacts that confirms high aggregative stability of synthesized copper suspensions.

Table 2. Morphological and dispersity characteristics of the dispersions (TEM data)

Dispersion	Particle size, nm	Average particle size, nm	Aggregate size, nm	Particle shape
WD-Cu	10 ... 40	38	70 ... 150	spherical
WDF-Cu	10 ... 40	26	100 ... 200	spherical
WAD-Cu	10 ... 20	20	100 ... 1000	spherical

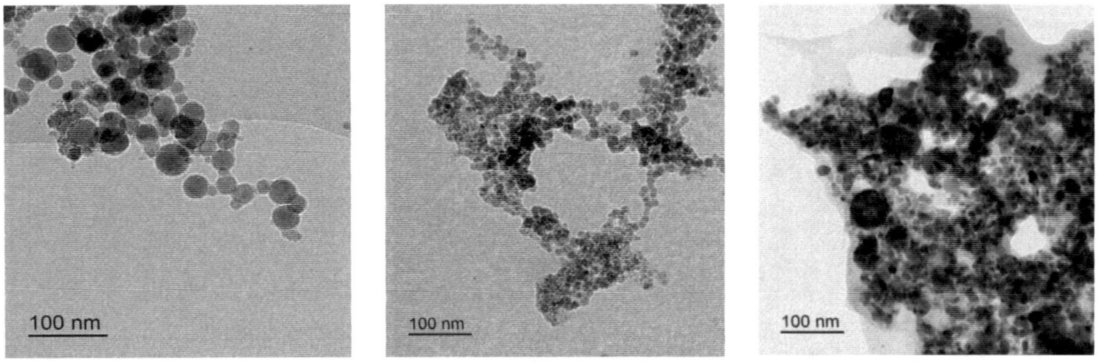

a) b) c)

Figure 1. TEM-images of the dispersions: WD-Cu (a), WDF-Cu (b), WAD-Cu (c)

Since zeta-potential value is proportionate to the aggregate stability of sols [15], WDF-Cu-dispersion has the highest stability (Table 3). The suspension was synthesized in a disperse phase with a higher degree of purification and, respectively, with a lesser ion power. Adding electrolytes is known to contribute to coagulation of hydrophobic colloids that causes surface charge neutralization. This fact explains lesser stability of the WD-Cu, received at the explosion of metal in distilled water. For nanoparticles of the WAD-Cu, synthesizing in the ethanol solution contributed to neutralization of the particle charge (Table 3).

Table 3. Average size and zeta-potential of the dispersions (DLS data)

Dispersion	Average size of agglomerates, nm	Zeta-potential of the particles, mV
WD-Cu	4780.2 ± 120.1	−12.00 ± 2.8
WDF-Cu	279.3 ± 22.3	−39.80 ± 3.1
WAD-Cu	435.0 ± 16.4	−1.74 ± 0.5

Evaluation of nanoparticles stability in suspensions. The figure 2 demonstrates a change of light transmission coefficient (%) of the synthesized water dispersions. WD-Cu-dispersion obtained at the explosion of copper granules in the distilled water had a minimal stability: it had the highest light transmission in the commence time (40 %), the particles were completely deposited within 24 hours. Whereas WDF-Cu suspension (synthesis in the additionally purified water) was characterized with a low light transmission (10 %), and sedimentation of particles started only in 15 hours.

It was experimentally proven that WAD-Cu-dispersion had the highest stability: change of the light transmission coefficient within 24 hours was 7 %. The most probable is chemisorptive stabilization by functional groups of ethanol. Due to the formation of stable micelles the aggregate stability of such a hydrosol may be kept for a long time.

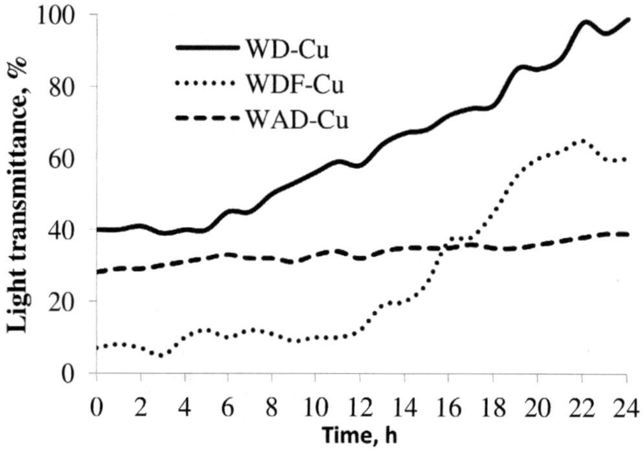

Figure 2. 24-h change of the light transmission coefficient of the dispersions

According to the experimental data a peak of the main fluctuations of hydroxyl group in the area of 3250 nm in the IR–spectra of the dispersions is presented (Figure 3). Thus, it was probably the formation of hydroxides of metals on the particle surface, which were not separated by centrifuging at a high speed.

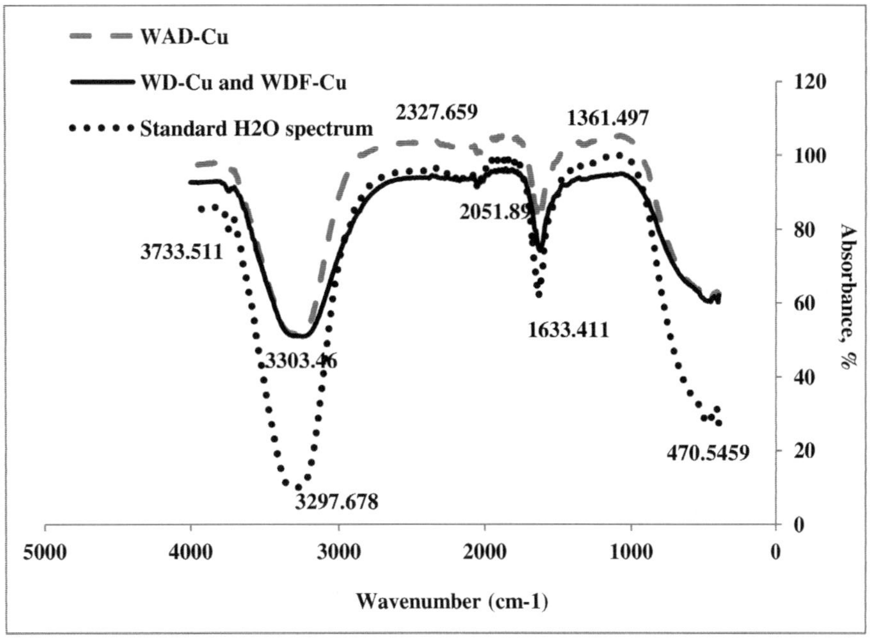

Figure 3. IR-spectra of the dispersions

Apparently, in suspension saturated with oxygen a partial dilution of copper nanoparticles is possible with formation of hydroxide in accordance with the data given in the article [16].

Antibacterial properties of copper nanoparticles.

Influence of nanoparticles dispersions on gram-negative microorganisms. For the line of G–-bacteria a tendency to a toxic action of the synthesized nanoparticles dispersions weakly depended on the type of the microorganism (Table 4). Thus, WD-Cu-dispersion was characterized with the highest DZGI value. The received result may be explained by the bigger size of dispersion particles (> 4 μm). The more the size of the particles was the more probable was the formation of cavities, hollows, and cracks between the disc penetrated with antibacterial particles and the surface of bacteria containing agar. As a result, it was quite possible to suppose the simplification of the diffusion of the smaller particles out of the limits of the disc that contributed to the increase of the DZGI (Figure 4).

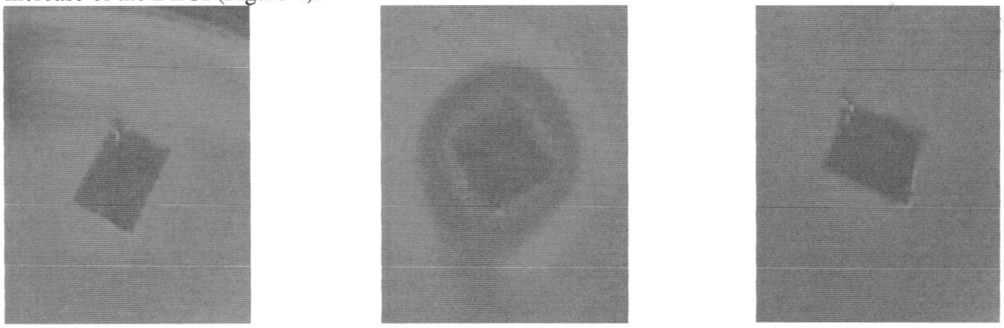

a) b) c)

Figure 4. Photos of the visual control of inhibiting zones of the *E. coli* C 600 exposed to WD-Cu (a), WFD-Cu (b), WAD-Cu (c)

Influence of nanoparticles dispersions on gram-positive microorganisms. It is known that the main difference of G–-bacteria from G+-ones is the presence for the first and absence for the second of the outer membrane that prevents penetration of antibodies inside the cell. That is why it was considered that due to its more capacious and impenetrable cell wall, G–-bacteria were more resistant to antibodies than G+-ones. We logically supposed that action of nanoparticles for G+-microorganisms would be the opposite. However, at the incubation of the cells in the dispersions there was no identical tendency: toxicity of nanoparticles depended not only on the method of their preparation but also on the type of the test organism.

We revealed that the G+ wax-like stick *B. Cereus* was instable to a larger extent in relation with WDF-Cu-dispersions than to WD-Cu. The increase of toxic effect may be related to the particle size and aggregation degree of the synthesized dispersions: the particle with less size are characterized with more probable migration activity, and, consequently, the probability of its penetration and impact on the cell may be increased. At the same time experimental results showed that the staphylococcus *S. aureus* was more sensitive to suspensions with a larger disperse phase as in WD-Cu-dispersion.

On the other hand, it was obvious that the maximal values of DZGI (4.2 mm) is typical for the WD-Cu-dispersions having the maximal size of the disperse phase (> 4 μm, Table 3). At that, this result was found for both G– (*E. coli*) and G+-bacteria (*B. Cereus*).

Thus, the studied Cu nanoparticles dispersions showed bactericidal action on all the test strains but the degree of such an influence depended on the method of preparation of the nanoparticles dispersion and the type of a test strain (Table 4).

Table 4. Diameter of diffusional zone of the cultures growth inhibition (DZGI) exposed to nanoparticles dispersions

Dispersion	Diameter of diffusional zone of growth inhibition of the test strain, mm			
	E. coli	*P. aeruginosa*	*S. aureus*	*B. cereus*
WD-Cu	4.2	3.0	3.1	Only below the disk
WDF-Cu	Only below the disk	Only below the disk	Only below the disk	4.4
WAD-Cu	Only below the disk	Only below the disk	Only below the disk	1.1

When describing the mechanism of toxicity of water nanoparticle dispersions of copper and its compounds it is important to note that copper-containing preparations cause breakage of ferments and disturbance of metabolism of microorganisms, i.e. they act as strong ferment poisons even in small concentrations (so called “oligodynamic action”). Either in the form of salts ($CuCl_2$), or in the form of organic compounds (such as a p-hydrozimercurybenzoat) cooper connects SH-groups and, thus, dramatically changes tertiary and quaternary structure of the ferment proteins.

Speaking about the toxicity of water-alcohol suspensions it is to be said that in addition to the action of copper the presence of alcohol molecules in suspension may not only cause breakages of the surface structures or layers of the cell, but also a coagulation of proteins. Consequently, presence of alcohol provokes bactericide effect of the suspensions.

4. Conclusion

The conducted studies allowed receiving dependencies which describe the link between the synthesis conditions, physicochemical characteristics and toxic properties of copper nanoparticles in water suspensions in the conditions *in vitro* in relation to strains *Escherichia coli*, *Pseudomonas aeruginosa*, *Staphylococcus aureus*, and *Bacillus cereus*. Water dispersions of copper nanoparticles were shown to inhibit the growth of test cells for both G+ and G- microbacteria but the degree of such an influence strongly depended on the type of a test strain. The authors have demonstrated that use of deeply purified water and alcohol-containing stabilizers at the synthesis of nanoparticles via metals electric erosion in the liquid prevents the copper nanoparticles coagulation and significantly influences on their physicochemical characteristics and, consequently, antibacterial properties.

Experimental data obtained using disc-diffusion method of strains sensitivity in relation to water-base dispersions of copper nanoparticles allowed making two main conclusions. First, when choosing antimicrobial agent for such cells a defining factor will be the dispersity of the synthesized suspensions. Secondly, bactericidal activity of copper nanoparticles increases due to a partial metal release and emissions of copper ions to the surface with their further diffusion into a cell which actively leads to a toxic effect (suppression of the strain growth).

Acknowledgments

The project was supported by the Ministry of Science and Education of the Russian Federation (project RFMEFI57814X0080).

References

[1] Veerapandian MS, Sadhasivam S, Choi J, Yun K 2012 *Chem Eng J* **209** 558-567
[2] Gunawan C, Teoh WY, Marquis CP, Amal R 2011 *ACS Nano* **5** 7214-7225
[3] Rackmetova AA 2011 *Study of Biological Activity of Cu Nanoparticles of Different Size and Phas Composition* PhD-thesis (Moscow: Moscow University)
[4] Hostynek JJ, Maibach HI 2003 *Rev on Env Health* **18**(3) 153-183
[5] Borkow G, Gabbay J 2004 *FASEB Journal* **18** 1728-1730
[6] Timofeev SS 2013 *Proc of the Xth Int Conf "Prospects of Fundemental Scinces"*, 23-26 Aprile, 2013. Available at: http://science-persp.tpu.ru/Previous%20Materials/Konf_2013.pdf
[7] Lopat'ko KG, Aftandilyants EG, Zaulichny YaV, Karpets MV 2010 *Proc of the Institute of Mater Sci Problems. Academy of Science of the Ukraine. Seria: Comosition, Layer, and Gradient Materials and Coatings. Electrical Contacts and Electrodes* **5** 232-243
[8] GOST RF 363979. *Etanol*. Technical Requirements
[9] *NIST Standart Reference Database*, available at: http://webbook.nist.gov/chemistry/
[10] He S, Feng Y, Gu N, Zhang Y, Lin X 2011 *Env Poll* **159** 3468-3473
[11] Baek Y, An Y 2011 *Science of the Tot Environ* **409** 1603-1608
[12] Practical Guideline RF 4.2.1890-04. Determination of the Microorganisms Sensibility to Antibacterial Substances (in Russian)
[13] Practical Guideline RF 2.1.4.1057-01. Organization of Inside Monitoring of the Water Quality for Sanitary-Biological Research (in Russian)
[14] GOST RF 12026-76. Paper for Laboratory Filtration
[15] Handy RD, von der Kammer F, Lead JR, Hassellov M, Owen R, Crane M 2008 *Ecotoxicology* **17** 287-314
[16] Godymchuk A, Savel`ev G, Gorbatenko D 2010 *Rus J of Gen Chem* **80**(5) 881-888

The silver ions contribution into the cytotoxic activity of silver and silver halides nanoparticles

A I Klimov[1], P M Zherebin[1], A A Gusev[2], A A Kudrinskiy[1, 3], Y A Krutyakov[1, 3]
[1] Department of Chemistry, M.V. Lomonosov Moscow State University, 1-3, Leninskie Gory, Moscow, 119991, Russia, tel./fax +7(495)9393549
[2] G.R. Derzhavin Tambov State University, 33, Internatsionalnaya str., Tambov, 392000, Russia
[3] National Research Centre "Kurchatov Institute", 1, Akademika Kurchatova pl., Moscow, 123182, Russia

E-mail: kudrinskii@petrol.chem.msu.ru

Abstract. The biocidal action of silver nanoparticles capped with sodium citrate and silver halides nanoparticles capped with non-ionic surfactant polyoxyethylene(20)sorbitan monooleate (Tween 80®) against yeast cells *Saccharomyces cerevisiae* was compared to the effect produced by silver nitrate and studied through the measurement of cell loss and kinetics of K^+ efflux from the cells. The cytotoxicity of the obtained colloids was strongly correlated with silver ion content in the dispersions. The results clearly indicated that silver and silver halides nanoparticles destroyed yeast cells through the intermediate producing of silver ions either by dissolving of salts or by oxidation of silver.

1. Introduction

At the beginning of the antibiotic era in the early 1930s the application of silver-based compounds in medical practice was of rare occurrence and the use of silver salts in the treatment of patients strictly diminished [1]. However the interest in silver has never entirely disappeared and nowadays it revived again with the development of nanosilver containing materials and composites mainly due to the rediscovery of high antibacterial activity of colloidal silver particles [2, 3]. The increased interest in nanosilver during the recent 10 years is mainly explained by the appearance and spread of pathogenic microorganisms with multiple drug resistance, including the resistance to antibiotics of the last generation. At the present time silver in the form of nanoparticles (NPs) has the highest degree of commercialization among all nanomaterials and became a promising alternative to traditionally used antimicrobial agents in the treatment of delayed wounds [4-6], burns [7, 8], ulcers, bladder inflammation and other diseases.

External use of silver-based antiseptic agents is known to be safe enough. In order to be toxic externally silver needs to be absorbed into the body in sufficient amounts [7, 8]. The most common change seen in the skin in response to percutaneous silver absorption is argyria – blue or bluish-grey skin discoloration [11]. At the same time it is well known that argyria is not serious disease and has only negative cosmetic effects.

Nevertheless in order to prevent possible side-effects in humans treated with nanosilver containing medications we need to obtain more accurate scientific information about colloidal silver action on

Content from this work may be used under the terms of the Creative Commons Attribution 3.0 licence. Any further distribution of this work must maintain attribution to the author(s) and the title of the work, journal citation and DOI.
Published under licence by IOP Publishing Ltd

living cells. Nowadays there is a lack of fundamental works aiming at the systematical study of the silver NPs action against living cells – neither bacterial nor mammalian.

This article is another step in the process of understanding of the mode of the silver NP's action on living cells. Three main concepts of antibacterial and antiviral action of nanosilver are commonly accepted and widely discussed. On the one hand, silver NPs themselves can penetrate through cell membrane and interact with organelles, thus leading to disorder of the cell functioning. This mechanism was well-proven with respect to antibacterial and antiviral activity of nanosilver [12, 13]. On the other hand, silver NPs can easily generate Ag^+-ions via oxidation of surface silver atoms by different oxidants dissolved in cytoplasm. So, the silver particles act as a buffer which maintains concentration of Ag^+ in the surrounding media at the approximately constant level. According to the third concept the bactericidal action of silver NPs is determined by the intermediates formed during the silver oxidation such as peroxides and free radicals.

Actually, the observed antibacterial activity of the nanosilver is a superposition of those three ways. The contribution of each way depends on many factors; that's why the detailed investigation of the mechanism of nanosilver action is necessary for the rational design of silver NPs-based pharmaceutical compositions. The aim of present research was to determine the contribution of silver NPs themselves and silver ions of NPs' origin into the activity of silver and silver halides nanoparticles against *Saccharomyces cerevisiae* cells.

2. Chemicals

Polyoxyethylene(20)sorbitan monooleate (Tween-80®) was purchased from Acros Organics and used without further purification, silver nitrate (99.9+%, Sigma-Aldrich), sodium citrate dihydrate (99+%, Sigma-Aldrich), sodium borohydride (Lancaster, 98+%), sodium bromide (99+%, Sigma-Aldrich), sodium iodide (99.9+%, Sigma-Aldrich), sodium chloride (99+%, Sigma-Aldrich), were of analytical grade and used without further purification. All aqueous solutions were prepared with doubly distilled water.

3. Synthesis of Tween-80® stabilized silver halides NPs

The preparation of Tween-80® stabilized colloidal silver halides was performed as follows. Fifty milliliters of $AgNO_3$ (0.0157 g, 9.2×10^{-5} mol) water solution was added dropwise to 50 ml of Tween-80® aqueous solution with vigorous stirring. The mixture was kept stirred for 15 min. The quantity of Tween-80® was adjusted to 0.05% (5×10^{-4} g ml^{-1}) total concentration in the reaction mixture. Then 100 ml of sodium halide (9.2×10^{-5} mol) water solution was added dropwise with intense stirring.

4. Synthesis of citrate-stabilized silver NPs

The preparation of citrate-stabilized Ag NPs was performed as follows. 50 ml of $AgNO_3$ (0.0063 g, 3.7×10^{-5} mol) aqueous solution was added dropwise to 50 ml sodium citrate solution with vigorous stirring. Then two 50 ml solutions with different concentration of $NaBH_4$ were added step by step (0.0008g, 2.1×10^{-5} mol and 0.0015g, 4×10^{-5} mol respectively). The quantity of sodium citrate was adjusted to reach 3×10^{-4} mol l^{-1} total concentration of citrate in the reaction mixture. Reduction process was carried out at room temperature.

5. Measurements

The UV-VIS absorbance spectra were recorded using a Shimadzu UV-1800 spectrophotometer (Japan), and 1 mm path length quartz cuvettes were used for the measurement of visible spectra.

The electronic images and diffractograms were made on Leo 912 AB Omega (Leo Ltd., Germany) transmission electron microscope (TEM) operating at 100 kV. The samples for TEM characterization were prepared by placing a drop of a colloidal solution on a formvar-coated copper grid which was dried at room temperature. All size distributions were calculated using Femtoscan Online v. 2.2.91 software (Advanced Technologies Center, Russia).

Zeta potential measurements of the obtained dispersions were carried out on Zetasizer Nano ZS analyzer with integrated 4 mW He-Ne laser, λ=633 nm (Malvern Instruments Ltd). Zeta potential was measured by applying an electric field across the dispersion of silver NPs using the technique of laser Doppler anemometry.

The concentration of silver ions in NPs suspensions was determined by Ag^+-ion selected electrode («ELIS-131Ag», Russia) in the thermostatic cell (3.0 ml) contained 0.01 M citrate buffer (pH 6.0) at 20°C with agitation. The calibration of electrode was conducted by fractional addition of 0.01 M $AgNO_3$ solution. The electrode ESr-10101/4,2 (Russia) was used as a comparative electrode.

6. Microbiological tests

Saccharomyces cerevisiae VKM Y-1173 cells were cultivated in the shaken flasks with Reader medium [14] with 2% glucose at 29°C until the culture has reached the logarithmic phase of growth [14]. After that yeast cells were washed with distilled water twice, harvested by centrifugation at 5000 *g* during 15 min and stored in a dense water suspension (0.9-1.2) $\times 10^9$ cells/ml throughout the experiment (3-4 hours) at 0°C.

For determination of *S. cerevisiae* survival rate after silver nitrate and citrate stabilized silver NPs treatment the primary cell suspension dissolved in distilled water (1:100). The studied agents (ionic form of silver (Ag^+): 0.01 M aqueous solution of $AgNO_3$ in concentrations from 0.75 to 18 μM was used. Following silver NPs were used: negatively charged citrate stabilized Ag-NPs in different concentrations from 1.8 to 18 μM were added to the cell suspension (~ $3 \cdot 10^7$ cells/ml) contained 0.5 ml 0.01 M citrate buffer (pH 6.0) and incubated during 30 min at 30°C. The suspension of non-treated cells was used as a control (100% of survival rate) sample. Different dilutions (from 1:10 to 1:1000) of yeast cells (0.1 ml) were inoculated on Petri plates with agar-agar and incubated during 3-5 days at 24-30°C, then the colony quantity was accounted.

For registration of the K^+ efflux from *S. cerevisiae* cells the ion selected electrodes («Orion», USA) and («ELIS-121K», Russia) [15] were used. The measurements were conducted in the thermostatic cell (3 ml) at 20°C with agitation. Cell suspension (60 μM) was added to measurement medium with 0.01 M citrate buffer (pH 6.0) to obtain the cell concentration ~ 3×10^7 cells/ml. The common level of *S. cerevisiae* intracellular K^+ was determined by addition of 20 μM Ag^+ to cell suspension or after heat treatment at 70°C during 15 min in the water bath. The K^+ level determined by the both methods was practically equal. Common K^+ level determined from native yeast cells was taken as 100% in the experiments with investigation of effects of nanosilver contained agents.

7. Results and discussion

NPs characterization

The manner of preparation of citrate-stabilized silver NPs was in the reduction of silver nitrate aqueous solution containing sodium citrate as non-covalent stabilizer by $NaBH_4$ solution. The optimal concentration of citrate-anion turned out to be in a good agreement with [16]. The as prepared silver NPs had an average diameter about 10 nm (figure 1A), were well dispersed and negatively charged, their zeta potential was -29.1 ± 0.5 mV at pH 7.8. Nanosilver dispersion was stable at least 6 month after synthesis and exhibited surface plasmon bands at 401 nm (figure 1B) corresponding to non-covalently stabilized Ag NPs.

The mechanism of the stabilizing action of Tween-80, apparently, is a classic for solubilizing surfactant. In our technique Tween-80 was used in concentrations one order greater than its own CMC (0.8 mM against CMC Tween-80 of 0.08 mM). Forming around the nanoparticle micelle acts as a wrapper that prevents coagulation, but allow silver ions go into solution and back.

Dispersions of silver halides (AgCl, AgBr, AgI) obtained by precipitation from aqueous solutions of silver nitrate in the presence of Tween-80 were characterized using UV-Vis spectroscopy (figure 2), transmission electron microscopy (figures 3-5)

The figure 2 clearly indicates the offset of the region of absorption to longer wavelengths area passing from silver chloride to bromide and then to iodide, which is consistent with the decrease of photolytic stability of halides in this series.

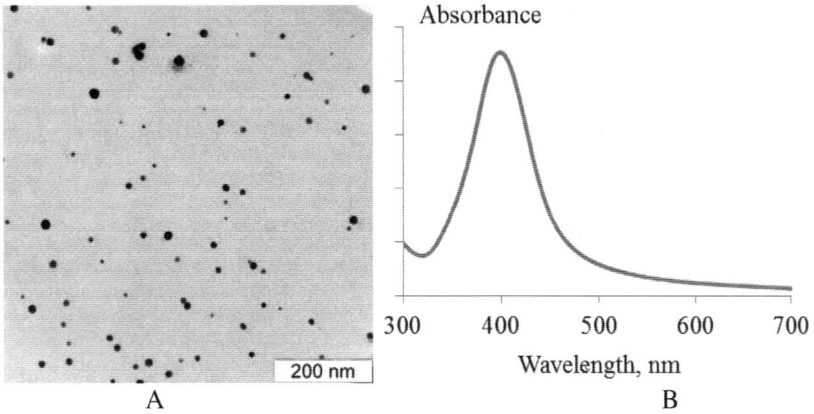

Figure 1. TEM image of citrate stabilized silver NPs (A), UV-visible absorbance spectrum of citrate stabilized silver NPs (B).

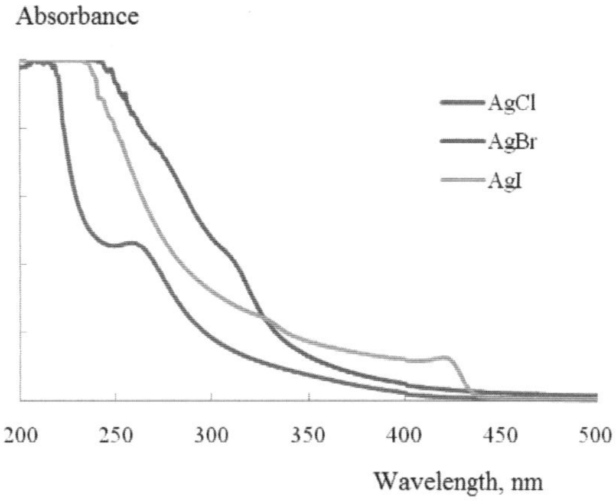

Figure 2. UV-visible absorbance spectra of silver halides NPs.

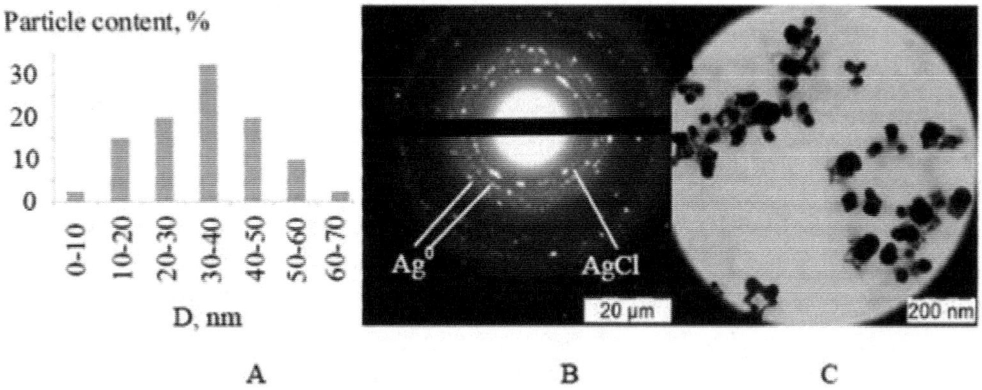

Figure 3. Histograms of the size distribution (A), electron microdiffraction data (B), electron micrographs (C), of silver chloride nanoparticles stabilized with Tween 80.

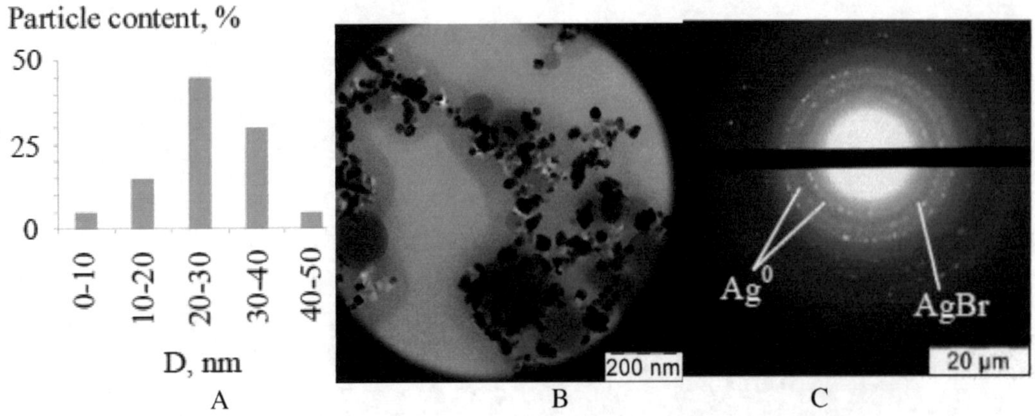

Figure 4. Histograms of the size distribution (A), electron micrographs (B), electron microdiffraction data (C) of silver bromide nanoparticles stabilized with Tween 80.

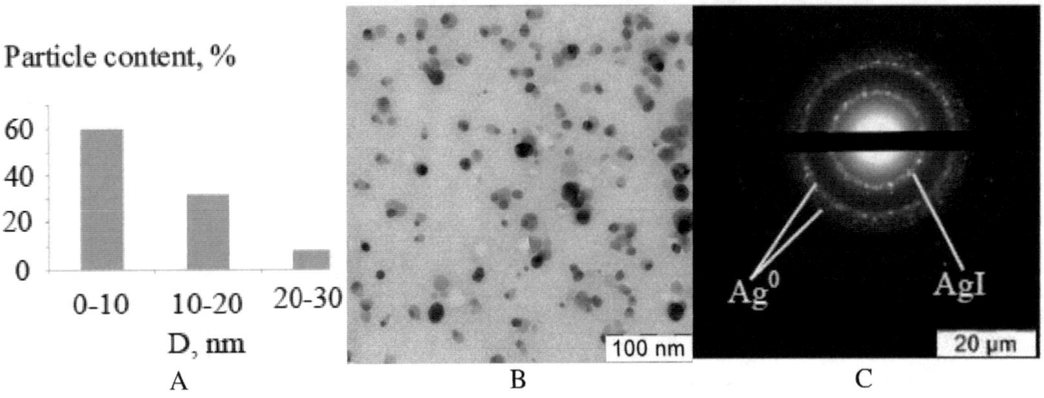

Figure 5. Histograms of the size distribution (A), electron micrographs (B), electron microdiffraction data (C) of silver iodide nanoparticles stabilized with Tween-80.

Microbiological activity of silver nanoparticles dispersions

The cytotoxic activity of silver nanoparticles was characterized by survival rate of yeast cells and by rate of K^+ efflux from the cells incubated with nanosilver dispersion. The data obtained by those two techniques are in good correlation therebetween (figures 6, 7). The rate of K^+ efflux and cell loss increase when concentration of $AgNO_3$ and Ag NPs increases. At that, the percentage of cell loss always exceeds the K^+ efflux at the end of an experiment, thus indicating that some dead cells still contain K^+ inside. This effect allows to suppose that cytotoxic activity of silver ions and nanoparticles against yeast cells does not relate completely to damage of cell membrane.

The determined rates of K^+ efflux from the cells incubated with Ag-NPs (figure 7A) were approximately 5 times lower as compared to the cells incubated with $AgNO_3$ of the same concentrations (figure 7B).

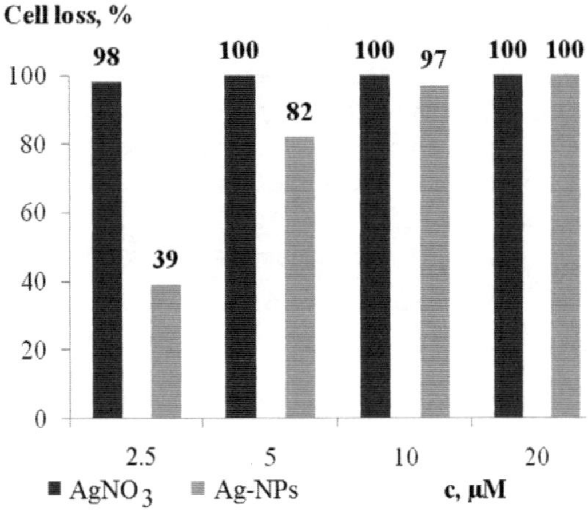

Figure 6. Survival rate of yeast cells incubated with $AgNO_3$ solutions and silver NPs colloids.

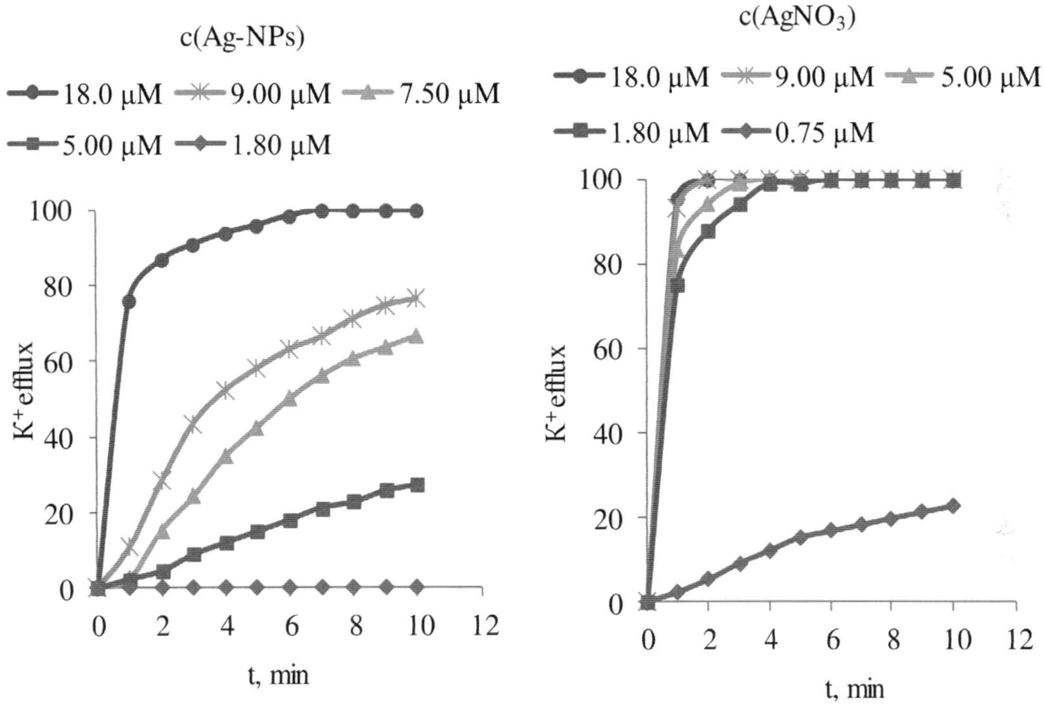

A B

Figure 7. The rate of K^+ efflux from the yeast cells incubated with silver NPs colloids (A) and with $AgNO_3$ solutions (B).

Addition of small amounts (up to the concentration of 50 μM) of $NaBH_4$ into the NPs sol before incubation resulted in dramatic decrease in the K^+ efflux (figure 8A). On the contrary, the addition of dilute H_2O_2 (up to 100 μM in the yeast cells suspension) to the cells incubated with silver NPs led to significant increase in the rate of K^+ efflux. It should be noted that $NaBH_4$ and H_2O_2 themselves in those concentrations did not influence on the cells' survival rate. Therefore, the additions of $NaBH_4$ and H_2O_2 does not directly influence on the yeast cell functioning. Those chemicals alter the rate of K^+

efflux indirectly by modifying the concentration of Ag^+-ions due to reduction of Ag^+ by $NaBH_4$ or at least partial oxidation of silver particles by H_2O_2. Since the reduction of Ag^+ led to total disappearance of the nanosilver activity it can be concluded that the activity of the studied silver NPs follows only from the activity of silver ions. On the basis of this finding and the obtained data concerning the rate of K^+ efflux induced by silver NPs and $AgNO_3$ the concentration of Ag^+-ions in the studied dispersion may be estimated as about 15% of the total concentration of silver. Nearly the same value was obtained from the direct measurements of the Ag^+ concentration in the synthesized nanosilver dispersion by Ag^+-ion selected electrode before the incubation with yeast cells.

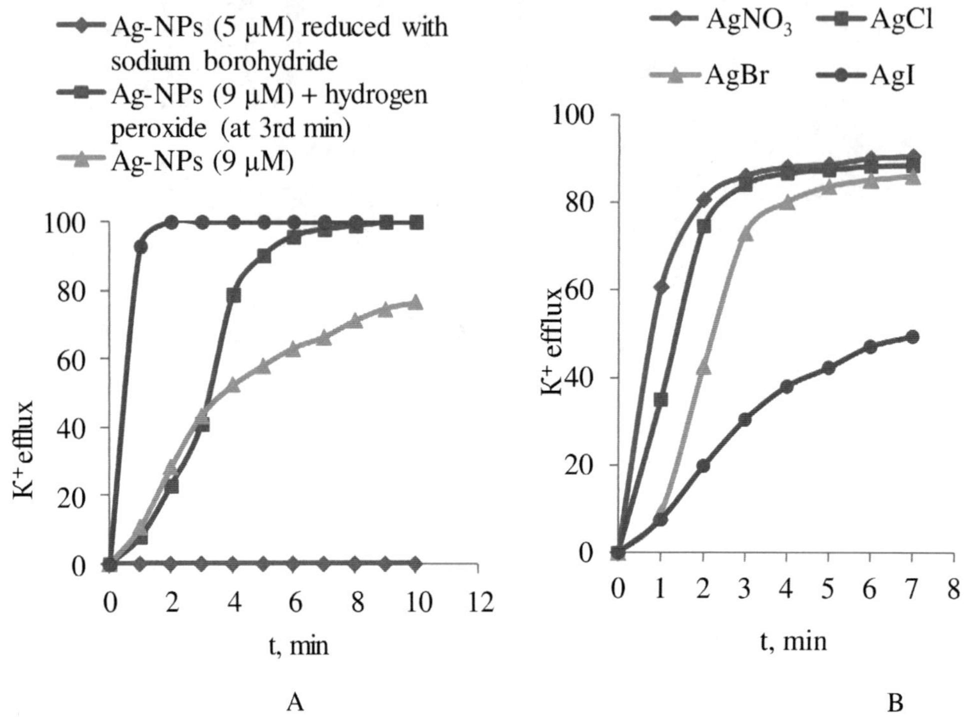

Figure 8. The rate of K^+ efflux from the yeast cells incubated with silver NPs colloids treated with $NaBH_4$ (50 µM) and H_2O_2 (100 µM), the corresponding curve for the yeast cells incubated with $AgNO_3$ is shown for comparison (A), the rate of K^+ efflux from the yeast cells incubated with silver halides colloids with silver concentration of 1.5 µM (B).

Short time after the reduction of $AgNO_3$ with the excess of strong reducing agents like sodium borohydride silver ions in dispersions of as obtained silver nanoparticles are of course absent. But after that oxygen or reactive oxygen species generated in living cells start to oxidize silver and produce Ag^+ ions. To estimate the rate of oxidation of silver nanoparticles we used well known and sensitive spectrophotometric technique with 1,10-phenanthroline and bromopyrogallol red. These organic substances formed an intensely blue ternary complex with ionic silver contained in silver NPs sol which then easily registered at the wavelength of 635 nm. Since the absorption spectrum of silver nanoparticles has a maximum at 405-415 nm, the presence of silver nanoparticles did not interfere with the spectrophotometric determination of ion Ag^+ at 635 nm. Intense bubbling of air for 30 minutes through silver nanoparticles dispersion does not resulted in the appearance of the absorption bands in the spectrum of the ternary complex, while adding sufficient amounts of hydrogen peroxide leads to a rapid (within 3 minutes) quantitative oxidation of nanoparticles with the formation of silver ions, which leads to a significant increase in activity against living cells.

Microbiological activity of silver halides dispersions

Taking into account that the biocidal activity of silver NPs against the *Saccharomyces cerevisiae* cells is mainly results from the silver ions action it could be assumed that slightly-soluble silver salts which can generate silver ions when slowly dissolving may provide prolonged biocidal action as strong as the action of silver nanoparticles. To provide a sufficiently intense and at the same time permanent and long-term release of silver ions, it is promising to use nanodispersed silver salts as a biocidal agents. To further understand the mode of action of poorly soluble silver salts dispersions on living cells, we also studied the effect of silver halides on the K^+ efflux rate from *Saccharomyces cerevisiae* cells. Figure 8B shows the dependence of the release of K^+ from the cells by incubation time with dispersion of silver halide.

Comparing the data from figure 7 it can be seen that the efficiency of biocidal effect of silver halide substantially exceeds the cytotoxic activity of silver nanoparticles and comparable to that of silver nitrate solution of the same concentration. However, at the initial stage of incubation the rate of release of potassium ions from the cells under the influence of nanoparticles of silver halides is several times less than the rate of release of potassium ions by the action of silver nitrate. This delay is due apparently to the slow dissolution of silver halides in citrate buffer. The solubility of silver halides in 0.01 M citrate buffer (pH 6.0), which can be calculated taking into account the possibility of the formation of complexes in solution $[AgHal]^0$, $[AgHal_2]^-$ и $[AgCit]^{2-}$ (Cit – citrate ion), equals 2.5 mM for AgCl, 133 μM for AgBr, 1.7 μM for AgI. Thus, the NPs dispersions with a silver halide concentration of 1.5 μM should be dissolved in a buffer solution. The rate of silver halides dissolution, as shown in [17] correlates with the silver salt solubility in the solution of complexing agent. This fact is in a good agreement with the observed decrease in the efflux rate of potassium ions from *Saccharomyces cerevisiae* cells in a row of AgCl-AgBr-AgI.

8. Conclusions

The discussion about the mechanism of the cytotoxic activity of the dispersions of silver nanoparticles lasts for a long time and reliable data concerning the nature of main active agent are hard to obtain. The results of the present work clearly indicate that the silver ions either remained in the dispersion of silver nanoparticles after their synthesis or generated afterwards by oxidizing silver particles make the crucial contribution into the cytotoxic activity of citrate-stabilized silver NPs against yeast cells. It was also shown that this activity most likely does not relate to damage of cell membrane. The same way the cytotoxic effect of silver halide NPs can be explained because of its attributing to generating Ag^+ ions through NPs dissolution.

References

[1] Klasen H J 2000 *Burns* **26**(2) 131
[2] Krutyakov Y A, Kudrinskiy A, Olenin A and Lisichkin G 2008 *Rus. Chem. Rev.* **77**(3) 233
[3] Vertelov G, Krutyakov Y A, Efremenkova O, Olenin A and Lisichkin G 2008 *Nanotechnology* **19** 355707
[4] Wu J, Zheng Y, Song W, Luan J, Wen X, Wu Z, Chen X, Wang Q and Guo S 2014 *Carbohydrate Polymers* **102** 762
[5] Habiboallah G, Mahdi Z, Majid Z, Nasroallah S, Taghavi A M, Forouzanfar A and Arjmand N 2014 *Modern Research in Inflammation* **3** 128
[6] Shankar S, Jaiswal L, Aparna R S L, Prasad R G S V, Kumar G P and Manohara C M 2015 *Mater. Express* **5**(2) 159
[7] Boonkaew B, Barber P M, Rengpipat S, Supaphol P, Kempf M, He J, John V T and Cuttle L 2014 *J. Pharm. Sci.* **103**(10) 3244
[8] Boonkaew B, Suwanpreuksa P, Cuttle L, Barber P M and Supaphol P 2014 *J. Appl. Polym. Sci.* **131**(9) 40215
[9] Lansdown A B and Williams A 2004 *J. Wound Care* **13**(4) 131
[10] Lansdown A B G 2006 *Current Problems in Dermatology* **33** 17

[11] Chang A L S, Khosravi V and Egbert B 2006 *J. Cutaneous Pathology* **33**(12) 809
[12] Morones J 2005 *Nanotechnology* **16** 2346
[13] Elechiguerra J L, Burt J L, Morones J R, Camacho-Bragado A, Gao X, Lara H H and Yacaman M J 2005 *J. Nanobiotechnology* **3** 6
[14] Vagabov V M, Trilisenko L V and Kulaev I S 2000 *Biochemistry* **65**(3) 349
[15] Vagabov V M, Ivanov A Y, Kulakovskaya T V, Kulakovskaya E V, Petrov V V and Kulaev I S 2008 *Biochemistry* **73**(11) 1224
[16] Henglein A and Giersig M 1999 *J. Phys Chem B* **103**(44) 9533
[17] Shiao D D F, Fortmiller L J and Herz A H 1975 *J. Phys. Chem.* **79**(8) 816

Bio-ecological consequences of crop seeds treatment with metal nano-powders

G Churilov
Division of General Chemistry with Course of Organic and Bio-Organic Chemistry, Ryazan State Medical University, Vysokovoltnaya Str., 9, Ryazan, 390026, Russia

E-mail: genchurilov@yandex.ru

Abstract. As a result of our investigations we have determined the optimal concentrations of ferrum, cobalt and cuprum nano-powders recommended to be used as micro-fertilizers increasing the yield and feed value of crops at the expense of accumulating biologically active combinations by 25-35 %. In unfavorable climate conditions, for example in a case of excess moisture or heat and drought, the plants development and ripening suffer. Our investigations have shown that the stimulating effect of nano-powders has lowered the effect of stress situations on plants development and simultaneously increased the rape seeds yield and quality. Treating the seeds with the drugs being studied has provided the high crop protection. If consider that the maximum efficiency of protectants Chinuk, SK (20 kg/t of seeds) and Cruiser, KS (10 kg/t of seeds) then for the same effect one needs nano-powders 0.1 g per hectare norm of seeds planting.

1. Introduction

The intensive agricultural production inevitably leads to the increased discharge of all soil mantle nutrients [1]. That is why compensation and support of the optimal balance of micro- and-macro-elements in agrocoenosis are most important problems when getting the high yield of the quality [2]. It is known that many micro-elements seriously influence the agricultural products yield and quality and the decline of fungal and bacteria diseases injuriousness for plants [3].

High deficit and cost of traditionally used micro-fertilizers lead to the search for their new types therefore nowadays they study new generation substances – metals nano-powders that are viewed as stimulators of biochemical processes [4,5]. Nano-powders possess great possibilities in inorganic nutrition and energy deposition [6]. At the expense of uncompensated links they easily form complex compounds with organic ones resulting in different ferments synthesis and activation influencing carbohydrate and nitrogen metabolism, amino acids synthesis and reactions of photosynthesis and cells breathing. The results of our investigation prove that. The powders themselves with their energetic effect stimulate the processes of adaptation and self-organization of biological systems to the environment. Nano-particles decline environment negative influence to some extent.

The proposed metals nano-powders allow using them in agriculture as inexpensive and non-toxic drugs partially solving the problem of expensive bio-additives deficit and increasing the rate of return [7,8,9].

2. Methods

Content from this work may be used under the terms of the Creative Commons Attribution 3.0 licence. Any further distribution of this work must maintain attribution to the author(s) and the title of the work, journal citation and DOI.
Published under licence by IOP Publishing Ltd

We have had phenological observations in field experiments, determined field germinating capacity and plants density. For this purpose we have detached permanent plots by the procedure of the State Crops Strain Test (1971) at every variant of the experiment. We have determined the plants height, herbage increment and accumulation of the dry matter collecting 100 plants from each variant. We have harvested at the plot discount areas, evaluated in metric quintal from 1 ha of the herbage.

We have chosen the plants to estimate the phytomass biological yield, to analyze the yield structure, to determine the dry matter and to conduct chemical analyses from the permanent plots (1 m^2) in the morning in dry weather after the dew drying (8-10 o'clock) and put them in poly bags with further temperature fixation for fermentative processes inhibition.

We have calculated the photosynthesis pure productivity or plants "net-assimilation" by formula [10]:

$$PhPP = \frac{M_2 - M_1}{1/2(L_1 + L_2) \cdot n},$$

where *PhPP* – the photosynthesis pure productivity ($g/m^2 \cdot$ deliquefaction);
M_1 and M_2 – plants dry matter at the beginning and the end of the period;
$(M_2 - M_1)$ – dry matter increase during *n* days;
L_1 and L_2 – leaves area at the beginning and the end of the period, m^2;
$1/2(L_1 + L_2)$ – average area of leaves during the experiment;
n – days between sequential periods of observation time.

We have determined total nitrogen by Kjeldahl tritrimetry (GOST R 50466-93) and evaluated "raw protein" using coefficient 6.25; phosphorus by photometry (GOST 26657-97), kalium by flame-photometry (GOST 30504-97), calcium by chelatometry, cuprum, ferrum and cobalt by atomic adsorption (GOST 30178-96) using the extract 1n of HNO_3 solution for cobalt (CINAO, 1981), 1n of H_2SO_4 solution for ferrum, 1n for cuprum, fiber by Shtoman, i.e. the mixture of concentrated nitrogen and acetic acids (1:20).

We have determined mobile phosphorus and kalium in the soil by Kirsanov, pH of salt extract by potentiometer method; the sum of the exchange bases by Kappen-Gilkovits; the hydrolytic acidity by Kappen; humus by Tyurin; absorbed bases by trilonometry; mobile forms of cuprum and cobalt by the atomic adsorption method (muriatic and nitrate extracts correspondingly).

3. Results and discussion

The rape seeds pre-plant treatment with ferrum, cuprum and cobalt nano-powders in doses 0.030 g, 0.048 g, 0.120 g per hectare seed application norm has promoted the vivid increase of seeds germination energy by 6-9 % and the true increase of the lab germination has achieved 10 %. Thus, high-reactive nano-metals have participated actively in enzymatic processes of splitting some extra nutrients, providing their active germination. It should be noted that the higher level of the experiment plantlets metabolism processes has promoted the true increase of the lab germination as compared with the control. It allows considering the seeds pre-plant treatment with nano-powders as an important agronomic method of preparing the seeds for planting.

The seeds pre-plant treatment with ferrum and cuprum nano-powders has had approximately the same increase of field germination as compared with the control and has been from 5 to 8 %. We have found out the highest effect when treating the seeds with ferrum and cuprum nano-powders in the dose of 0.048 – 0.12 g/he.

The plants linear growth is an important parameter implicitly characterizing the intensity of cells division or expansion. We have discovered in our investigations that the seeds pre-plant treatment with ferrum, cobalt and cuprum nano-powders has promoted the increase of the rape plant height. In 2010 with extreme weather conditions (heat and drought) the plants had the highest height in a case with nano-cuprum and that was 21-27 cm or 35.6-45.8 % higher than the height of the control ones (0.12 g/he). In the case with nano-ferrum only the dose of 0.48 g has given the height increase by 7 cm or 11.8 %. The pre-plant treatment with nano-cobalt has stimulated the rape growth promoting this parameter increase by 10-12 cm or 16.9-20.3 %.

In 2012 and 2013 being more favorable for growth and development all the experiment variants had activation of the plants linear growth. It has been most significant with nano-cobalt, not so much and approximately the same with nano-ferrum and nano-cuprum. On the average, for 3 years of investigations the highest excess of the linear growth as compared with the control has been discovered when using the nano-metals in the dose of 0.048 g. The rise to the control has been from 8 to 13 cm or 11-18 %.

Studying the influence of the seeds pre-plant treatment with nano-metals has shown that they can also influence positively the assimilating apparatus – leaf area fulfilling the leading role in photosynthesis. The difference in the control variant in these years has achieved about 7 thousand m^2/he. We have seen the highest excess of the leaf area as that of the control in the variant with nano-ferrum in all three doses and that has been from 5 to 7 thousand m^2/he.

We have got the maximum rise in a case with seeds treatment with nano-ferrum in the dose of 0.048 g and nano-cuprum in the dose of 0.03 g per hectare seed application rate (Table 1). It is known that even the short unfavorable influence of this or that environment already at the beginning of the seeds germination period causes considerable biological changes influencing later the yield. Therefore, one can suppose that seeds treatment with nano-metals can change the metabolism process and its intensity increasing its activity during all ontogeny. For 4 years of investigations the herbage maximum yield increase has been 65.0-56.7 c/he or 30.4-27.0 % as affected by nano-ferrum and nano-cuprum in the dose of 0.03-0.048 gr/hectare seeding rate (chosen as optimal). The nano-metals have not only changed the dynamics of the herbage increase but also promoted the plants chemical composition change. The seeds pre-plant treatment with all the metals in doses of 0.03 g and 0.048 g has promoted the protein increase by 40-47 % depending on the metal. The variants with nano-cobalt treatment have had the highest protein. It shows that cobalt participates in nitrogen and protein metabolism promoting the protein synthesis.

We have discovered that the variants have not had considerable differences in phosphorus and calcium. At that the content of microelements used for treating the seeds has not practically changed. So ferrum has varied in the range of 51.0–55.5 mg/kg of the dry matter in the experiment variants and 53.4 mg/kg in the control one. Cobalt in the experiment variants has varied in the range of 0.12–0.17 mg/kg of the dry matter and 0.11 mg/kg in the control one. The content of ferrum and cobalt in the soil before the harvest has been 17.3 mg/kg and 1.33 mg/kg correspondingly and 17.36 mg/kg и 1.35 mg/kg after harvesting that has not exceeded the control figures (17.30 mg/kg for ferrum and 1.23 mg/kg for cobalt) and ODC.

Table 1. Influence of seeds pre-plant treatment on rape herbage yield (c/he)

Variants	2009	2010	2011	2013	Average Increase for 4 years	
					c/he	%
Control	181	152	257	249	-	-
Nano Fe, 0.012 g	258	157	288	271	33.7	16.0
Nano Fe, 0.03 g	252	235	308	299	63.7	30.4
Nano Fe, 0.048 g	273	224	303	299	65.0	30.9
Nano Co, 0.012 g	237	225	270	268	40.2	19.2
Nano Co, 0.03 g	249	214	288	297	52.2	24.9
Nano Co, 0.048 g	247	214	285	295	50.5	24.1
Nano Cu, 0.012 g	236	200	283	266	36.4	17.4
Nano Cu, 0.03 g	247	231	296	292	56.7	27.0
Nano Cu, 0.048 g	245	215	290	289	49.9	23.8

$HCP_{0.95}$	18.9	22.7	26.7	16.0

Investigation of the Prolonged Effect. The seeds harvested in 2009 were treated with nano-ferrum in the dose of 0.03 g per hectare seeding rate and later were planted in 2010 without nano-powders treatment. The yield was 203 c/he that exceeded the control by 14 % but was 27 % less than the yield got from the seeds treated with ferrum nano-powder in 2010. The yield increase is explained by higher quality seeds having more nutrients. The seeds harvested in 2010 were planted in 2011 without any nano-powders treatment. The yield has not differed from that of the control that shows reducibility of the ecosystem (Le Shatelye principle) to preserve homeostasis.

Later we have studied the higher concentrations nano-particles influence on the rape as we have planned to use the nano-metals to protect the plants decreasing herbicides to be used [11]. At such concentrations we have specially controlled microelements in the soil.

At the experiments with the rape seeds pre-plant treatment with metals nano-powders we have not discovered any authentic change of ferrum and cuprum as well as other chemical elements in the soil. These data prove ecological safety of the agricultural seeds pre-plant treatment with nano-powders (Table 2).

Table 2. Soil macro- and microelements when rape seeds treatment with metals nano-particles

Parameters	Control	Experiment
Before planting		
pH	5.5±0.05	5.6±0.05
P (phosphorus), mg/100 g of soil	12.2±0.04	12.0±0.04
K (kalium), mg/100 g of soil	14.1±0.03	14.3±0.04
Humus, %	3.2±0.01	3.1±0.08
Fe (ferrum), %	5.0±0.5	4.8±0.04
Cu (cuprum), mg/kg of soil	17.0±0.04	16.7±0.02
Co (cobalt), mg/kg of soil	1.2±0.04	1.3±0.05
Zn (zinc), mg/kg of soil	0.9±0.03	0.92±0.004
Pb (lead), mg/kg of soil	14.1±0.06	14.0±0.5
Cd (cadmium), mg/kg of soil	2.1±0.04	2.0±0.3
After rape harvesting		
pH	5.6±0.05	5.7±0.02
P (phosphorus), mg/100 g of soil	12.2±0.04	12.3±0.05
K (kalium), mg/100 g of soil	14.0±0.03	14.4±0.08
Humus, %	3.3±0.05	3.1±0.03
Fe (ferrum), %	4.8±0.08	4.7±0.05
Cu (cuprum), mg/kg of soil	16.8±0.04	16.9±0.06
Co (cobalt), mg/kg of soil	1.3±0.05	1.4±0.04
Zn (zinc), mg/kg of soil	0.91±0.005	0.93±0.003
Pb (lead), mg/kg of soil	13.9±0.08	13.5±0.09
Cd (cadmium), mg/kg of soil	2.0±0.5	2.1±0.06

Note: -$P \leq 0.05$

At that there is quantity and quality increase of plants vegetative and generative organs (Table 3 and grain yield.

Table 3. Mass correlation change between rape organs depending on seeds pre-plant treatment

Parameter	Control	Nano Fe	Nano Cu

		(g/he. seeding rate)				(g/he. seeding rate)			
		0.05	0.1	1	10	0.05	0.1	1	10
Leaves, c/he	57.5	65.3	68.0	73	69.8	76.4	87.9	90.8	88.2
Stems, c/he	87.0	89	97	99	98	97	101	102	99
Pods and inflorescences, c/he	70.5	76.4	79.0	80	78.5	84.3	90.4	89.3	80.0
Total phytomass, c/he	206	204	244	267	265	232	279	280	278
Pods per 1 plant, items	23.3	24.0	25.0	28.5	28.3	24.8	25.8	31.3	30.8
Pods length before harvest, cm	3.5	3.5	3.9	4.1	3.7	3.6	3.6	4.3	4.2

Agro-ecological investigation of metal nano-particles influence on biological objects, plants in particular, is not possible without analysis of biologically active substances including vitamins: ascorbic acid and carotene. Nano-particles toxic affect on plants could have led to their synthesis in tissues decline that could have proved the nano-materials suppressing affect on agrobiocenosis.

Rape seeds treatment with optimal concentrations ferrum and cuprum nano-particles has promoted the increase of vitamin C in the herbage by 25.8 % and 26.9 % correspondingly in reference to the control. Carotene accumulation in the herbage has increased under the treatment up to 28-29 % as compared with the control figures.

The rape herbage chemical composition has not had any considerable changes except nitrogen-free extractable substances. Their content under metals nano-particles influence has increased by 10-14 % as compared with the control (Table 4).

The ability of ultra-fine metals to accumulate both in plants and the soil is very important while analyzing agro-ecological aspects of nano-technologies in energy crops growing. Table 4 presents the mineral composition of rape seeds.

Table 4. Mineral substances in rape seeds, mg/kg

Elements	Control	Variants	
		Nano Fe (0.1 g/he. seeding rate)	Nano Cu (0.1 g/he. seeding rate)
Ferrum	55.40±0.09	54.20±0.01	56.70±0.02
Cuprum	7.10±0.01	7.40±0.06	7.0±0.05
Cobalt	0.15±0.003	0.16±0.002	0.12±0.004
Cadmium	0.03±0.001	0.02±0.001	0.04±0.002
Lead	0.40±0.03	0.70±0.03	0.90±0.06
Arsenic	0.17±0.004	0.11±0.004	0.13±0.006
Zinc	51.8±0.02	47.40±0.05	46.70±0.05
Nickel	1.40±0.06	1.70±0.03	2.0±0.6

Note: -$P \leq 0.05$

Microelements and heavy metals in the rape seeds got with the use of ferrum and cuprum have not differed greatly from the control ones. We have not also seen any ferrum and cuprum accumulation in seeds of the experiment variants that proves the absence of cumulative properties of the studied nano-materials.

Comparative evaluation of efficiency of spring rape different dressers and nano-powders has been conducted field tests data-driven got at the farm "Stenkino". The rape seeds have been pre-plant treated with nano-powders having the concentration of 0.1 g per hectare seeding rate. In the control we have soaked the seeds in d water. We have used not only cuprum nano-powder but copper oxide one having high antibacterial and fungicide activity.

High efficiency of rape protection in variants with dressers Chinuk, SK and Cruiser, KS is well known [2,11]. Therefore the possibility of nano-powders to act has been compared with dresser Cruiser, KS.

Table 5. Degree of crops damage by most dangerous pests and rape yield with dressers

Variant	Number of Most Dangerous Pests, spec./ 100 plants			Total damages, tems per 100 plant	Yield, c/he	Saved yield %
	Cruciferous flea beetles	Cabbage moth	Rape weevil			
Nano Cu, 0.1 g	11.1	9.1	8.0	28	23.0	16.6
Nano CuO, 0.1 g	10.0	8.8	7.8	24	22.6	15.7
Cruiser, KS, 10 kg/t of seeds	10.5	8.9	7.8	26	19.5	14.7
$HCP_{0.5}$					2.2	

Treating the seeds with all the drugs being investigated has provided the crop total protection. Cruiser, KS has been a little bit more efficient against the rape weevil though on the whole there have been few differences. We should point out that biological efficiency in 15 days of using Cruiser, KS has declined a little in comparison with the effect of nano-powders but in spite of the high level of pests' development they have saved the yield considerably.

In the process of observing the season dynamics of harmful fauna population in rape and other agrocenosis we have discovered that the pests' population has changed considerably at phonological phases of crops development [12]. The crops colonization with pests has begun in the third decade of April. Cruciferous flea beetles have appeared first. At gradual temperature increase their population has been increasing for two weeks before the rosette formation. From the budding phase their population density has declined considerably at the expense of hibernate species dieback. We have noticed the largest number of pests and pollinators in the period of cabbage blooming.

Cruciferous flea beetles have turned to be most harmful. They have damaged the plants at the most expose phase, i.e. sprouts that could lead to their loss. The rape yield has depended on the used insecticide and nano-powders (Table 6).

Table 6. Spring rape yield depending on insecticide

Yield, c/he			
Control	Kinmix 0.2 l/he	Nano Cu, 0.1 g	Nano CuO, 0.1 g
14.2	18.9	23.5	19.6

The decline of total damages has happened mainly at the expense of cruciferous flea beetles decline.

Kinmix has proved to have a fast solid action and high efficiency. But it has got a serious drawback – resistance of organisms to this class of combinations after several repeated applications. The high biological efficiency of the seeds pre-plant treatment with nano-powders has remained till the middle of the crop blooming so we have not needed the repeated treatment with insecticides.

Slow development of the spring rape aerial portion in the first period of growth has declined its competiveness to weeds. In the case with the seeds pre-plant treatment with nano-powders the powerful development of the aerial portion lets rape dominate over weeds. The weeds species composition at the experiment area during the experiments has been widely presented mainly by annuals.

The perennial weeds infestation has been presented by rhizome plants – quitch (Elytrigia repens), field horsetail (Equisetum arvense); burr plants – Canadian thistle (Cirsium arvense), field milk or yellow thistle (Sonchus arvensis), field bindweed (Convolvulus arvensis), dandelion (Taraxacum officinale) and great plantain (Plantago major). We have pointed out that if herbicides have had their better effect at temperature higher than 10° C, high relative air and soil humidity, then nano-powders have been especially efficient even in dry periods.

The nano-powders have not reduced the weeds but increased the yield and the first sprouts sustainability and the weeds affect has been minimal. The rape crop density and sustainability has been 233-240 species/m^2. There has also been some increase in the crop density, pods number per plant, mass of 1000 seeds and photosynthetic parameters.

The use of nano-materials has promoted formation of the powerful and active leaf apparatus making possible to increase the amount of accumulating PhAR that has been a good condition for crops yield increase.

Table 7. Influence of herbicides on spring rape yield

Yield, c/he				
Without herbicides (Control)	Klocet 1.5 l/he	Trophy 1.0 l/he	Nano Cu, 0.1 g	Nano CuO, 0.1 g
18.6	21.7	22.5	23.3	22.1
Saved average yield, %	15.3	13.4	16.0	17.1

Thus, in today ecological situation full of problems we propose native highly-efficient nontoxic nano-drugs of wide application spectrum possessing the ability to protect rape, get good yields and quality seeds.

4. Conclusions

Nano-crystal metals having unique properties can be used as new generation bio-drugs. Besides they are cost-effective and influence the increase of crops yield. High efficiency of nano-powders as growth stimulators has been shown in this article by the example of rape. We have shown that nano-powders having optimal concentrations up to 0.1 g/he seeding rate are ecologically safe as they do not accumulate in neither plants nor soil and have no mutagenity. The maximum increase caused by their use with spring rape as compared with the control has been up to 25 %.

When growing rape it is very important to protect the plants. Developing integrated systems of rape protection means minimal negative environment impact of protective measures and maximum mobilization of natural biotic factors to suppress hazardous organisms' development. We have noticed high efficiency of the spring rape seeds treatment with nano-materials against the pollen beetle and other pests and the yield average increase has been 4.4-6.9 c/he as compared with that of the control. The use of herbicides and nano-powders treatment speeds up the crop ripening for 1-2 days, increases parameters of the yield structure, promotes powerful and active leaf apparatus functioning, increases accumulating PhAR and as a result serves a background for the crops yield increase and this rape yield increase when the pre-plant treatment with nano-powders can be up to 22 % with other parameters considerable increase.

The adoption of resource-saving technologies with nano-materials use will help to escape the arable lands physical properties deterioration, soil degradation caused by the fertilizers active effect, pesticides and agricultural engineering. They must become additions to agricultural methods including science based rotations, choosing cultivars resistant to diseases, systems of tillage, weed control and timely harvesting. The peculiarity of nano-particles chemical cooperation with liquid fluid is one of determining factors in stimulating plants development that gives possibility to use nano-powders as growth stimulators and plants protectors.

References

[1] Bullock D G, and Sawyer J E, 1991 Nitrogen, *J. of Production Agriculture Vol.* **4** 4550
[2] Lapushkin V M and Kidin V V 2011 *Soil Fertility* **5** 5
[3] Graham R D and Webb M J 1991 Micronutrients and Disease Resistance and Tolerance in Plants
[4] Folmanis G A and Kovalenko L V 1990 Ultrafine metals in agriculture (*Moscow, IMETRAN)* **80**
[5] Polishchuk S D, Nazarova A A, Churilov D G, Kutskir M V, Churilov G I and Ivanycheva J N 2013 *Nano-techniq*ues **3(35)** *67*
[6] Glushenko N N, Bogoslovskaya O A and Olhovskaya I P 2002 *Chemical physics* **4** 148
[7] Churilov G I 2009 *Bulletin of Orenburg State University* **12** 148
[8] Sushilina M M 2004 *Soil Fertility* **5** 26
[9] Polishchuk S D, Nazarova A A and Kutskir M V 2012 *The Bulletin of RSATU* **3** 68
[10] Kuzmin N A, Korneev T V and Shevchenko V V 1998 *Voronezh* 127
[11] Mazin V V, Popova T A, Hadeeva N V, Burutin A B and Maisuryan A N 2007 *Agricultural Biology* **1** 86
[12] Popova T A and Petrova N I 2014 *Processing of Agricultural Academy Named after Timiryazev K A* **1** 125

Perspectives of Using Ultra-Fine Metals as Universal Safe Bio-Stimulators to Get Cattle Breeding Quality Products

S Polishchuk
Division of chemistry, Ryazan State Agrotechnological University, 1 Kostycheva st., Ryazan, 390044 Russia

E-mail: svpolishuk@mail.ru

Abstract. We have conducted investigations of ultra-fine metals biological activity with lab non-pedigree white rats, rabbits breed "Soviet chinchilla" and cattle young stock of the black and white breed as the most widely spread in the central part of Russia. One can see the possibility of using microelements of ultra-fine iron, cobalt and copper as cheap, non-toxic and highly effective biological catalyst of biochemical processes in the organism that improve physiological state, morphological and biochemical blood parameters increasing activity of the experimental animals' ferment systems and their productivity and meat biological value. We have proved the ultra-fine powders safety when adding them to the animals' diet.

1. Introduction

Modern vet and livestock technologies are closely connected with nano-technologies providing potential vector of solving many actual problems of mankind, mainly food industry. The aim of agriculture and livestock breeding in particular is providing people with high quality full value food products. Healthy animals' products consumption is the basis for human diseases prevention, improving lifetime and its quality.

As far as transferring the livestock sector to the industrial ground and increasing production it is necessary to pay more attention to full value, balanced animal nutrition and increasing the index of fodder efficiency. Being the necessary component of many biologically active compounds like proteins, ferments, hormones, vitamins and pigments or influencing their functions microelements participate in different processes of animals' life activity and metabolism. Bio-drugs of new generation, i.e. microelements in the form of metals ultra-fine powders (MUFP) arouse great interest. Ultra-fine powders of iron, cobalt, copper, manganese, selenium and others have higher biological activity. Investigations conducted in last years have shown their efficiency in crop breeding, fodder production and livestock raising [1, 2, 3, 11].

2. Methods

We have considered the multi-factorial variant of using nanomaterials, i.e. ultra-fine metals as highly active bio-stimulators of animals' physiological processes. This optimizes the hormone, immune and bio-chemical status of animals and increases the young stock livability.

We have studied these bio-drugs in different regions of Russia (Moscow, Ryazan, Tula and other oblasts, Krasnodar and Stavropol territories). We have been studying ultra-fine powders at Ryazan State

Content from this work may be used under the terms of the Creative Commons Attribution 3.0 licence. Any further distribution of this work must maintain attribution to the author(s) and the title of the work, journal citation and DOI.
Published under licence by IOP Publishing Ltd

Agrotechnological University Named after P.A. Kostychev since 1997. We had experiments in 2001-2014 and they included lab and farm tests.

3. Results and Discussion

3.1. Complex influence of iron, copper and cobalt ultra-fine powders in the system Plant – Animal

We have shown the possibility of their use in agriculture as cheap and non-toxic drugs partially solving the problem of expensive micro-fertilizers deficit for feed crops. In a case of adding these drugs to the animals' diet there is improvement of rabbits' physiological state (live weight, survival and offspring livability and reproduction) and morpho-biochemical parameters of their blood.

Based on the experimental data connected with metallic nano-powders use for vetch seeds cultivar Lygovskaya-28 pre-plant treatment we have estimated that the optimal concentration of iron and cobalt ultra-fine powders is 0.03 g per 12 kg of seeds. At that the vetch green mass yield has increased by 25.5% and 32.1 % correspondingly. Ascorbic acid in vetch green mass has 1.5 times increased when seeds pre-plant treatment with iron nano-powder and 3-4 times when using cobalt nano-powder as compared with the control. Iron and cobalt nano-powders have increased carotene in vetch green mass 1.5 and 4-5 times correspondingly as compared with the control. When treating the seeds with nano-powders there has been increase of water-miscible polysaccharides by 68 %, protein by 45 % and lectin has declined by 24 % with iron nano-powder and by 59 % with cobalt nano-powder that has increased the vetch fodder value. Adding the vetch grass grown with iron and cobalt nano-powders to the rabbits' diet has increased live weight gain on the average by 9 % and 18 % correspondingly as compared with the control due to the increase of polysaccharides and decline of lectin fraction.

Adding vetch grass grown with iron and cobalt nano-powders has stimulated the hematogenesis function that manifested in erythrocytes and hemoglobin increase. There has also been a change of leucocytes formula percentage to lymphocytes increase. One could also see some true increase of crude protein in blood serum at the expense of globulin-rich fraction that has testified the increase of immune-biological reaction (Table 1). The increase of LDG has presupposed intensively going glycolysis, i.e. carbohydrate metabolism strengthening and ALT increase has shown protein metabolism stimulation.

Table 1. Morpho-Biochemical Blood Parameters of Rabbits that Received Metal Nanopowders

Blood Parameters	Control	Fe Nanopowder	Cu Nanopowder
Erythrocytes, *1012/l	5.4±0.03	5.9±0.02	5.8±0.04
Hemoglobins, g/l	110±2	120±.3	118±0.5
Trombocytes, n.103/l	205±6	200±7	210±.3
Leukocytes, *109/l	4.1±0.04	5.3±0.05	5.1±0.06
Leukocytic formula, %			
Lymphocytes	43.0±0.3	51.0±0.4	50.0±0.4
Monocytes	6.0±0.2	6.0±0.3	6.1±0.01
Granulocytes	51.0±0.5	43.0±0.4	44.0±0.4
Total serum proteins, g/l	52.5±0.06	58.0±0.07	56.5±0.05
Protein fraction, %			
α1 - globulins	3.4±0.05	4.0±0.5	4.2±0.06
α2 - globulins	11.2±0.03	6.5±0.04	9.5±0.02
β - globulins	8.0±0.4	10.6±0.05	13.0±0.1
γ - globulins	10.0±0.3	12.5±0.04	12.0±0.2
Abumins	67.4±0.08	66.4±0.07	61.3±0.08

Note: - $P \leq 0.05$

The degustation commission has discovered that organoleptic parameters (taste, aroma, delicacy, juiciness and others) of broth, fried and boiled meat of the experimental and control groups have not differed greatly and can be eaten without any restrictions.

3.2. Influence of nano-crystal metals on physiological state of lab and agricultural animals and safety of livestock products when adding them to the diet

Nano-powders are in meta-stable state and therefore have higher chemical activity subjected both dimensioned and structural factors and changed mechanical, electrical, optical and other properties of substances. Nano-particles possess more energy than traditional powders. So they have got a number of unique physical and chemical properties. Nano-particles are digested and produce physiological action with the organism several times stronger than familiar natural mineral substances or their chemical analogues. Biological activity of nano-powders is due to small sizes of particles and their possibility to penetrate into the organism. As opposed to metal ions nano-particles before their penetration into the plant and animals cells do not have any charge therefore they do not form large complexes with transmitting protein and can easily pass through pores of plasmodesma in a bi-lipid layer of cell membranes sized 50 nm that is comparable with the size of nano-particles (25-35 nm) and get into cytoplasm. One can see the nano-powders effect on the organism in nano-particles on the cell and micro-molecular levels [4, 5].

For full development and safe implementation of biologically active drugs based on metals nano-powders sized 25-45 nm in livestock raising we have developed their veterinary and toxicological characteristic. As a result of the conducted experiments we have estimated that nano-powders of iron, cobalt, copper and copper oxide do not possess any cumulative properties and differ by lower toxicity than inorganic salt of these metals. This allows using them in premixes as a source of micro-elements [6, 7]. We have tested their efficiency for prophylaxis of cows' postnatal complications and newborn calves' enteric infections. The primary evaluation of safety and efficiency has shown high preventive properties of the drug and its safety.

As a result of some preventive measures the cows' lochial period has passed without any complications. The cows have been hogging timely that has proved the normal revival of the reproductive system after calving. 80 % of animals getting the drug have not got any postnatal complications (endometritis). 20 % of the experiment cows have had postnatal endometritis in a benignant catarrhal form. More than 60 % of cows have been inseminated in one hog without being stimulated by hormones.

The newborn calves' prophylaxis has left out young stock mortality from enteric infections. The use of nano-materials having high biological activity can make a considerable contribution to solving this problem and offer an efficient system of diseases prophylaxis by activating own protective strength of the organism.

We have studied the influence of metals ultra-fine powders on immune, bio-chemical and hormone parameters of the organism so that physiologically justify the use of metals nano-particles as non-medicinal drugs of animals' prophylaxis and correction.

We have chosen animals to the groups for the experiment according to the principle of the balance group-analogues considering gender, age, breed, live weight and the same feeding and housing conditions. We have taken blood from the side ear vein four- timely at interval of 10 days before morning feeding. We have determined all blood parameters (hemoglobin, erythrocytes, leucocytes, leucocytic formula and blood sedimentation rate) according to unitized methods of clinical diagnostics [9]. We have analyzed the biochemical parameters of blood with spectrophotometer "Spectrum" firm "Abbot" according to unitized methods of clinical lab investigations.

3.3. Determining optimal doses of metals nano-powders when adding them to the diet of lab animals

We have had experiments with rabbits breed "Soviet chinchilla" aged 30 days. We have got 10 experiment groups 16 rabbits weighing 800±0.5 g each. We have used the mixed fodder treated with the drugs suspension to add metals nano-powders to the diet. We kept the rabbits in recommended housing and 2 times feeding conditions. The influence of different doses of nano-crystal metals on live weight in the control and experiment groups of animals is shown in figure 1.

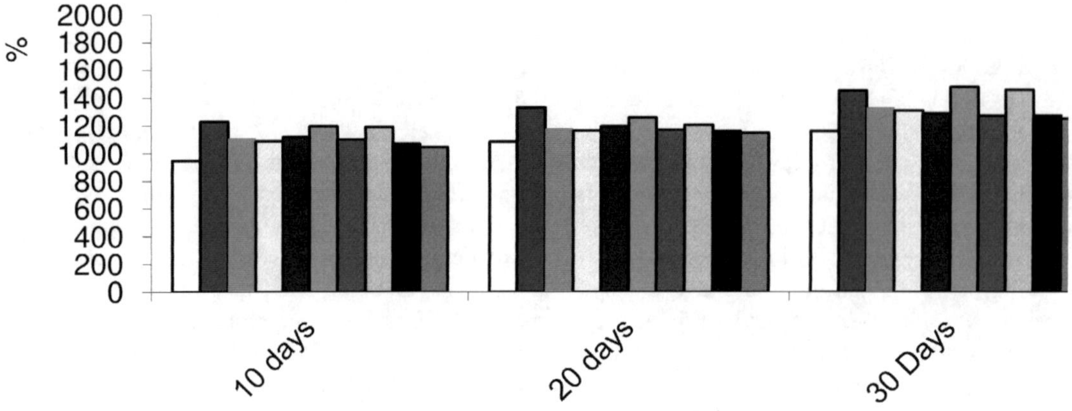

Figure 1. Live Weight of Lab Rabbits, g

One can consider 0.08 mg/kg of live weight as the optimal dose of iron nano-powder. The increase of the animals' weight has been 25% after 30 days of giving the drug. The optimal cobalt nano-powder concentration has been 0.02 mg/kg that has allowed increasing the live weight by 27 %. The corresponding data for copper have been 0.04 mg/kg and 26 %. The cancellation of MUFP has not led to decline of the experiment animals' weight. Moreover the tendency of mass increase has remained even after 10 days after the cancellation of metals ultra-fine powders (Figure 1). We have studied the influence of these concentrations nano-powders on rabbits' physiological parameters and morpho-biochemical parameters of blood. The experiment lasted for 60 days and the age of the animals at the beginning of the experiment was 30 days. The animals of all 4 groups have regularly gained the weight. They have been active and got a good appetite. Adding iron nano-powder to the diet has increased the rabbits' live weight by 11.7 % by the end of the experiment as compared with that of the control, cobalt nano-powder by 17.8 % and copper by 16.3 %. We have also investigated morphological and biochemical blood parameters of the experiment animals.

All parameters of blood serum when adding nano-crystal metals have been within the limits of the physiological norm. But there have been authentic increase of erythrocytes and hemoglobin, crude protein at the expense of the globulin fraction and α_1 - and β-globulins in particular. Increasing these fractions leads to strengthening the active transfer of carbohydrates and lipids to tissues that influences the globulins number and activity that activates metabolism. One can see the same character of changes with γ-globulins. At the expense of γ-globulins increase as compared with the control there has been an increase of immune-biological reactivity as they are defensive antibodies (immunoglobulins) and are responsible for specific immune reply. The immune system that is structurally and functionally organized complex of lymphoid cells interacting with each other and accessorial cell elements on some stages of immunogenesis can serve an indicator reflecting the nano-materials affect on the organism. If consider that microelements in ultra-fine state activate the immune, ferment and humoral systems of the organism promoting the increase of metabolism and better digestion and nutrition then one can use them as bio-additives [10, 11].

3.4. The influence of nano-crystal iron, cobalt and copper on the physiological state of the young stock (heifers)

We have had experiments with black and white cattle as most widely spread in Ryazan oblast at farm JSC "Starozhilovo stud farm".

We have treated the mixed fodder with water powders suspension before feeding the animals. Suspension consumption has been 1 l per 1 t of mixed fodder. They have added the treated mixed fodder to the diet of the control and experiment animals on the basis of 2 kg of mixed fodder per 1 animal a day during the whole experiment. The age of experiment and control animals at the beginning of the experiment has been approximately 4 months. The experiment lasted for 8 months. We weighed all animals once a month. The first group (control) did not get ultra-fine powders, the second one (experiment) got nano-crystal iron (0.08 mg/kg of live weight a day), the third experiment group has got nano-crystal cobalt (0.02 mg/kg), the forth experiment group has got nano-crystal copper (0.04 mg/kg).

The experiment animals during the whole experiment have regularly exceeded the control animals in live weight beginning from the 1st month of the experiment. At that, every next month the weight gain of the live weight of the experiment heifers has increased. Adding nano-crystal iron to the heifers' diet has promoted the live weight increase in 8 months by 22.4 as compared with the control, cobalt by 13.7 % and copper by 10.7 % as compared with the control. Daily average gain has also increased (Figure 2).

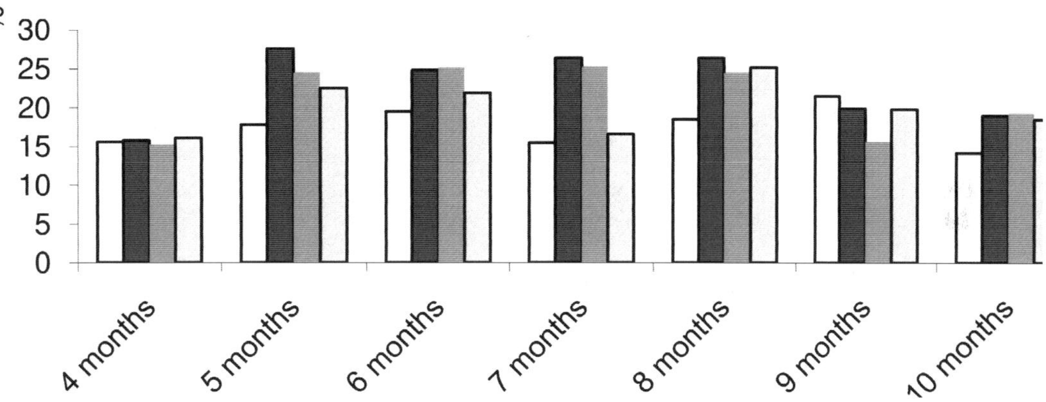

Figure 2. Influence of Iron, Cobalt and Copper Nano-Powders on Gross Increase of Experimental Animals, kg

By the 8th month of the experiment the gross gain in the group getting iron nano-powder has exceeded the control by 53.8 %, the group getting cobalt nano-powder by 25.4 % and the group with copper nano-powder by 23.1 %.

The increase of live weight, daily average and gross gain in a case of adding nano-crystal metals proves catalytical action of these bio-drugs that manifests in metabolism strengthening on the whole and protein metabolism in particular.

3.5. Influence of nano-crystal iron, cobalt and copper on morphological parameters of heifers' blood

Table 2. Morphological Parameters of Heifers Blood when Adding Iron, Cobalt and Copper Nano-Powders to Their Diet (8 month of experiment)

Blood Parameters	Control	Fe Nano-Powder	Co Nano-Powder	Cu Nano-Powder
Erythrocytes, $*10^{12}/l$	5.60±0.08	6.70±0.07	6.60±0.02	6.20±0.03
Hemoglobins, g/l	105.0±0.5	123.0±0.5	120.0±0.4	118.0±0.6
Trombocytes, $10^9/l$	338±0.9	310±0.8	315±0.4	330±0.5
Leucocytes, $10^9/l$	7.9±0.05	8.5±0.03	8.3±0.04	8.0±0.06
Leucocytic formula, %				

Leucocytes	61	68	66	64
Monocytes	5	4	5	5
Granulocytes:	34	26	28	25
Stab neutrophils	3	2	2	1
Segmented neutrophils	30	22	22	28
Basophils	-	1	1	1
Eosinophils	1	1	1	3

Note: - $P \leq 0.05$

Copper nano-powder influence on blood morphological content (Table 2) has manifested in little rise of erythrocytes (by 10.7 %) and hemaglobin (by 12.4 %) as compared with the control. All other blood parameters have remained within the norm and differed inconsiderably from the control.

One could see leucocytes increase in the 8th month of the experiment by 7.6 % (iron) and 5.1 % (cobalt) and the change of the leucogram, i.e. lymphocytes increase for iron cobalt and copper has been 11.4 %, 8.2 % and 5% higher than the control. These changes prove the increase of the immune system of animals as the main function of B-lymphotcytes is producing antibodies or protective immunoglobulins. The lymphocytes increase has caused the significant decline of granulocytes and neutrophils lower than the control in particular.

The lymphocytes increase in the leucogram has proved strengthening protective functions, whereas the increase of hemoglobin and erythrocytes has proved the improvement of the hemopoietic system of the organism. With nano-powders there is growth and renewal of blood cells and strengthening of oxidation processes that is one of the criteria in diagnosing and evaluating the nano-materials safety.

3.6. Influence of nano-crystal iron, cobalt and copper on bio-chemical blood parameters of the experiment animals

Bio-chemical blood parameters reflect metabolism of protein, fat, carbohydrates vitamins, hormones and water-mineral characteristics of the organism. They let interpret the growth and development of the organism, help to discover latent diseases.

Using ultra-fine iron has considerably activated much ferment. So by the end of the experiment ALT content in blood serum has increased by 12.5 % and AST by 7.6 % as compared with the control. At that de Rittis coefficient (correlation of AST to ALT) has changed from 1.65 in the control to 1.40 in the first experiment group of animals that got nano-iron and up to 1.55 in the 2nd and 3rd experiment groups that got nano-cobalt and nano-copper correspondingly. This parameter in the experiment groups has reached the physiological norm (1.2 – 1.4) that proves stabilization of animals' metabolisms. These ferment activation in the blood of the experimental animals is connected with strengthening the protein metabolism that, in its turn, leads to intensive gain of the live weight (Table 5). This is also connected with the increase of γ-glutamyltransferase in blood serum by 14.2 - 21.4 % when using nano-metals as compared with the control. This ferment participates in amino acids metabolism and its activity also reflects protein metabolism intensity. The uric acid has increased by 13.3-23.3 %, urea by 13.6 - 18.2 %, creatinine by 5.5 - 9.5 % higher than the control one. These substances are products of protein metabolism and the fall under residual nitrogen and the increase of their amount are connected with strengthening protein metabolism intensity in animals.

Table 3. Biochemical Parameters of Heifers Blood when Adding Iron, Cobalt and Copper Nano-Powders to Their Diet

Blood Parameters	Control	Fe Nano-Powder	Co Nano-Powder	Cu Nano-Powder
1. Alanine Aminotransferase (AAT), mmol/l	0.80±0.01	1,0±0.06	0.9 ± 0.04	0.9 ±0.03
2. Aspartate Aminotransferase (AST), mmol/l	1.30±0.01	1.4±0.02	1.6 ±0.01	1.4±0.05
3. α-amylase, mg/sec*l	3.5±0.01	4.5±0.02	4.2±0.03	3.8±0.05
4. Alkiline Phosphatase, u/l	310±1	401± 1	345±2	325±3
5. γ- glutamiltranspeptidase, u /l	14.0±0.3	16.0±0.5	17.0±0.2	16.0±0.2
6. Uric Acid, mkmol/l	30.0±0.4	35.0±0,6	37.0±0,2	34.0±0.1
7. Urea, mmol/l	2.2±0.02	2.5±0.04	2.6±0.03	2.4±0.04
8. Creatinine, mkmol/l	73.0±0.1	78.0± 0.5	77.0±0.2	80±0.4
9. Cholesterine, mmol/l	3.4±0.03	2.5± 0.02	2.2±0.04	2.4±0.03
10. Crude Protein, g/l	68.5±0.09	74.5±0.04	66.8± 0.01	69.6±0.03
	11. Protein Fractions:			
α_1 – globulins, %	4.0±0.4	5.0±0.1	5.0±0.3	6.0±0.5
α_2 – globulins, %	16.0±0.8	8.0±0.1	10±0.6	9.0±0.5
β – globulins, %	12.0±0.7	13.0±0.4	13.0±0.2	14.0±0.9
γ – globulins, %	30.0±0.2	38.0±0.7	33.0±0.6	31.0±0.7
Abumins, %	38.0±0.4	42±0.03	40±0.8	40±0.6

Note: - $P \leq 0.05$

Cholesterol in blood of the experimental animals as compared with the control has declined by 30% that positively influences the animals' physiological state. This is proved by the increase of α-amylase by 17.1 – 28.6 % as compared with the control. The intensive growth of the organism also proves the activity of alkaline phosphatase. Its content is 24.2 - 29.3 % higher than the control. This ferment participates in regulating the processes of bone tissue formation and catalyzes phosphorus metabolism in animals and the increase of alkaline phosphatase activity is particularly important for the growing organism.

Such a considerable change of ferment system activity is connected with using some inconsiderable amount of nanocrystal metals that has presupposed some additional increase of micro-elements at the expense of these elements better digestion in animals' digestive duct.

3.7. Nano-metals influence on mineral substances in animals' blood, muscle tissue and liver

The results of our investigations have proved that nano-metals stimulate accumulation of other macro- and micro-elements (Table 6). That has been proved by the analysis of blood, muscles and liver mineral composition.

Table 4. Mineral Substances Content in Heifers Blood Serum when Adding Iron, Cobalt and Copper Nano-Powders to Their Diet

Blood Parameters	Control	Fe Nano-Powder	Co Nano-Powder	Cu Nano-Powder
K (Kalium), mmol/l	4.5±0.08	4.8±0.06	5.0 ±0.1	4.8±0.04
Na (Natrium), mmol/l	138.0±0.4	144.0±0.3	142.0±0.7	145.0±0.1
Ca (Calcium), mmol/l	2.1±0.05	2.5±0.06	2.4±0.06	2.4±0.04

P (Phosphorus), mmol/l	2.3±0.05	2.4±0.05	2.6±0.05	2.4±0.04
Cl (Chlorine), mmol/l	98.0±0.7	106.0±0.9	103.0±0.4	104.0±0.9
Fe (Iron), mcmol/l	16.9±0.06	17.9±0.03	17.1±0. 01	17.3±0.05
Cu (Copper), mcmol/l	22.0±0.1	24.0±0.5	21.0±0.9	24.0±0.7

Note: - $P \leq 0.05$

After 8 months of the experiment there has been authentic increase of kalium by 6.6-11.0 %, natrium by 3.5 -5.1%, calcium by 14-19 % and phosphorus by 4.4-13.0 % in blood of animals that have got ferum nano-powder. The content of chlorine ions has increased by 5.1-8.8 %. There has also been 6.0 % increase of iron and 9.0 % of copper as compared with the control.

In this case the iron increase has promoted copper accumulation. This proves that we have selected correct doses as iron surplus in the organism could have led to copper and calcium deficit.

Table 5. Influence of Nano-Powders in the Diet on Mineral Substances in Muscle Tissue and Liver of Heifers, mg/kg

Substance	Muscle Tissue				Liver			
	Control	Fe Nano-Powder	Co Nano-Powder	Cu Nano-Powder	Control	Fe Nano-Powder	Co Nano-Powder	Cu Nano-Powder
Barium (B)	27±0.1	29±0.1	26±0.1	27±0.1	25±0.1	30±0.1	27±0.1	27±0.1
Calcium (Ca)	210±0.2	370±0.3	280±0.8	210±0.6	40±0.9	80±0.3	20±0.4	00±0.2
Cobalt (Co)	0.10±0.03	0.13±0.005	0.12±0.002	0.12±0.005	1.1±0.03	1.3±0.05	1.2±0.02	1,4±0.05
Chrome (Cr)	1.5±0.04	1.7±0.03	1.2±0.06	1.4±0.05	<0.1	<0.1	<0.1	<0.1
Copper (Cu)	2.5±0.07	3.0±0.1	2.9±0.03	2.7±0.04	52±0.5	54±0.3	57±0.3	52±0.4
Iron (Fe)	250±0.2	255±0.3	264±0.3	282±0.4	190±0.2	200±0.3	204±0.6	200±0.1
Kalium (K)	$2.7*10^3$	$3.4*10^3$	$3.0*10^3$	$3.0*10^3$	$2.6*10^3$	$3.8*10^3$	$2.8*10^3$	$3.0*10^3$
Magnesium (Mg)	170±0.5	190±0.7	210±0.9	200±0.4	140±0.5	170±0.7	150±0.9	150±0.4
Manganese (Mn)	5.3±0.04	6.9±0.06	5.4±0.06	7.1±0.08	8.8±0.04	9.3±0.06	9.4±0.06	8.1±0.08
Molybdenum (Mo)	0.2±0.05	0.3±0.02	0.2±0.03	0.2±0.02	0.7±0.05	0.9±0.02	0.8±0.03	0.6±0.02
Natrium (Na)	350±0.3	360±0.4	430±0.5	380±0.4	530±0.3	600±0.4	680±0.5	700±0.4
Nickel (Ni)	<0.1	<0.1	<0.1	<0.1	<0.1	<0.1	<0.1	<0.1
Plumb (Pb)	<0.1	<0.1	<0.1	<0.1	<0.1	<0.1	<0.1	<0.1
Selenium (Se)	0.5±0.03	0.6±0.04	0.4±0.02	0.5±0.03	0.7±0.03	0.9±0.04	0.5±0.02	0.8±0.03
Strontium (Sr)	<0.1	<0.1	<0.1	<0.1	<0.1	<0.1	<0.1	<0.1

Note: - $P \leq 0.05$

Ultra-fine cobalt and copper have also contributed to kalium, calcium and phosphor increase in blood. Other substances content has not changed (Table 5).

Iron nano-powder in the diet has increased the content of calcium, copper, kalium, magnesium and manganese in muscles 20%-30 % higher than those of the control. Iron in muscle tissue has not exceeded the control one that shows the absence of cumulative properties of the drug and its ecological safety. Strengthening digestion and metabolism processes has not led to heavy metals accumulation in muscles that has influenced positively the general physiological state of the animals. Cobalt and copper nano-powders have also changed the content of mineral substances in animals' muscles but the level of cobalt and copper themselves has not exceeded the control one. At that heavy metals have not accumulated. One can find similar results when analyzing the mineral composition of the control and experiment animals' liver. We have not seen any metals including iron, copper and cobalt accumulation in blood, liver and muscles of the animals.

Table 6. Nano-powders Influence on Vitamins in Heifers Tissues, mg/100 g

Substance	Muscle Tissue				Liver			
	Control	Fe Nano-Powder	Co Nano-Powder	Cu Nano-Powder	Control	Fe Nano-Powder	Co Nano-Powder	Cu Nano-Powder
Vitamin A	1.2± 0.02	1.4± 0.03	.5± .04	1.7± 0.03	3.5± 0.02	4.8± 0.04	3.9± 0.03	4.5± 0.04
Vitamin C	8.4±0.03	9.7±0.05	8.8±0.07	9.3±0.06	17.7±0.1	20.9±0.2	19.5±0.1	21.3±0.02
Vitamin E	0.27± 0.004	0.28± 0.005	0.30± 0.003	0.31± 0.006	0.21± 0.003	0.22± 0.004	0.28± 0.005	0.26± 0.003

Note:- $P \leq 0.05$

3.8. Biochemical investigations of internal tissues and organs

We have determined the protein value of meat as for the correlation between essential and non-essential amino acids. The number of essential amino acids in meat protein of animals that have got iron nano-powder has increased by 3.9 %, in case with cobalt by 4.2 % and copper by 4.8 % that means the increase of the meat food value of the experiment heifers. There has also been some increase of glycine, histidine, methionine and serine, in particular stimulating anti-oxidants accumulation.

The content of fatty acids in subcutis of the experiment animals has not differed from that of the control one that proves the absence of pathological biochemical processes in the organism and the normal development of lipid metabolism. There has been an increase of vitamin A up to 11-35%, vitamin C up to 10-20 % and E (up to 33 %) in the liver and in muscles (A 17-42 %; C 8-16 %; E to15 %) of the experiment animals that has also increased meat value as a food product.

If at the beginning of the experiment the amino acid composition has been approximately the same in the control and experiment groups, by the end of the experiment there has been a considerable common increase of amino acids content in blood of the animals getting cobalt nanopowder by 26.8 % and copper by 21.9 % as compared with the control (Figure3,4).

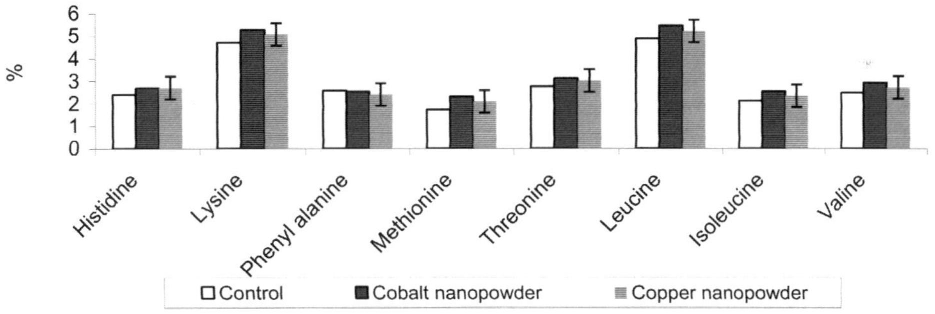

Figure 3. Essential Amino Acids in Blood after Feeding Cobalt and Copper Nano-Powders

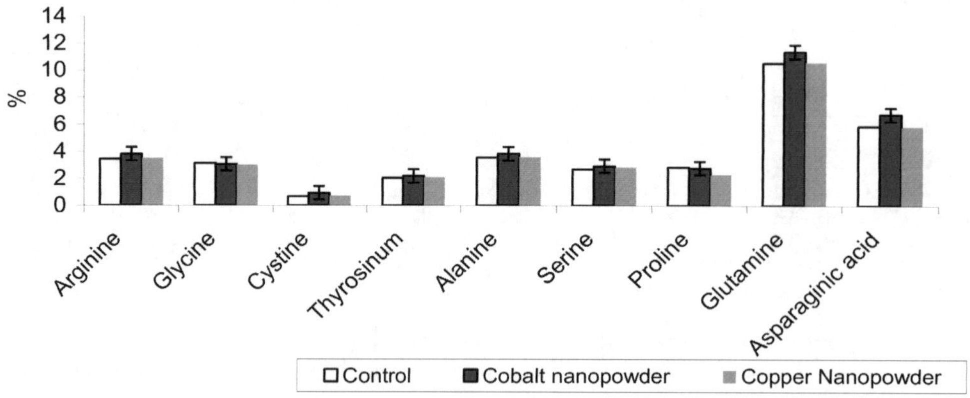

Figure 4. Contiguous Amino Acids in Blood after Feeding Cobalt and Copper Nano-Powders

We have got similar conclusions when studying the influence of cobalt and copper nano-particles on physiological and productive parameters of black and white bull calves (2008-2010). We have used the suspension on the basis of 1 liter per 1 ton of the mixed fodder. The metals nano-powders activity has remained during 1 month. The animals getting metals nano-powders have had higher digestive activity than the control ones. The dry matter digestion index has been 3.9 % and 2.5 % higher than the control for cobalt and copper nano-powders. The organic matter content has been 6.8 % and 3.5 % higher, protein 5.6 % and 3.8 % higher, fat 4.4 % and 3.8 % higher, fiber 3.6 % and 4.0 % higher and anazotic extractive matters 3.8 % and 4.0 % higher than those of the control for cobalt and copper nano-powders correspondingly. Adding cobalt and copper nano-powders has caused the increase of nitrogen and calcium balance in the organism and the decline of their fecal and urinary excretion.

There have been some changes of the experiment animals' meat productivity parameters. As a result of the slaughter the cows having nano-cobalt in the diet have had the hot carcass mass increased by 18.7 %, the dead weight by 16.9 % and the slaughter yield by 4.0 % as compared with the control. For copper the hot carcass mass has increased by 14.0 % and the slaughter yield by 3.5 %. At that the meat energetic value has increased by 10.1 – 12.5 %.

3.9. Immunity increase under the influence of iron, copper and cobat nano-particles

We have conducted investigations during the postpartum period characterized by the organism immune depression. In many instances one can consider it to be the reason for most cows having complications as a result of necessary protective strengths absence. In comparison with ill animals the cows getting metals nano-particles safely escaped complications of the postpartum period. At that they have had T indexes, total lymphocytes, considerably lower, only 38.3 % as opposed to 53.1 % of ill animals, whereas healthy cows have had this index equal to 44 %. This shows the absence of total lymphocytosis and low specific immunity in the experiment group. T-killers of the tested cows have been considerably higher (24.7 %), especially under the influence of nano-particles as compared with sick animals having 3.7 %. This fraction provides activation of the specific cell immunity. As the animals having such high exponents under nano-particles have not fallen ill we can consider it a favorable factor. B-lymphocytes have been vice versa lower in the experiment group that also shows the absence of complications development threat. In this situation the organism of the animal with the help of nano-particles copes with the activation of only the cell link not including the humoral one that is, as you know, increases when acute bacterial infections. One can also connect with this T-helpers lowering of cows having got nano-particles and stable high level of T- suppressors. We suppose stimulating the immunity is connected with the immune protective action of metals nano-particles. Copper and iron are important bio-metals participating in immune reactions. Their deficit weakens the function of the immune system: the decline of antimicrobial activity of macrophages, tissue saturation with granulocytes, cell eating suppression and anti-bodies formation. Production of lactoferrin that is an immune response modifier and an important component of supporting homeostase of the organism systems, participating in transporting iron and possessing antimicrobial, antitoxic and antiphlogistic activity is also connected with iron. All these facts prove the immune stimulating action of iron, copper and cobalt nano-particles in cows' bodies.

3.10. Milk productivity

We have studied milk productivity and milk quality of the first calf black and white heifers having iron ultra-fine powder in winter housing. We have formed 2 groups of animals numbering 20 animals in each of them. While doing that we have considered the animals' age, live weight and the cows' pregnancy period. All the experiment animals have weighed 470... 475 kg being 5…6 months pregnant and been kept in the same housing and feeding conditions. As for feeding difference the experiment first calf heifers got not only the balanced diet (according to norms) but the mixed fodder with 0.009 mg of iron per 1 kg of the mixed fodder a day.

Table 7. Milk Productivity of First Calf Cows in the First Phase of Lactation

Exponent	Group (n =20)		
	Control	I Experimental	II Experimental
Milk Yield, kg:			
Daily Average Milk Yield, kg	14.0	15.9	15.8
Natural Milk within 100 Days of Lactation	1402	1599	1508
Fat Content, %	3.90±0.06	4.2±0.01	4.3±0.02
Milk Fat within 100 Days of Lactation, kg	58,0	63.6	63.5
Protein, %	3.40±0.06	3.90±0.05	3.56±0.05
Density, °A	29.1±0.01	29.5±0.01	30.0±0.01
Acidity, °T	18.5±0.02	18.3±0.01	18.0±0.50
Energy Value of 100 g of milk, kJ	290.0 ±0.28	292.0±0.49	296.0 ±0.51
Calcium in Milk, mg	98.2±0.08	106.0±0.2	108.0±0.5
Phosphorus in Milk, mg	83.7±0.06	86.8±0.08	89.1±0.04

Note: $P \leq 0{,}05$;

Adding iron ultra-fine powder do the diet of the experiment animals has increased the daily average milk yield of the first calf heifers by 13.3 % as compared with the control (Table 7). For the first 100 days of the experiment group first calf heifers' lactation we have got 147 kg more milk as compared with the control group. Fat in the milk of the experiment animals has been 4.25 % that is 9.0 % higher than that of the control group. Milk fat in the milk for 100 days of lactation in the experiment group has been considerably higher (5.6 kg) than that in the control group as they have got more milk from the experiment cows and fat mass percentage in it has been a little higher.

Organoleptic estimation of the experiment first calf heifers' milk has not shown any differences. It has met the requirements of GOST R 52054-2003 in taste, color and smell.

All parameters under study of the experiment group have to some extent exceeded the analogues of the control group. Calcium and phosphorus in milk of the experiment group first calf heifers has been 9.0% higher than those of the control group (Table 7).

Physical-chemical properties of milk determined by concentration and the dispersiveness degree of the compounds are of great importance for the processing industry. To physical-chemical properties of milk one can attribute density and acidity. Milk density of the experiment first calf heifers has been

0.1-0.9° A higher that that of the control. The base-titratable acidity of all experiment animals' milk has not practically differed and corresponded to the norm. Thus, adding ultra-fine iron to the diet has increased milk yield, milk fat and improved milk chemical composition.

4. Conclusions

Ultra-fine microelements affect the organism mainly indirectly changing activity of ferments, hormones, protein, vitamins and other biologically active substances including metals or sensitive to their concentration change in the environment. Adding nano-crystal metals especially iron and cobalt promotes intensive accumulation of the animals' live weight due to the positive influence of protein, carbohydrates and mineral substances digestion. This supposition has been proved by morphological and biochemical investigations of blood and the content of mineral substances in blood serum. Ultra-fine microelements activate immune, enzymatic and humoral systems of the organism promoting metabolism increase and better digestion and fixation of the diet.

The increase of protein, carbohydrates and mineral metabolism leads to considerable increase of the experiment bull calves' and black and white heifers' meat productivity and the quality parameters of meat (chemical and biochemical composition, vitamins, essential amino acids, macro- and micro-elements) and milk.

References

[1] Nazarova AA, Zheglova TV, Polishchuk SD, Churilov GI, Ivanycheva YN and Folmanis GE 2008 *Rabbit Breeding and Fur-Farming* **6** 8

[2] Ovsyannikov AI 1976 Basis of Experimental Work in Animal Breeding (*Moscow: Kolos)* p. 304

[3] Churilov GI, Polishchuk SD and Nazarova AA *Journal of Materials Science and Engineering B* **3** 379

[4] Ilyin AP, Korshunov AV and Astankova AP 2007 *International Science and Practice Conference "NanoTechnologies and NanoMaterials for Biology and Medicine". Novosibirsk,* 43

[5] Glushchenko NN, Bogoslovskaya OA and Olkhovskaya IP 2002 *Chemical Physics* **21** 79

[6] Kureneva VP, Egorov IA, Glushchenko NN, Fatkullina LD and Fedorov YI 1985 *Collection of Scientific Papers Borovsk* **31** 80

[7] Nazarova AA and Polishchuk SD 2012 Cattle Metabolism Peculiarities when Adding Cobalt NanoParticles to Ration Materials of VII International Symposium Fundamental and Applied Scientific Problems (*Moscow: RAS)* p.257

[8] Kovalsky VV and Rish MA 1970 Biological Role of Copper in Animals Organism Biological Role of Copper *Symposium. Moscow, April 4-6* (*Moscow: "Science")* p. 113.

[9] Delektorskaya LN and Dobryanskaya LD 1974 Unitized Methods of Clinical Lab Investigations *Moscow* 5

[10] Zenova N, Nazarova A and Polishchuk S 2010 *Milk and Meat Animal Breeding* **1** 30

[11] Nazarova AA , Polishchuk SD, Stepanova IA Churilov GI, Nguyen HC and Ngo QB 2014 *J. Advances in Natural Sciences: Nanoscience and Nanotechnology* **5** 1

IOP Conf. Series: Materials Science and Engineering **98** (2015) 012037 doi:10.1088/1757-899X/98/1/012037

Toxicological characterization of bio-active drugs on basis of Iron Fe, Co, and Copper Cu nanopowders

S Polishuk, A Nazarova and I Stepanova
Division of chemistry, Ryazan State Agrotechnological University, 1 Kostycheva st., Ryazan, 390044, Russia

E-mail: svpolishuk@mail.ru

Abstract. The article presents investigations of toxicological parameters (acute and chronic toxicity, cumulative coefficient) of iron, cobalt, copper and copper oxide nanoparticles with white rats in labs. We have estimated the optimal concentrations of the above mentioned substances with rabbits. We have also studied morphological, physiological and biochemical parameters of the animals when adding the optimal doses to the diet for a long term.

1. Introduction

The authors have studied the biological and catalyst properties of nanomaterials and developing bio-drugs on the basis of metal nanoparticles to be used in livestock-raising and veterinary at the Centre of nanomaterials and nanotechnologies for AIC in Ryazan Agrotechnological University Named after P.A. Kostychev. Investigations have shown that one can use ultra-fine iron, cobalt and copper as non-expensive, nontoxic and highly efficient biological catalysts of biochemical processes in the organism improving physiological state, morphological and biochemical blood parameters increasing activity of the experiment animals' ferment systems and increasing their productivity and biological value of meat [1,2,3,4].

The bio-drugs study was carried out in different regions of Russia (Moscow, Tula and other oblasts, Krasnodar, Stavropol Territories). We have been studying ultra-fine powders at Ryazan Agrotechnological University Named after P.A. Kostychev since 1997. We have studied the influence of nano-crystal metals on the physiological state of lab and agricultural animals, their productivity and products safety in a case of adding metals to the diet.

For effective use of nano-bio-drugs it is necessary to characterize them toxicologically and determine toxic and optimal concentrations of metals nano-particles for lab animals [5,6].

2. Methods

The aim of the investigations has been determining toxic and optimal doses of nano-bio-drugs on lab animals and morphological-physiologic and biochemical characteristics of animals in a case of long-term adding to the diet the optimal doses of the drugs being studied.

For the investigations we have chosen lab non-pedigree white rats and rabbits, breed "Soviet chinchilla".

We conducted our experiments in 2006-2009 and they included lab and farm tests. [4]

In first experiments we have studied:

1. Metals nano-powders toxicity (acute and chronic toxicity, LD_{50}, cumulative properties).
2. Optimal doses of nano-crystal metals being added to the diet.

Content from this work may be used under the terms of the Creative Commons Attribution 3.0 licence. Any further distribution of this work must maintain attribution to the author(s) and the title of the work, journal citation and DOI.
Published under licence by IOP Publishing Ltd

3. The effect of nano-crystal metals on the physiologic state, blood morphological-biochemical parameters, health and reproduction of rabbits in the lab.

We have chosen the experiment animals into groups according to the principle of balanced groups-analogues taking into account their sex, age, breed, live weight. The animals have been in the same food and housing conditions. We have taken blood from the side ear vein four- timely at interval of 10 days before morning feeding. We have determined all blood parameters (hemoglobin, erythrocytes, leucocytes, leucocytic formula and blood sedimentation rate) according to unitized methods of clinical diagnostics [7]. We have analyzed the biochemical parameters of blood with spectrophotometer "Spectrum" firm "Abbot" according to unitized methods of clinical lab investigations [8].

We have determined the number of alanine and aspartate transaminase by Wrightman-Frenkel's colorimetric method; the number of lactate dehydrogenase by the reaction with 2.4-dinitrophenyl diamide (method of Sevel & Tovarec); the number of alkaline phosphatase by the reaction with p-nitrophenyl phosphate (method of Bessey, Lorie & Brock); the amount of uric acid by the reaction with phosphatotungstic reagent and the number of cholinesterase by acetylcholine chloride hydrolysis with the help of the colorimetric method.

For true development and safe implementation of biologically active drugs based on metals nano-powders in livestock-raising we have developed their vet and toxicological characteristic.

On the first stage of investigations we have determined acute, chronic toxicity and cumulative properties of iron, cobalt and copper nano-powders. We have conducted our investigations in determining acute toxicity in lab conditions. The objects of the investigation have been non-pedigree male rats weighing 180-210 g. The feeding diet is presented in table 1.

Table 1. Feeding diet of lab rats during experiment (in grams per 1 animal a day)

Fodder and Additives	Amount, g
Grain mixture (oats, millet, wheat, sunflower)	15
Wheat-bread	4
Grits	3
Mixed fodder	10
Milk	8
2nd category meat	5
Succulent fodder (carrot, beetroot, cabbage)	10
Greenery	10
Fish oil	0.1
Feed yeast	0.2
Bone flour	0.2
Salt	0.2

We have got 8 experiment groups 10 animals each and have chosen the doses of each metal being studied taking into account some peculiarities (table 2).

Table 2. Doses of studied drugs for experiment groups

Groups	Iron Nano-Powder	Copper Nano-Powder	Copper Oxide Nano-Powder	Cobalt Nano-Powder
Control	-		-	
Experiment 1	100 mg/kg	10 mg/kg	10 mg/kg	10 mg/kg
Experiment 2	500 mg/kg	50 mg/kg	20 mg/kg	50 mg/kg
Experiment 3	1000 mg/kg	100 mg/kg	50 mg/kg	100 mg/kg
Experiment 4	1500 mg/kg	150 mg/kg	80 mg/kg	150 mg/kg
Experiment 5	2000 mg/kg	200 mg/kg	100 mg/kg	200 mg/kg
Experiment 6	2500 mg/kg	250 mg/kg	250 mg/kg	250 mg/kg
Experiment 7	3000 mg/kg of live	300 mg/kg of live	300 mg/kg of live	300 mg/kg of live

weight	weight	weight	weight

We have estimated the average mortal doses for each metal nano-powder a posteriori:

Iron LD_{50} - 2130 + 80 mg/kg
Copper LD_{50} - 380 + 30 mg/kg
Copper oxide LD_{50} - 300 + 80 mg/kg
Cobalt LD_{50} - 145 + 10 mg/kg

The groups getting the maximum concentration of the drug being studied have demonstrated considerable increase of leucocytes in animals' blood (up to 30 % higher than the control). It is connected with rats' protective reaction to getting the drug because the leucocytes level as immune factors shows the immune response toxic substances. One can also see considerable increase of bilirubin and its fractions during the experiment (up to 30 %) that is connected with liver cells damage because of the toxic effect of maximum doses of the drug being studied. The increase of crude protein (up to 24 %) and protein metabolism products (uric acid, creatinine) proves the possible damage of the excretory system and kidneys in particular. The experiment animals' autopsy has proved the biochemical results of the experiment. We have seen liver hyperemia and asphyxic heart.

We have had the experiment in determining sub chronic toxicity and cumulative properties in 3 series during 40 days. On first 5 days every rat has got 1/10 from the existing single dose LD_{50}. Then we have increased the dose by 1.5 times every 5 days. During the experiment we have registered cases of toxicosis and animals' death and prosected the dead animals.

Based on the experiments we have estimated the chronic median lethal dose:

for iron nano-powder chronic LD_{50} = 25 743.8 mg/kg of live weight;
for copper nano-powder chronic LD_{50} = 4 120.5 mg/kg of live weight;
for copper oxide nano-powder chronic LD_{50} = 1 543.8 mg/kg of live weight;
for cobalt nano-powder chronic LD_{50} = 1 780.0 mg/kg of live weight.

During the experiment we have noticed in animals a considerable increase of leucocytes (up to 30-40 %) and determined the toxic dose of the drug being studied.

The cumulative coefficient ($K_{cum)}$ for nano-powders has been as follows:

iron – 12.09; copper – 10.84; copper oxide – 5.15; cobalt – 12.27.

According to the standard classification some cumulative coefficients prove that the accumulation of all the metals being studied has been mild.

Taking into account the results we have got one can say that nano-powders of iron, cobalt, copper and copper oxide do not posses any cumulative properties and according to GOST 12.1.007-76 "Hazardous substances" can be attributed to the 4th class of hazard, i.e. inconsiderably hazardous substances.

3. Results and Discussion

To determine the metabolism state of the experiment animals in a case of feeding them with copper nano-powder we have estimated the dynamics of some blood biochemical parameters characterizing this or that metabolic process (table 3).

We have carried out our investigations with non-pedigree male rats in the lab. According to the principle of animals-analogues we have chosen 20 rats. We have taken the experiment animals' blood sampling and autopsy in the lab together with some specialists.

The experiment groups animals have got an increase of crude protein and the level of residual nitrogen during the whole experiment as compared with initial rates on the average by 5.4-9.1 % and 9.2 – 20.6 % ($P<0.05$) correspondingly.

The blood of the experiment rats has not practically differed by morphological parameters as compared with the control groups. The average results of erythrocytes, hemoglobin, leucocytes, color index and leukogram have been according to the norm. The study of main biochemical parameters has not educed any negative effects of the nano-powders usage. On the contrary there has been some positive shift in the dynamics of glucose concentration and the level of cholesterol and urea total bilirubin and creatinine has declined. Therefore, the metabolic state of the experiment animals when feeding them with nano-powders has improved.

Table 3. The Experiment animals' blood biochemical parameters

Group	Time of Investigation, days				
	Initial	10	20	30	60
		Crude Protein, g/l			
1st (1/40 of LD_{50})	67.0±0.5	68.7±0.05	68.5±0.03	70.5±0.05	71.1±0.04
2nd (1/10 of LD_{50})	65.7±0.02	66.6±0.03	69,4±0.07	71.7±0.07	72.1±0.06
3d (1/4 of LD_{50})	66.7±0.09	67.6±0.03	70,2±0.03	72.0±0.9	71.1±0.08
Control	67.0±0.6	66.7±0.05	99.7±0.06	65.4±0.05	67.0±0.7
		Residual Nitrogen, mg %			
1st (1/40 of LD_{50})	33.3±0.04	34.5±0.07	36.6±0.06	39.2±0.04	38.4±0.03
2nd (1/10 of LD_{50})	36.2±0.02	34.8±0.04	37.6±0.04	39.5±0.04	38.3±0.04
3d (1/4 of LD_{50})	32.6±0.08	36.2±0.09	38.3±0.03	39.3±0.06	39.0±0.3
Control	33.0±0.3	35.5±0.07	35.7±0.03	35.0±0.2	36.1±0.04
		Glucose, mol/l			
1st (1/40 of LD_{50})	7.7±0.01	8.0±0.3	8.4±0.01	8.9±0.02	9.1±0.01
2nd (1/10 of LD_{50})	7.7±0.02	8.1±0.03	8.5±0.02	8.9±0.01	9.2±0.05
3d (1/4 of LD_{50})	7.0±0.3	8.0±0.1	8.6±0.04	9.1±0.01	8.3±0.02
Control	7.0±0.4	7.6±0.01	8.0±0.9	7.8±0.02	7.8±0.04
		Cholesterol, mol/l			
1st (1/40 of LD_{50})	1.80±0.05	1.84±0.009	1.84±0.009	1.81±0.009	1.81±0.003
2nd (1/10 of LD_{50})	1.80±0.09	1.84±0.005	1.82±0.002	1.80±0.02	1.80±0.04
3d (1/4 of LD_{50})	1.80±0.09	1.84±0.003	1.80±0.07	1.78±0.025	1.79±0.009
Control	87.0±0.5	88.0±0.2	90.0±0.2	1.9 ±0.02	1.87±0.009
		Total Bilirubin, mcmol/l			
1st (1/40 of LD_{50})	3.18±0.02	3.00±0.06	2.92±0.004	2.83±0.002	2.71±0.001
2nd (1/10 of LD_{50})	3.12±0.07	3.10±0.07	2.95±0.003	2.87±0.002	2.22±0.001
Group	Time of Investigation, days				
	Initial	10	20	30	60
3d (1/4 of LD_{50})	3.23±0.02	3.00±0.05	2.87±0.007	2.87±0.007	2.85±0.003
Control	3.20±0.13	3.23±0.02	3.18±0.001	3.18±0.001	3.10±0.001
		Urea, mol/l			
1st (1/40 of LD_{50})	5.0±0.01	4.7±0.08	3.0±0.3	3.5±0.04	2.1±0.03
2nd (1/10 of LD_{50})	5.1±0.03	4.9±0.04	4.0±0.02	3.1±0.03	2.1±0.02
3d (1/4 of LD_{50})	4.9±0.04	4.8±0.05	3.9±0.02	3.0±0.03	2.3±0.01
Control	5.0±0.02	5.2±0.05	5.1±0.03	5.0±0.02	4.5±0.02
		Creatinine, mcmol/l			
1st (1/40 of LD_{50})	51.0±0.1	50.8±0.02	50.2±0.01	51.2±0.06	50±1
2nd (1/10 of LD_{50})	51.2±0.01	51.0±0.1	50.6±0.02	50±2	50.1±0.02
3d (1/4 of LD_{50})	51.0±0.2	50.9±0.02	50.4±0.01	50.2±0.02	50.2±0.03
Control	51.3±0.02	51.0±0.2	51.0±0.02	50.9±0.02	50.8±0.02

Note: - $P \leq 0.05$

We have determined the optimal doses of metals nano-powders when adding them to the diet of the lab animals with rabbits breed "Soviet chinchilla" aged 30 days. We have got 10 experiment groups each having 6 rabbits.

To add metals nano-powders to the feeding diet we have used the mixed fodder treated with the drugs suspension. Every 10 days of the experiment and 10 days after it we have weighed the experiment animals and had morphological and biochemical tests of the experiment animals' blood and supervised the animals' general development, appetite and activity. The rabbits have been kept in conditions recommended for their management and breeding in a case of twice-a-day feeding. The influence of different doses of ultra-fine metals on the live weight of the control and experiment animals is presented in table 5.

One can consider 0.08 mg/kg of the live weight to be the optimal dose of the iron nano-powder. This concentration has promoted the increase of the animals' weight up to 7.5 % in 30 days after giving the drug. The rest concentrations have had less effect: the dose of 0.16 mg/kg of the live weight has given 5.5 % weight increase and the dose of 0.24 mg/kg has brought a 3.9 % increase as compared with the control.

The optimal dose for cobalt nano-powder has been 0.02 mgr/kg of the live weight that let increase the live weight by 4.8 % higher than the control. For copper nanopowder the optimal dose has been 0.04 mg/kg of the live weight a day that has increased the rabbits live weight by 3.8 as compared with the control. The increase of copper nanoowder dose up to 0.12 mg/kg of the live weight has negatively influenced the animals' weight gain that has been 0.8 % lower than that in the control group in 30 days of the experiment. Moreover, the animals in this group have been less active in comparison with the control ones. Therefore, high concentration copper has oppressed the rabbits' health.

Table 4. The rabbits' feeding diet when using the combined feeding type (g per 1 animal)

Fodder	Age (days)		Replacement older than
	30-60	61-120	120 days
	Live weight (kg)		
	1.0 – 1.5	1.5 – 2.5	2.5 – 3.5
Grain (oats), g	60	80	80
Mixed fodder, g	10	20	30
Sunflower coarse meal, g	5	15	10
Succulent fodders, g (carrot, beetroot, cabbage, potato)	50	120	150
Meadow hay, g	80	100	150
Meadow grass, g	250	400	450
Fish oil, g	0.3	0.5	0.5
Bone flour, g	2.0	2.0	2.0
Feed yeast, g	0.2	0.4	0.4
Salt, g	0.6	1.0	1.2
Fodder units (EFU)	0.11-0.15	0.16-0.20	0.21-0.23
Metabolic energy, mJ	1.10-1.45	1.61-2.00	2.09-2.30
Dry matter, g	93-130	130-195	200-220
Crude protein, g	18-27	27-37	34-37
Digestible protein, g	14-21	21-28	26-29
Crude fiber, g	17-23	20-32	35-39
Calcium, g	0.8-1.0	1.1-1.4	1.5-1.7
Phosphorus, g	0.6-0.7	0.7-0.9	0.9-1.1
Carotene, mg	0.8-1.4	1.5-2.0	2.4-2.6
Vitamin D, mg	100-170	170-240	320-380
Vitamin E, mg	2.0-3.4	3.4-4.8	6.4-7.6

Table 5. Live weight of the lab rabbits, g

Groups	Before Experiment	In 10 Days	In 20 Days	In 30 Days	10 Days after Experiment
Control	845	1010	1145	1260	1510
Group 1 Fe NP 0.08 mg/kg	840	1130	1230	1355	1660
Group 2 Fe NP 0.16 mg/kg	830	1110	1180	1330	1625
Group 3 Fe NP 0.24 mg/kg	840	1090	1165	1310	1610
Group 4 Co NP 0.01 mg/kg	850	1120	1195	1290	1620
Group 5 Co NP 0.02 mg/kg	840	1125	1210	1320	1630
Group 6 Co NP 0.03 mg/kg	835	1100	1170	1275	1630
Group 7 Cu NP 0.04 mg/kg	830	1090	1185	1305	1590
Group 8 Cu NP 0.08 mg/kg	840	1070	1160	1275	1560
Group 9 Cu NP 0.12 mg/kg	845	1045	1150	1250	1505

Note: - $P \leq 0.05$

The cancellation of copper NP has not led to the weight decline of the experiment animals. On the contrary the tendency of increasing the body weight has remained even 10 days after the cancellation of metals ultra-fine powders (table 5).

During the investigations all morphological parameters of the experiment rabbits' blood have been within the physiological norm.

3.1. Metals nano-powders influence on physiological parameters of rabbits and blood morphological biochemical indexes.

For the experiment we have formed the following groups 8 animals each: group 1 – control; group 2 – experiment – 0.08 mg/kg iron ultra-fine powder; group 3 – experiment – 0.02 mg/kg cobalt ultra-fine powder; group 4 – experiment – 0.04 mg/kg copper ultra-fine powder a day. The experiment has lasted for 60 days. Every 10 days we have weighed the control and experiment animals. The age of the animals at the beginning of the experiment has been 30 days. The animals in all 4 groups have steadily gained the weight, been active and had good appetite. We have seen some positive results of nano-crystal metals usage in 10 days of the experiment. Adding iron nanopowder to the diet has increased the rabbits' live weight by 8.8 % in 30 days and by the end of the experiment by 11.7 % higher than the control. Cobalt nano-powder has also promoted the rabbits' live weight increase by 5.6 % in 30 days and by 7.8% in 60 days. The increase in a case with nano-copper has been 6.3 % higher than the control at the end of the experiment.

We have carried out a simultaneous investigation of morphological and biochemical parameters of the experiment animals' blood. We have chosen 4 animals from each experiment group. Blood sampling has taken place every 10 days before morning feeding. Before the experiment all blood parameters have been within the physiological norm. Table 6 presents the results of investigating the blood of the control and experiment animals in 60 days of the experiment.

Table 6. Blood morphological-biochemical parameters of rabbits got the nano-crystal metals additive

Blood Parameters	Control	Iron	Cobalt	Copper

		Nano-Powder	Nano-Powder	Nano-Powder
Erythrocytes, 10^{12}/l	5.4±0.03	5.9±0.03	5.7±0.04	5.8±0.03
Hemoglobin, g/l	110.0±0.2	120.0±0.3	119.0±0.4	118.0±0.2
Blood platelets, thousand	205.0±0.6	200.0±0.7	206.0±0.6	210.0±0.5
Leucocytes,*10^9/l	4.1±0.04	5.3±0.05	4.5±0.05	5.1±0.06
Leukogram, %				
Lymphocytes	43.0±0.3	51.0±0.4	53.0±0.5	50±1
Monocytes	6.0±0.02	6.0±0.03	6.0±0.05	6.0±0.05
Granulocytes	51.0±0.5	43.0±0.04	41.0±0.03	44.0±0.04
Total protein of serum, g/l	52.5±0.16	58.0±0.17	61.0±0.19	56.5±0.05
Protein fractions, %				
α_1 - globulins	3.4±0.05	4.0±0.5	4.2±0.03	4.2±0.06
α_2 - globulins	11.2±0.03	6.5±0.04	4.6±0.05	9.5±0.02
β - globulins	8.0±0.04	10.6±0.05	15.0±0.09	13.0±0.01
γ - globulins	10.0±0.03	12.5±0.04	13.0±0.01	12.0±0.02
Albumins	67.4±0.08	66.4±0.07	64.0±0.05	61.3±0.08

Note: - $P \leq 0.05$

The rabbits that have got ultra-fine metals as an additive have shown the true increase of erythrocytes and hemoglobin.

The iron nano-powder has increased erythrocytes by 9.3 % and hemoglobin by 9.1 % in 60 days of the experiment as compared with the control that has positively influenced the animals' physiology. By the end of the experiment there has been a 29.3 % increase of leucocytes and the leukogram change, i.e. the increase of lymphocytes (by 8 %) and the decline of granulocytes by 8 %. There has also been an increase of total protein in serum (by 10.5 %) and α_1 - globulins (by 2.5 %) that has caused the decline of α_2 - globulins and albumins. The rest parameters have had inconsiderable deviations from the control.

One can see similar changes in other experiment groups as well. Cobalt nano-powder has increased erythrocytes by 5.5 %, hemoglobin by 10 %, leucocytes by 9.8 %, lymphocytes by 10 % and total protein by 16.2 %.

The copper nano-powder has increased the level of erythrocytes and hemoglobin (by 7.4 % and 7.3 %), leucocytes and lymphocytes (by 24.4 % and 7 %). All serum parameters in a case of nano-crystal metals have been within the physiological norm. But there has been an increase of total protein at the expense of the globulin fraction and α_1 – and β - globulins in particular. It is known that α_1 – and β - globulins csynthesize in the liver and serve active transmitting agents of different blood substances. The increase of these fractions leads to strengthening the active transfer of carbohydrates and lipids to tissues that influences the number and activity of globulins. This in its turn activates metabolism. γ - globulins have had the same character of changes. At the expense of γ -globulins increase as compared with the control there has been an increase of the immune-biological reactivity as they are protective anti-substances (immune globulins) and are responsible for the specific immune response. These parameters have changed most vividly in a case of adding Iron and cobalt nanopowders.

The immune system that is structurally and functionally organized combination of lymphoid cells interacting with each other and some secondary cell elements on different stages of immune genesis can be an indicator reflecting nanomaterials effect on the organism. If one takes into account that ultra-fine microelements activate immune, ferment and humoral systems of the organism increasing metabolism and better digestion of nutrients then we can use them as bio-additives.

Female and male rabbits after getting the offspring have been dissected to see the state of their internal organs. At that we have discovered the total absence of internal organs pathologies: the liver, the kidneys, the heart, the lungs, the bowels have been in norm, without any external changes, had normal size. The structure of each organ has been in agreement with the norm.

References

[1] Kulikova OV, Nazarova AA and Polyschuk SD 2012 *Bull. Ryazan State Agrotechnol. Univ.* **2** 70
[2] Nazarova AA , Polishchuk SD, Stepanova IA ,Churilov GI, Nguyen HC and Ngo QB Adv. 2014 *Nat. Sci.: Nanosci. Nanotechnol.* **1** 1
[3] Churilov GI, Nazarova AA and Polyschuk SD 2010 Scientific and practical basis of application of nanometal powders in feeding farming animals Monograph (Ryazan City: Russia) p. 144
[4] Churilov GI, Polishchuk SD and Nazarova AA 2013 *J. Mater. Sci. Eng.* **3** 379
[5] Karkischenko NN 2006 *Biomed.* **1**(1) 5
[6] Onishenko GG, Bikotko BG, Pokrovsky V.I., Potapov A.I. 2007 The Concept of toxicological studies, risk assessment methodology, methods for identifying and quantifying nanomaterials (in Rissian)
[7] Zupenetc IA, Mis Misyureva SV and Propisnova VV 2005 Clinical Laboratory: Research Methods: the Manual NFaU "Zolotye Straitsy"
[8] Ilyin L and Sanotsky IV 1990 Methods for Determining the Toxicity and Hazards of Chemicals *(Moscow: Medicine).*

Colloidal metal oxide nanoparticle systems: the new promising way to prevent antibiotic resistance during treatment of local infectious processes

Y Karasenkov[1], G Frolov[2], I Pogorelsky[3], N Latuta[4], A Gusev[2], D Kuznetsov[2], V Leont'ev[2]
[1] Rosdent Ltd., 9-4 Michurinsky ave., Moscow, 119192, RF
[2]National University of Science and Technology "MISIS", 4 Leninsky pr., Moscow, 119049, RF
[3]Vyatka State University, 36 Moscovskaya str., Kirov, 610000, RF
[4]First Moscow State Medical University, 8-2 Trubetskaya str., Moscow, RF

[1]E-mail: rosdent@mail.ru

Abstract. New bactericidal containing nanoparticles colloids for application in dentistry, maxillofacial surgery, urology, obstetrics, gynaecology, ENT, proctology have been developed. The various water colloidal nanodispersive systems of metals and oxides have been obtained by means of electric impulse – condensation (electroerosion) method. These systems are based pure elements and alloys of argentum (Ag), titanium dioxide (TiO_2), iron oxide (Fe_2O_3), tantalum oxide (TaO), vanadium oxide (VO_2), cobalt oxide (CoO), tantalum dioxide TaO_2, zinc oxide (ZnO), copper oxide (CuO) and mixed suspensions of titanium, aluminium and molybdenum oxides. The research has been made on culture of dentobacterial plaque and mixed culture issued from gingival spaces. The composition of culture was identified with S.aureus, S.epidermidis and nonfermentable kinds of E.coli. The observation period lasted more than nineteen days. All solutions showed highly prolonged bactericidal activity in dilutions from the whole solution 1-20 mg/L. The bactericidal activity of powder specimen of silica containing Ag and Fe_2O_3 nanoparticles used as dental filling material and disintegrates of composite materials (produced by "StomaDent" CJSC) have been studied. Tested materials have long (up to 19 days and more) bactericidal activity.

1. Introduction

The principally new drugs with high antibacterial activity, which would significantly reduce the scale of using antibiotics in medical practice and thereby reduce further spread of antibiotic resistance are in the spotlight now. A lot of different antibiotics which make parts of materials for topical treatment of odontogenic infections are used in dentistry and oral surgery. However the activity of antibacterial components in these medicines is short-term. In addition an attempt to add resistant antibacterial properties to cements, restorative materials and adhesives has so far failed. And after all it has long been observed that the use of light-cure composite materials, even those of the latest generation, in practice does not guarantee further protection of dental hard tissues from appearance of secondary caries. Therefore, current research is related to the future use of colloidal solutions of metal nanoparticles as an antibacterial component of dental materials.

Content from this work may be used under the terms of the Creative Commons Attribution 3.0 licence. Any further distribution of this work must maintain attribution to the author(s) and the title of the work, journal citation and DOI.
Published under licence by IOP Publishing Ltd

The high pH of oral fluid, a great variety of biochemical processes, a wide range of natural and pathogenic oral microorganisms, peculiarities of structure and chemical composition of tooth tissues, etc., determine the following requirements for filling materials and adhesive systems:

1. A wide range of bactericidal action.
2. Prolonged bactericidal effect.
3. Absence of local and systemic toxic effects.
4. Adhesion to dental hard tissues and prosthetic surfaces.
5. Insolubility in oral fluid.
6. Preservation of physicochemical properties when interacting with other dental materials.

The possibility of using nanomaterials in medicine has been studied recently by many researches. The bactericidal activity of silver nanoparticles on various individual strains of microorganisms is studied well enough. From a practical point of view it is necessary to analyze the antimicrobial properties of aqueous dispersions of silver and other metals directly with plaque microflora.

2. Materials and methods

Water dispersions of Ag, ZnO, FeO, TaO_2, TiO_2, CuO, VO_2 nanoparticles were produced by means of as called electro-erosion method. was filled into the arc Glass chamber was filled with distilled water and electric impulses of the capacity of 6.5 kW were fed to the metal electrodes made of Ag, Zn, Fe, Ta, Ti, Cu, VO_2) accordingly to type of nanoparticle systems. As a result of evaporation of electrodes the formation of colloidal solution of metals in water occurs. This method allows to obtain the stable dispersions of oxide nanoparticles. The nanoparticle systems are stable due to formation of hydrate multilayer shells on the surface of nanoparticles.

Information about dislocation microstructure of nanoparticles was obtained with using of transmission electronic microscopy (TEM) using a microscope JEM 2100 (Jeol, Japan). The suspensions were dropped on TEM copper grids with amorphous carbon substrate. All the measurements were conducted under voltage of 200 kV; cathode LaB6 was used as a source of electrons. The microbiological samples were dehydrated prior TEM experiments by means of treatment with acetone.

The culture of coliform bacterium E.Coli was used to evaluate antibacterial activity. Solid nutritional medium (agar) was made on the base of Hottinger broth (protein base used for nutritional media preparation for microorganisms. PanEco, Moscow.). Combination of the medium components provides the necessary nutritional elements for the growth of cultures in the form of certain colonies on the solid medium surface. Hottinger agar was melted on a boiling water bath, after that it was cooled to 45…50 °C and was dispensed into sterile 100 mm Petrie dishes by 25 ml with the resulting layer of 4.0 ± 0.5 mm. After setting the Hottinger agar was dried at 37 ± 1 °C for 45-60 min. Two days prior to researches the test culture was reinoculated from the storage medium onto slant nutritional agar in test tubes. The next day the obtained agar culture was used to produce suspension of the test strain employing turbidity standard.

To control dilution 0.1 ml (100 mkl) of suspension from 6th and 7th dilutions of culture was plated to the dishes with nutritional agar by method of direct surface inoculation. Three such inoculations were made out of each culture. After inoculation test tubes with culture were immediately placed in a refrigerator. Dishes with cultures were incubated in a thermostat at 37 ± 1 °C within 18…24 hours. On the day of the research the average number of colonies grown in the three dishes was calculated for each series of inoculations. All laboratory glassware was treated with 3-%water solution of phenol or 2% chloramine solution for disinfection.

Disk diffusion test was used to assess antibacterial properties of the studied dispersions. It consists of measuring inhibition zone around the paper carrier of of antibacterial preparation. The inhibition zone forms as a result of the preparation diffusion from the carrier into the media. Within certain limits the inhibition zone is reversely proportional to the minimum suppressing concentration (MSC). The study results in attributing the microorganism to one of the sensitivity categories (sensitive, intermediate or resistant). For the research small batches of disks (wafers) were made of filtering paper.Laboratory glassware was washed twice in 2% HNO_3 solution and rinsed 7 times with distilled water. All the surfaces including inner surfaces of air hoods and tables were treated with disinfecting compositions of sodium bicarbonate. Application of the discs was implemented with the help of sterile tweezers. The distance from the disc to the edge of the dish and between disks was 15…20 mm.

3. Results and discussion

There are two ways to obtain hydrosols of metals and their oxides: the chemical method of producing nanocolloids based on reduction of cations and physical method based on evaporation of metal in different liquid media. The first method has a various significant shortcomings. Aqueous dispersion of metal and oxide nanoparticles obtained by equilibrium chemical reduction have relatively low surface energy level, these colloids have some toxicity due to the presence of ions that makes its applications to living organisms undesirable [1-4].

Therefore the most promising methods of production of such nanosystems of metals and oxides for biological applications are physical. One of these methods is electrical erosion of metal electrodes by a voltaic arc in liquid media. A method is based on electrical erosion of electrodes under the influence of spark arc discharge occurring between them. The resulting low-temperature plasma in the voltaic arc is cooled by liquid (water or another liquid) which condenses nanoparticles at ambient temperature. The result of huge temperature difference is condensation of electrode material in liquid with formation of polydisperse colloidal system which consisted of nanosized metal or metal oxides in amorphous or crystalline phase homogeneously dispersed in liquid. Thus obtained colloidal solutions usually are stable to precipitation during several months or several years, The concentrations of nanoparticle depends of production conditions and has a range of 0.1 - 30 mg/l.

Analysis of transmission electron microscopy photos of as obtained various nanosystems on the base of Ag, ZnO, FeO, TaO_2, TiO_2, CuO, VO_2 (Figures 1-4) indicates that colloidal systems vary in different distribution of particles and the ratio of crystalline and amorphous phases, , what allows to suggest a high potential surface energy of particles in these systems. The potential energy will be transferred to the dispersion medium for a long time as a result of the processes of crystallization of the amorphous phase, coagulation and coalescence of the particles, recrystallization of the crystalline phase and redox processes associated with changes in the chemical composition of the dispersed phase.

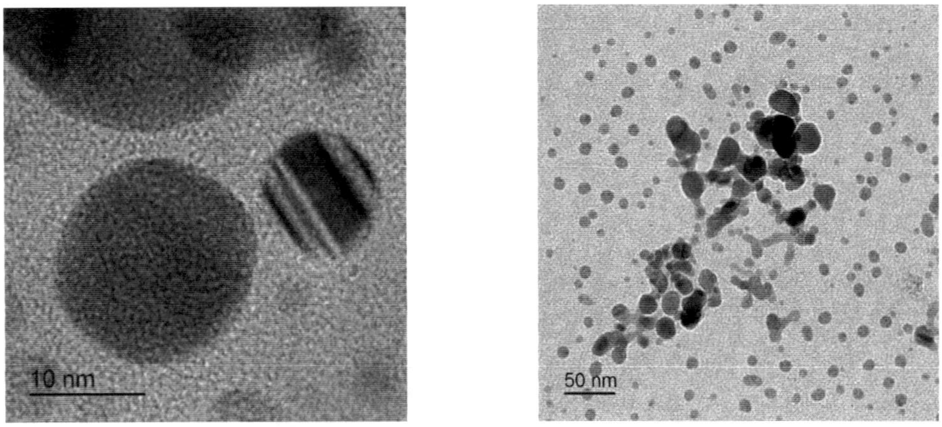

Figure 1. TEM-images of silver nanoparticles obtained.

IOP Conf. Series: Materials Science and Engineering **98** (2015) 012038 doi:10.1088/1757-899X/98/1/012038

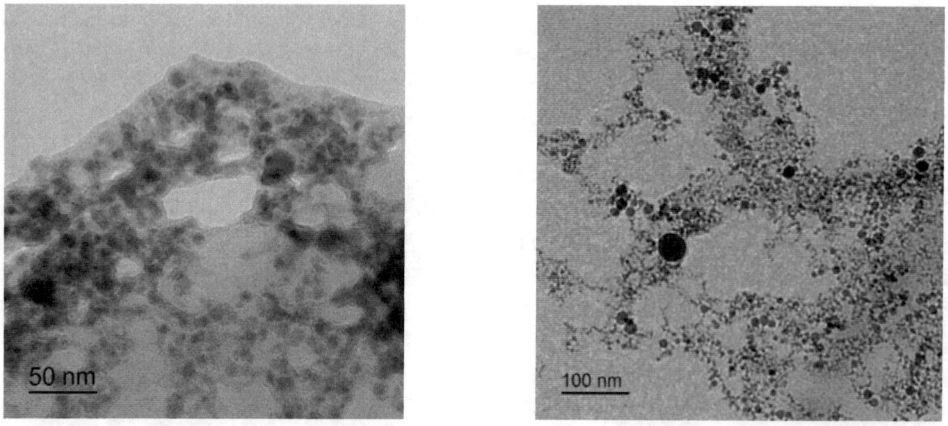

Figure 2. TEM-images of copper oxide nanoparticles (I) and zinc oxide (left), iron oxide (II) (right).

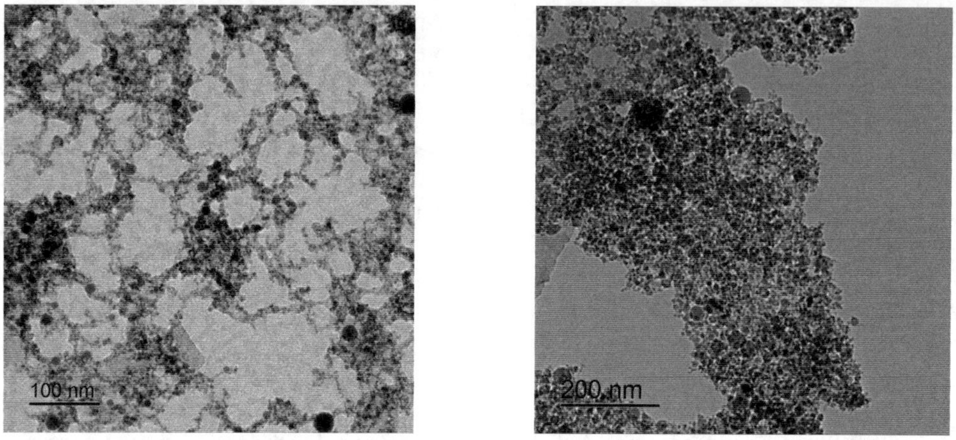

Figure 3. TEM-images of tantalum dioxide nanoparticles (left) and titanium dioxide (right).

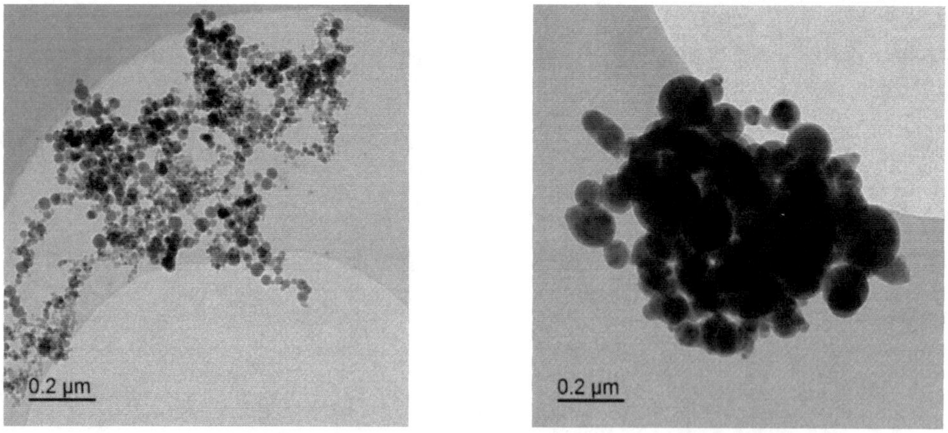

Figure 4. TEM-images of copper oxide nanoparticles (I) (left) and vanadium dioxide (right)

Antibacterial properties of silver, titanium, copper, cobalt, nickel and zirconium nanoparticles are caused by the presence of double electric layer around the nanoparticles that has high reaction activity and interact with adsorption centers of cell membrane peptidoglycans [2, 3]. Breaking the integrity of microbial wall and cytoplasmic membrane, metal nanoparticles penetrate into the cell and react as a catalyst in oxidation processes with the release of free radicals and destruction of cell structure [1, 4-8]. For example, such high antibacterial activity of Ag-ZnO, Ag-CuO composite nanoparticles composite against microorganisms P.gingivalis; F.nucleatum and P.intermedia was shown [9].

In our study the bactericidal properties of colloidal solutions of metals and metal oxides have been analyzed by means of disk diffusion method. We obtained the clearly defined bactericidal effect for solution concentrations from 1 to 10 x 10^{-7} mg/l. For evaluating the antimicrobial activity a mixed culture of dental deposit biofilm was used. The calico tests – objects previously sterilized and impregnated with test solutions – were placed in Petri dishes with the seeding of indicator bacteria, on the surface of FT – agar based on Hottinger's broth. The incubation of Petri dishes with the seeding of indicator culture and test samples was carried out at a temperature of 370 °C for 24 hours.

The investigated aqueous colloidal nanoparticles solutions of silver, iron oxide (II), nickel oxide (II), titanium dioxide, tantalum dioxide, mixed solution of titanium dioxide and molybdenum dioxide with corresponding concentrations of 12.2 mg/l, 12.4 mg/l, 8.2 mg/l, 10.3 mg/l, 7.5 mg/l, 6.8 mg/l were multiply diluted to 10, 10^2, 10^3, 10^4, 10^5, 10^6 and 10^7 times. During incubating no bacterial medium was found under the test objects for all solutions and all dilutions.

The results of bactericidal activity researches of calico tests impregnated with colloidal solution of silver and iron oxide (II) nanoparticles against dental deposit on the second day are presented by Figures 5 and 6. Figure 7 presents the results of research after nine days of incubations, one can see that silver and iron oxide nanoparticles have prolonged bactericidal effects.

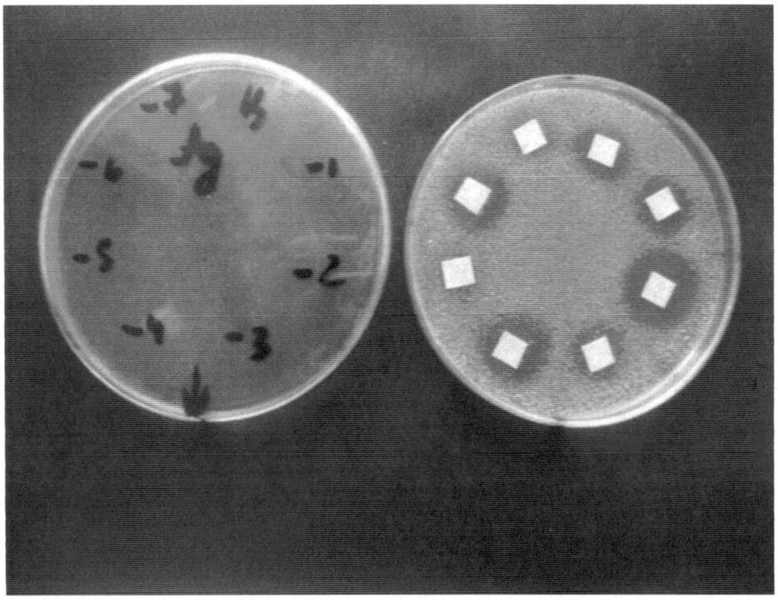

Figure 5. Bactericidal activity of calico tests impregnated with colloidal solution of silver against dental deposit on the second day.

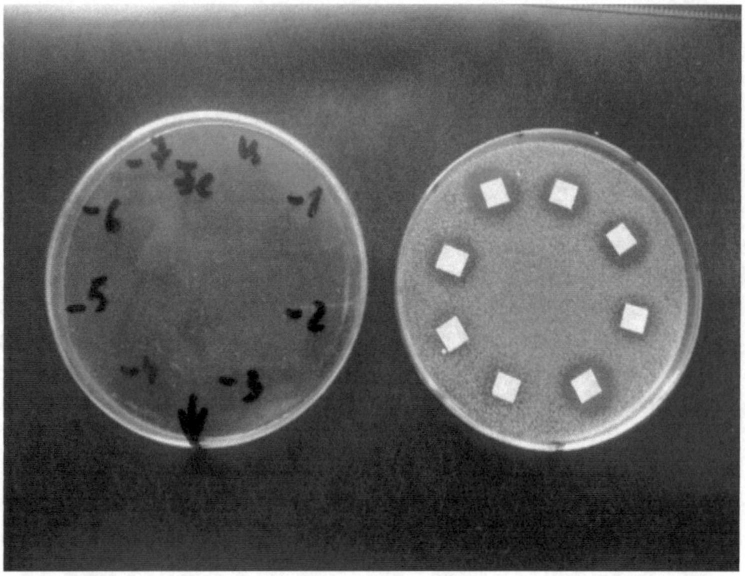

Figure 6. Bactericidal activity of calico tests impregnated with colloidal solution of iron oxide (II) against dental deposit on the second day.

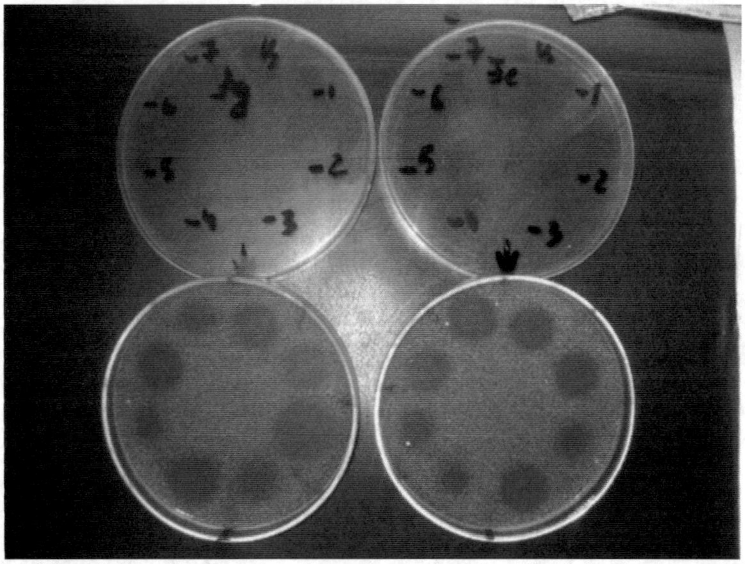

Figure 7. Bactericidal activity of calico tests impregnated with colloidal solutions of silver and iron oxide (II) against dental deposit on the ninth day

It was established that the maximum values of antibacterial activity against dental bacterias were shown by following water solutions of nanoparticles:

- silver with the concentration from 1 to 10^{-5} mg/l;
- iron oxide (II) with the concentration from 1 to 10^{-6} mg/l;
- nickel oxide (II) with the concentration from 10 to 10^{-5} mg/l;
- titanium dioxide with the concentration from 10 to 10^{-6} mg/l.

The results of microscopic researches of model bacterial system (Bacillus cereus) with colloidal solution of titanium dioxide nanoparticles have shown the destruction of bacterial cell wall and spores membrane with the release of organelles into the environment (Figures 8 and 9).

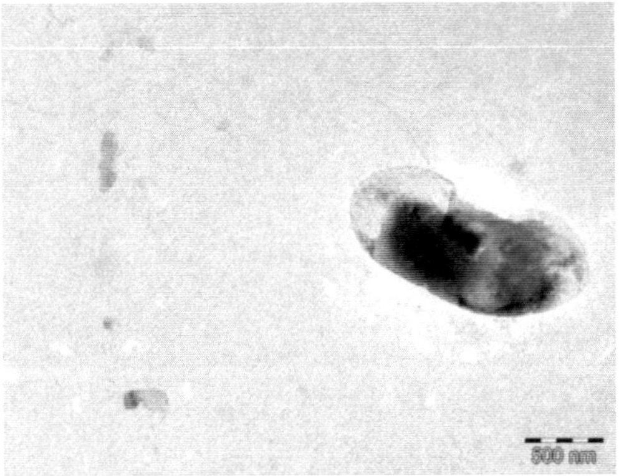

Figure 8. Initial cell Bacillus cereus.

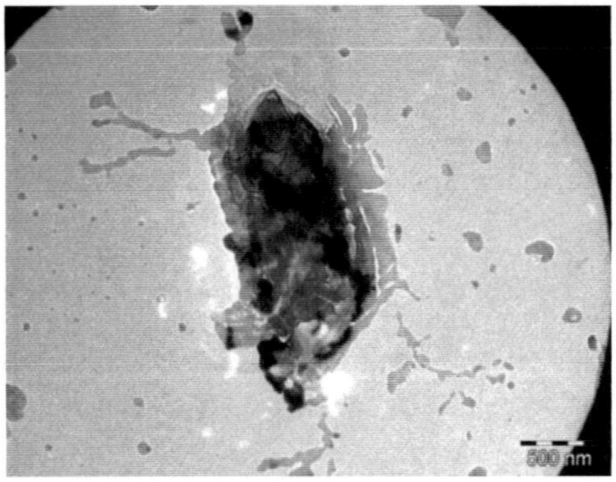

Figure 9. Cell B. cereus after interaction with the nanoparticles of titanium dioxide.

Therefore the long-term bactericidal effect of metal nanoparticles against pathogenic germs of dental deposit proven by our and earlier bacteriological studies [8-13] makes possible the introduction of these nanoparticles in the composition of dental restorative filling materials.

As a result of our research we can stipulate that obtained by means of electroerosion method metal and metal oxide nanoparticles could be potencially new alternative to traditional antibiotics in the treatment of local dental infectious processes. Such nanoparticles colloids have a prolonged bactericidal effect and wide range of antibacterial activity. Therefore it is possible to use these colloidal solutions as an antibacterial component of dental composite filling materials, adhesive systems and pickling gels. It would allow to improve the results of therapeutic treatment and help to prevent dental caries.

4. Acknowledgments

The project was supported by the Ministry of Science and Education of Russian Federation (project RFMEFI57814X0080).

References

[1] Bondarenko VM, Chubunov VF and Bondarenko VM 1987 *Microbiol. J* **3** 118-126
[2] Pogorelskiy IP, Frolov GA and Gurin KI 2012 *Disinfection Affairs* **4** 37-40
[3] Pogorelskiy IP, Frolov GA and Gurin KI 2014 *Disinfection Affairs* **2** 35-37
[4] Polyakov VA, Burachevskiy II, Morozova SS, Frolov GA, Karasenkov YN, Dmitrieva EV and Domanskaya DS 2013 *Food Industry* **2** 24-27
[5] Kalinkin MN, Rumyantsev VA, Zhmakin IA, Davydov BN, Bakanova KB, Tetarchuk VV and Artamonov AA 2014 *Nanotechnology in Dentistry* in the Proc. of All-Russian Conf. "The 60th Anniversary of the Faculty of Dentistry of Tver State Medical Academy" 175-177
[6] Yakovlev GV and Stekhin AA 2008 *Hygiene and Sanitation* **6** 21-26
[7] Denyer SP and Stewart GS 1998 *Int Biodeterioration and Biodegradation* **41** 261-268
[8] Fan W, Wu D and Tay FR 2014 *Int J Nanomedicine* **12** 217-230
[9] Vargas-Reus MA, Memarzadeh K and Huang J 2012 *Int J Antimicrobial Agents* **40** 135-139
[10] Zhang K, Li F and Imarato S 2013 *J Biomed Mater Res B Appl Biomater* **101** 929-238
[11] Godymchuk A, Yunda E, Kosova N, Karepina E, Kuznetsov D, Gusev A and Arzamastseva E 2015 *Adv Mater Res* **1085** 363-369
[12] Yunda E, Godymchuk A, Kosova N, Kuznetsov D and Senatova S 2014 *Adv Mater Res* **872** 263-270
[13] Karepina E, Godymchuk A, Kuznetsov D and Gusev A 2014 *Adv Mater Res* **872** 248-256

The method of radioactive tracer for measuring the amount of inorganic nanoparticles in biological samples

Yu Buzulukov[1,2], A Antsiferova[1,2], V A Demin[1,2], V F Demin[1,2], P Kashkarov[1,2,3]
[1]National Research Centre “Kurchatov Institute”, AcademikaKurchatova sq. 1, Moscow, 123182, Russia
[2]Moscow Institute of Physics and Technologies, Institutski lane, Dolgoprudny Moscow Region, 141700, Russia
[3]Lomonosov Moscow State University, GSP-1, Leninskie Gory 1, Moscow, 119991, Russia

E-mail: buzulukov@inbox.ru

Abstract. The method to measure the mass of inorganic nanoparticles in biological (or any other samples) using nanoparticles labeled with radioactive tracers is developed and applied to practice. The tracers are produced in original nanoparticles by radioactive activation of some of their atomic nuclei. The method of radioactive tracers demonstrates a sensitivity, specificity and accuracy equal or better than popular methods of optical and mass spectrometry, or electron microscopy and has some specific advantages. The method can be used for study of absorption, distribution, metabolism and excretion in living organism, as well as in ecological and fundamental research. It was used in practice to study absorption, distribution, metabolism and excretion of nanoparticles of Ag, Au, Se, ZnO, TiO_2 as well as to study transportation of silver nanoparticles through the barriers of blood-brain, placenta and milk gland of rats. Brief descriptions of data obtained in experiments with application of this method included in the article. The method was certified in Russian Federation standard system GOST-R and recommended by the Russian Federation regulation authority ROSPOTREBNADZOR for measuring of toxicokinetic and organotropy parameters of nanoparticles.

1. Introduction

The production of nanotechnology-based consumer goods and commercial applications of nanoparticles (NP) develops quite quickly [1-3]. Data base on consumer goods (made with using of NPs) includes 1824items up to beginning of 2015. During 2014 the number of names of consumer goods increases by 10 % and continue to grow. Almost all NPs in consumer goods are inorganic ones. The most popular are NPs of Ag (24% of all goods), Ti and TiO2 (10%), C (4%), Si and SiO2 (3%), ZnO (2%), Au (1%). Other NPs are used sufficiently less [4].

Potential health and environmental risks of engineered nanoparticles (NPs) should ensure environmental and health safety (EHS) of nanotechnologies, especially of nanotechnology-based consumer goods. It is necessary to control mass or concentration of NPs inside living organism, plants, food, agriculture row materials, consumer goods etc.[5]. In order to estimate potential danger and EHS of NPs it is important to study processes of absorption, distribution, metabolism and excretion (ADME) of NPs in living organism. The ADME process studies are performed usually by model experiments with administration of engineered NPs in living organism and measuring of their

Content from this work may be used under the terms of the Creative Commons Attribution 3.0 licence. Any further distribution of this work must maintain attribution to the author(s) and the title of the work, journal citation and DOI.
Published under licence by IOP Publishing Ltd

mass or concentration in organs and biological liquids. Thus it is obvious that developing the enhanced methods to measure mass/concentration of engineered NPs in biological samples is actual problem for estimation of safety of products containing NPs as well as for other EHS problems.

2. Methods to measure mass of inorganic NPs in biological samples

There are not universal methods to measure mass of any NPs in biological samples nevertheless there are a lot of specific methods which are applicable for any specific NPs [5]. As shown in section 1, consumer goods containing NPs are developed on base of short set of inorganic NPs mentioned above, and so they are the most actual for practice applications of EHS research. The most popular methods to measure in samples mass of these NPs are elemental analysis, e.g. optical spectroscopy (atomic absorption or atomic emission spectroscopy) and mass spectrometry, as well as transmission electron microscopy (TEM) and methods of labeled NPs.

TEM method is quite expensive and has a low representativity. Methods of elemental analysis are applicable only to non-biophile elements (i.e. elements with low natural mass concentration in organism) and need laborious preparing of samples before measuring. One of the most efficient alternative methods is one of labeled particles, which is applicable to variety of model experiments, e.g. to study ADME of NPs administered to laboratory animals [6], or transportation of NPs along trophic chains in ecological systems, or transportation of NPs from packing with NP into the food [7, 8]. Fluorescent labels may visualize NPs in biological samples, e.g. by using of confocal microscopy. The uptake of fluorescently labeled NPs in cells may be measured by flow cytofluorometry. Labels which produce a characteristic signal in the Raman spectra also may be used to measure mass of engineered NPs in samples or just in the body of a laboratory animal [9]. The main requirements to labels are minimal influence on chemical and physical properties of NPs under investigation and simple way to register labeled NPs in samples. Hence the ideal labels are isotopic tracers, as they produce very low changes in chemical/physical properties of NPs and don't influence biological and surface properties of labeled NPs [5, 12]. Isotope tracers may be stable or radioactive. One can measure an amount of isotopes in samples using mass-spectroscopy (for stable isotopes) or gamma- or beta spectrometry (for radioactive ones). In specific situations the other methods of registration of isotope labels may be available, e.g. nuclear magnetic resonance spectroscopy. Generally radioactive isotopes may be registered in living organism (e.g. by methods similar to positron-emission tomography) without disturbing the vital activity of organism. Really radioactive tracers are popular in biology, e.g. many experiments were performed with ^{3}H tracer [10, 11], experiment with labeled NPs of ZnO described in [12].

3. Incorporation of radioactive tracers in NPs

Radioactive tracer method (RTM) in applications aimed at measuring of mass/concentration of NPs in biological samples implies administration to laboratory animals NPs labeled with radioactive tracers.

The simplest way to incorporate isotope tracer into NP is transformation (activation) of some stable nuclei in original NPs to their radioactive isotopes by means of neutron irradiation. The transformation may be performed by irradiation of original NPs by neutrons of very low energy in the course of nuclear reaction of neutron capture, e.g. ^{109}Ag (stable) $(n, \gamma) \rightarrow {}^{110m}Ag$ (radioactive).

The other way to incorporate radioactive tracers into original NPs is transmutation of stable nuclei in original NPs to radioactive ones of other elements, e.g. transmutation of stable isotopes of Ti to radioactive ^{48}V or ^{47}Sc by irradiation of original Ti in NPs with fast protons or neutrons. The main parameters of isotopes which can be used as radioactive tracers in experiments with Ag, Au, Se, ZnO are shown in table 1 [5,14,16].

Table 1.Main parameters of isotopes of Zn, Ag, Au, Se used as tracers.

Nano-material	Original stable isotope in NP composition		Corresponding radioactive isotope			
	isotope	Natural abun-dance [%]	isotope	$T_{1/2}$ [days]	type of radiation	E_γ [MeV]
Silver	^{109}Ag	48.2	^{110m}Ag	249.8	β^-,γ	0.6577 0.8847
Selenium	^{74}Se	0.9	^{75}Se	119.8	γ	0.1360 0.2647
Gold	^{197}Au	100	^{198}Au	2.7	β^-,γ	0.4118
Zinc	^{64}Zn	48.6	^{65}Zn	244.3	β^+,γ	1.115

Note.Designations: $T_{1/2}$is half-life of radioactive decay,$E_\gamma$is energy of main γ - radiation spectral line.

4. Measurements of mass of inorganic NPs in biological samples by radioactive tracer method

Measurements of the mass of NPs labeled with radioactive tracers may be performed by measurements of nuclei radiation of radioactive tracer. Nuclei radiation from the tracer can be gamma-radiation and beta radiation, both of them are caused by radioactivity (activity) of tracer. Usually isotopes have both kinds of radiation and power of radiation is strongly proportional to activity of tracers in the sample and so to mass of labeled NPs in the sample. In principle activity of tracers may be measured by measuring gamma or beta radiation, but in practice usage of gamma radiation has sufficient advantages. The main advantage is good penetration ability of gamma-rays and very low level of absorption and reducing in matter of samples. Beta radiation reduces sufficiently inside a sample and an operator should perform special treatment of the samples and calculate correction on self-absorbtion of beta radiation inside sample. Also modern gamma-spectrometer equipment is less expensive in comparison to beta-radiometrical equipment. High sensitivity and selectivity of modern gamma-spectrometers needs very low activity of samples, which is comparable with natural activity of human body (about 7 kBq). It is absolutely safe and sufficiently lower than level of maximum significant activity nominated by International Committee on Radiological Protection.

The method of RTM was certified in RF standard system GOST-R for measurements of mass concentration of NPs of Ag and ZnO in biological samples and recommended by RF regulation authority ROSPOTREBNADZOR for measuring of toxicokinetic and organotropy parameters of engineered nanomaterials in experiments with laboratory animals[15]. Some experiments on *in vivo* study of ADME of commercially popular NPs were performed with using of RTM [5, 14, 16] and results are briefly reported briefly in next section.

5. Results and discussion

The RTM was used for ADME measurements of NPs of Ag, Au, Se, ZnO and TiO_2 [13] as well as for study of transportation of NPs with milk of lactating rats and through placenta barrier of rats [16].Sensitivity of RTM method is limited by parameters of radiation spectrometer equipment and radiation background in area of spectrometer detector. It is limited also by mass and content of the samples. In particular sensitivity depends on the value of minimal detection activity (MDA)in the form of quantitative measurements limit L_q which is described and recommended by ISO [17]. In experiments with single peroral administration of NPs of Ag, Au and ZnO [5] sensitivity was different for different samples, but in the worst case it was better than one ng per sample. Radioactive tracer ^{75}Se has low energy of γ - radiation (see Table 1) and consequently has large background in energy area of measurements, which caused by Compton scattering of γ-rays. Thus sensitivity for NPs of Se was lower and was better than 6 ng per sample. In agree with the research described in [14] accuracy of RTM method in application to NPs of Ag is better than ±15% under P=0.95, the methods for measuring of NPs Au, Se and ZnO are similar and accuracy of measurements assumed to be the same. Some examples of data obtained in experiments with using of RTM are shown in Tables 2, 3.

Data obtained in experiments with study of ADME of NPs of Ag in rats after single peroral administration are shown in Table 2.

Table 2. Distribution of Ag NPs in organs of rats after single intragastric administration. Data are shown in percent of administered dose of 1 mg per animal.

Biological specimen	Time after administration [hours]		
	24	48	72
GIT+feces (the sum)	>98	>98	>99
Carcass	0.36±0.17	<0.6	0.23±0.09
Liver	0.6±0.2	0.8±0.3	0.18±0.10
Kidneys	0.014±0.002	0.029±0.008	0.007±0.003
Blood	0.13±0.05	0.20±0.05	0.05±0.02
Lungs	0.009±0.003	0.016±0.003	0.006±0.003
Heart	0.004±0.002	0.006±0.002	0.0032±0.0007
Pancreatic gland	0.008±0.002	0.012±0.005	0.004±0.001
Spleen	0.05±0.02	0.06±0.03	0.010±0.004
Gonads	0.016±0.003	0.033±0.007	0.010±0.004
Brain	0.003±0.001	0.012±0.002	0.005±0.002
Urine	0.012±0.002	0.032±0.009	0.048±0.037

Note. Designation GIT means gastrointestinal tract.

RMT was also used in an experiment on study of transportation of Ag NPs with milk of lactating rat females to suckling mammary pups of rats [6]. The single peroral dose for lactating female was about 1.8 mg of Ag NPs per body weight. Time of lactation was 48 hours. The total accumulation of Ag NPs into the milk exceeded 1.94 ±0.29 % of the administered dose over 48 hours of lactation. Main data are shown in Figure 1. It was the first time experimental evidence of the transfer of NPs from mother to offspring through the breast milk.

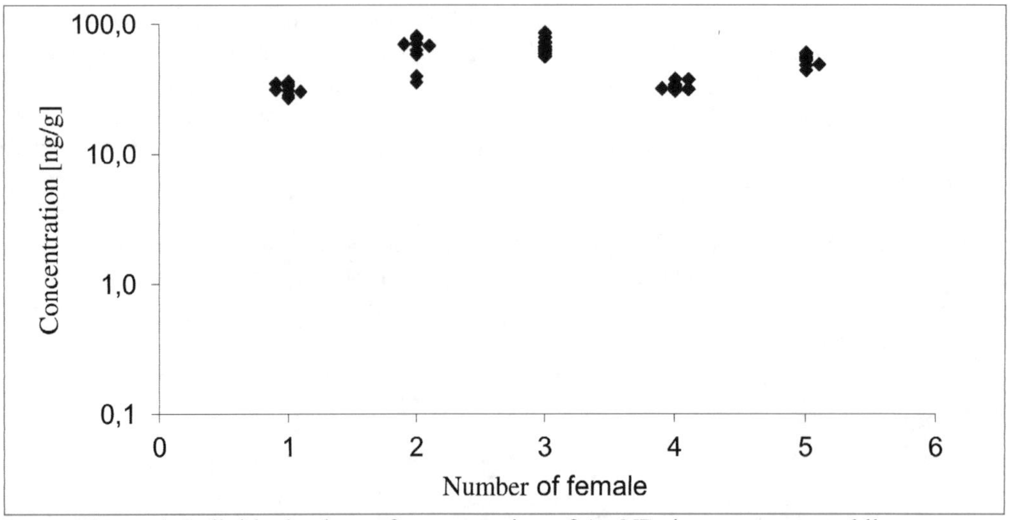

Figure 1. Individual values of concentration of Ag NPs in mammary suckling rat pups of lactating females

In precise experiments there is always the risk of influence of external contaminations of samples. In the experiment with mammary infant rats there was a risk of external contamination of rat pup body from feces of mother female. And therefore additional experiments were performed to verify data shown at Figure 1 against external contaminations of infant rat bodies with Ag NPs by feces of lactating female. As shown in Table 3, feces contain about 98% of administrated dose of Ag NPs (after peroral administration). Thus additional measurements of several rat pup internal organs and carcass were performed to clarify the localization of main part of NPs in rat pup bodies. The data obtained are shown in the Table 3.

Table 3. Distribution of NPs of Ag in the body of mammary infant rats, which were drinking milk from lactating mother rat exposed with NPs of Ag.

Biological sample	Mass of NPs of Ag in organs of mammary infant rats (in percent of dose of 1.8 mg administered to lactating female rat)							
	1	2	3	4	5	6	7	Ave-rage
Carcass	0.0155	0.0101	0.010	0.0119	0.0144	0.0081	0.0135	0.0134
GIT	0.095	0.141	0.164	0.111	0.132	0.098	0.161	0.129
Liver	0.0245	0.0251	0.0259	0.0239	0.0283	0.0305	0.0245	0.0266
Kidneys	0.0009	0.0010	0.0011	0.0010	0.0014	0.0016	0.0017	0.0013
Sum of sample weights	0.1102	0.1510	0.2010	0.1478	0.1761	0.1366	0.2007	0.1605
Wholebody	0.149	0.1520	0.166	0.1451	0.146	0.152	0.164	0.153
Difference	0.0388	0.0010	-0.0350	-0.0028	-0.0301	0.0154	-0.0367	-0.0071
Difference %	26.0	0.7	-21.1	-1.9	-20.6	10.1	-22.4	-4.6

Note: specimens were obtained in24 hours after intragastric administration of 1.8 mg of Ag NPs to mother rat.

Data of table 3 shows that main part of NPs registered in the infant rats is localized inside body and surface contaminations are negligeable. Also the sum of masses of NPs detected in organs and carcass was compared to the mass of NPs registered in the whole rat pup bodies. The differences were shown in table 3 (last two rows). The average deviation does not exceed measurement uncertainties of the RMT [15] which confirms the reliability of the obtained data.

6. Conclusion

It can be concluded that RTM is quite appropriate to model biological experiments with inorganic NPs on ADME study as well as for broad variety of EHS investigations. It has several serious advantages in comparison with conventional methods of determination mass/concentration of NPs in biological samples: by the side of optical and mass-spectrometry methods of elemental analysis the RTM can measure mass/concentration of NPs of biophile elements at their natural background in organism, RTM does not need any preparing of samples before measuring and thus is sufficiently less labor-consuming. Also by contrast to conventional methods (including TEM) RTM determinates mass of NPs in whole sample and interpretation of obtained data is more transparent. The experimental data obtained in experiments with ADME study demonstrate sensitivity and accuracy comparable or better than conventional methods. The RTM was used successful in the model experiments on *in vivo* study of ADME of some commercially actual NPs, for the first time were obtained experimental evidences of the transfer of Ag NPs from mother to her off-spring through placenta and breast milk.

Acknowledgements

The authors are sincerely grateful to I.V. Gmoshinski, S.A. Khotimchenko, A.A. Shumakova, V.A. Tutelyan (Federal State Scientific Research Institute of Nutrition, Moscow, Russia) for strong support and cooperation in biological part of the experiments mentioned above.

References

[1] J. Tian, K. K. Y. Wong, Ch. M. Ho, Ch. N. Lok, W. Y. Yu, Ch. M. Che, J. F. Chiu, and P. K. H. Tam, 2007 *J. Med. Chem.* **2** 129
[2] Kujda M, Oćwieja M, Adamczyk Z, Bochenska O, Bras G, Kozik A, Bielanska E and Barbasz . *J. of Nanoscience and Nanotechnol.* 2015 **15**(5) 3574
[3] Bondarenko O, Juganson K, Ivask A, Kasemets K, Mortimer M and Kahru A *Archives of Toxicology* 2013 **87**(7) 1181
[4] *Projecton Emerging Nanotechnologies*, http://www.nanotechproject.org, Woodrow Wilson International Centre for Scholar, Washington, USA
[5] Gmoshinski I, Khotimchenko S,Popov V, Dzantiev B, Zherdev A, Demin V and Buzulukov Yu2013 *Russian Chemical Reviews* **82** (1) 48
[6] Raspopov R, Arianova E, Trushina E, Mal'sev G, Kuz'min P, Shafeev G and Khotimchenko S 2011 *Vopr. Pitan.* **80** (4) 36
[7] Balbus J, Manyard A,Colvin V, Castranova V, Daston G, Denison R, Dreher K, Goering P, Goldberg A, Kulinowski K, Monteiro-Rivere N, Oberdorster G, Omenn G, Pinkerton K, Ramos K, Rest K, Sass J, Silbergeld E and Wong B 2007 *Environ. Health Persp.* **115** 1654
[8] Draft Opinion of the Scientific Committee on the Potential Risks Arising from Nanoscience and Nanotechnologies on Food Feed Safety 2008, *European Food Safety Authority* 35
[9] Keren S, Zavaleta C, Cheng Z, Zerda A, Gheysens O, and Gambhir S 2008 *Proc. Natl. Acad. Sci. USA* **105** 5844
[10] Chambers E and Mitragotry S 2007 *Exp. Biol. Med. (Maywood)* **232** 958
[11] Nurmi J, Tratniek P, Sarathy V, Baer D, Amonette J, Pecher K, Wang C, Linehan J, Matson D, Penn R and Driessen M 2005 *Environ.Sci.Technol.* **39** 1221
[12] Chen J, Shih M,Peir J, Liu C, Chou F,Lai W, Chang L, Lin P, Wang M, Yang M and Yang C 2010 *Analyst* **135** 1742
[13] Kotenko K, Belyaev I,Buzulukov Yu, Bushmanov A, Demin V, Gmoshinski I, Zhorova E, Kalistratova V, Marchenkov N, Nisimov P, Raspopov R and Soloviev V 2012 *Meditzinkaia radiologiia i radiatsionnaya bezopasnost* **57**(5) 5

[14] Buzulukov Yu, Gmoshinski V, Raspopov R, DyominV, Solov'yev V, Kuzmin P, Shafev G and Khotimchenko S 2012 *Meditzinkaia radiologiia i radiatsionnaya bezopasnost* **57**(3) 5
[15] Demin A, Demin V, BuzulukovYu, Kashkarov P and Levin A *Nanotechnologies in Russia* 2013 **8** (5–6) 347
[16] Melnik E, Buzulukov Yu, Demin V, Gmoshinski I, Tyshko N and Tutelyan V *Acta Naturae* 2013 3(18) 48
[17] ISO 11929: 2010. Determination of characteristic limits (decision threshold, detection limit and limits of the confidence interval) for measuring of ionizing radiation. Fundamental and application // ISO standard, 2010

IOP Conf. Series: Materials Science and Engineering **98** (2015) 012040 doi:10.1088/1757-899X/98/1/012040

Activators of Biochemical and Physiological Processes in Plants Based on Fine Humic Acids

G Churilov[1], S Polishuk[2], M Kutskir [2], D Churilov[2] and S Borychev[3]
[1] Division of general chemistry with course of organic and bioorganic chemistry, Ryazan State Medical University 9 , Vysokovoltnaya st., Ryazan, 390026, Russia
[2] Division of chemistry, Ryazan State Agrotechnological University, 1 Kostycheva st., Ryazan, 390044, Russia
[3] Division of building of engineering constructions and mechanics,Ryazan State Agrotechnological University, 1 Kostycheva st., Ryazan, 390044б Russia

E-mail: wiclipucli@hotmail.com

Abstract. This article describes the application of ultrafine humic acids as growth promoters and development of crops, for example corn. During the study we determined the optimal concentration of humic acids in ultrafine state for presowing treatment of seeds of maize. An analysis of laboratory and field tests was presented. We showed the relationship between physiological changes and biochemical processes.

1. Introduction

Humic acids are biogenic origin components of the soil. They have high molecular organic structures that may bind metal ions found in soil solutions. It is necessary for plants to subsequent absorption of metals into the root system. Metals perform a wide range of the roles and functions in organism. Also they are an essential part of coenzymes, so they maintaining balance of metals in plants which depends on the presence of humic acids in soil. Humic acids are used in agricultural technology widely, in agricultural practice they are used as a fertilizer mainly. But usage of humic acids depends on many physical and chemical properties of the soil (pH, humus content) and artificial changing the values of these parameters can lead to disruption of homeostasis of soil [1].The problem of studying is the mechanism of action and use of substances in the dispersed state as stimulators of growth and the development stimulators of agricultural crops has been widely dicussed in modern Russian science [2,3,4]. Main results confirming the action of ultra-dispersed forms of substances on plants are reflected in the works of Russian scientists. Application of humic acids in the ultrafine state with présowing treatment of seeds allows acting on separate plants. So, it may to improve the quality and quantity of crop production significantly reduced production costs.

The experimental research was carried out from 2009 to 2013. The objects of study were the seeds of corn cross-breed "Obsky 140". All seeds used in the study were consistent with one class. Under laboratory conditions, seeds were germinated in a thermostat (27 ^{0}C) in Petri dishes (50 pcs. seeds in each) in the liquid growth medium, which is a distilled water solution with addition of ultra fine humic acids in a given concentration. Laboratory and field studies were carried out at the Center of Nanotechnology and nanomaterials for agriculture at Ryazan State Agrotechnological University named after P. A. Kostychev. Energy of growth and germination was determined in accordance with GOST 12038-84 [5]. The enzyme activity of peroxidase and superoxide dismutase was determined by method of spectrophotometry in the laboratory of the Ryazan State Medical University. To determine the productivity of photosynthesis they used mass the Nichiporovich method (1961).

Content from this work may be used under the terms of the Creative Commons Attribution 3.0 licence. Any further distribution of this work must maintain attribution to the author(s) and the title of the work, journal citation and DOI.
Published under licence by IOP Publishing Ltd

2. Results of the research.

At the first stage of our research it was investigated the concentration range (from 0.1 to 10.0 grams) of active ingredient per hectare of seed rate to determine the concentration of stimulating the development and growth of maize seedlings (g/h.s.r.). The conducted experiments provide following results.

Table 1. Vital indexes of corn seeds treated with fine humic acids

Variant	Germinative energy , %	Germinating power, %
Control	88.40±0.03	96.60±0.03
HA - 0.1	92.40±0.02	98.70±0.02
HA - 0.5	90.30±0.02	94.20±0.02
HA - 1.0	89.60±0.03	92.40±0.02
HA - 5.0	90.30±0.02	95.20±0.03
HA - 10.0	91.60±0.02	94.20±0.03

The maximal value of index of germinative energy was observed only at a concentration of 0.1 g of active substance g/h.s.r. and was relative to a control by 3.4% under the influence of humic acids. In use of humic acids at a concentration of 0.1 g/h.s.r. laboratory germination exceeded the control one by 2.1%.

Table 2. Mass indexes of corn plantlets treated with biological active ultrafine humic acids

Variant	Mass of sprouts, g	Variant	Mass of roots,gr
Control	4.20±0.09	Control	1.40±0.06
HA - 0.1	3.70±0.06	HA - 0.1	4.50±0.09
HA - 0.5	2.50±0.07	HA - 0.5	4.40±0.05
HA - 1.0	2.80±0.06	HA - 1.0	3.80±0.08
HA - 5.0	3.20±0.09	HA - 5.0	5.70±0.03
HA - 10.0	3.70±0.08	HA - 10.0	4.70±0.08

Weight of root seedlings in all experimental versions exceeded the control average of 3.1 times (Table 2). Maximum weight of seedlings (in 4 times) in the application of humic acids in the ultrafine state was observed at a concentration of 5.0 g/h.s.r. The test results confirmed that biologically active humic acid in the ultrafine state promote the development of seedling roots of maize seeds, which is a prerequisite for the survival of the next growing season. The laboratory research allowed them to determine the concentrations promote the maximum stimulating effect on seedlings of corn seeds (0.1 g/h.s.r., 0.5 g/h.s.r. 1.0 g/h.s.r.). These concentrations have been tested in the field research.

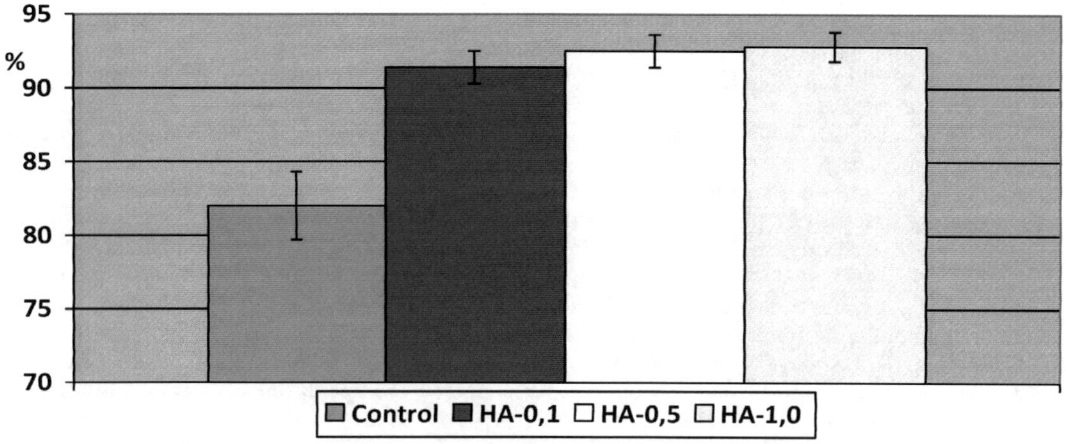

Figure 1. The field emergence of corn plants treated with ultrafine humic acids.

The field emergence of seeds of the test plants significantly exceeded the control, at a concentration of 0.1 g/h.s.r. by 9.8%, and at a concentration of 1.0 g/h.s.r. by 10.8% (figure 1). Increasing the field emergence increases the number of plants per plot which subsequently enhances crop productivity.

Table 3. Height and weight of corn plants in phase of 3-4 leafs.

Variant	Height of one plant, cm	Variant	Weight of one plant, g
Control	46.0±0.2	Control	154.0±0.5
HA – 0.1	49.50±0.02	HA - 0.1	160.0±0.2
HA – 0.5	51.30±0.07	HA - 0.5	164.0±0.2
HA – 1.0	48.40±0.03	HA - 1.0	169.0±0.2

Phase of 3-4 leaf for corn plants is decisive during the growing season, because in this period the plants are most subjected to weed invasion. The higher plants in a given period, the greater the probability of preserving the harvest. As one can see from table 3 the use of biologically active humic acids in the ultrafine state contributed to the increase of plant height. The highest elevation (by 11.5%) of the control values was observed at a concentration of 0.5 g/h.s.r. of active ingredient. Mass of maize plants treated with the humic acids in ultrafine state was higher than control in all experimental embodiments: at the concentration of 0.1 g/h.s.r. by 4.0%, at 0.5 g/h.s.r. by 6.5%, at 1.0 g/h.s.r. by 10.6%. The increase in weight of the acceleration confirms synthesis processes that subsequently had a positive impact on the formation of high-quality crop. Tendency of growth stimulate and development of the test plants was maintained throughout the growing season (figure 2.).

Figure 2. Corn plants of control and tested groups in phase of 7-9 leaf.

There were determined leaf area and photosynthetic productivity of corn plants during the growth period (Table 4).

Table 4. Leaf area and photosynthetic activity.

Variant	Leaf area, thousand .m^2/ha	Variant	Photosynthetic productivity, $g/m^2 \cdot$ day
Control	11934.00±0.04	**Control**	5.010±0.003
HA - 0.1	11984.70±0.04	**HA - 0.1**	5.270±0.003
HA - 0.5	12542.50±0.04	**HA - 0.5**	5.680±0.004
HA - 1.0	12469.20±0.03	**HA – 1.0**	5.350±0.003

The leaf area of corn plants seed treated with humic acids at a concentration of 0.1 g/h.s.r. did not differ from the controls, but beginning with concentration of 0.5 g/h.s.r. the positive difference to the control was 5.1%, and at a concentration of 1.0 g/h.s.r. - 4.5%. (Table 4). The value of index of photosynthetic productivity of corn plants was maximum at concentration of 0.5 g/h.s.r. of active substance. The value of the index of experimental plants exceeded the control by 13.4%. The positive correlation between leaf area and plant mass determines the photosynthetic productivity which confirms stimulation processes of synthesis.

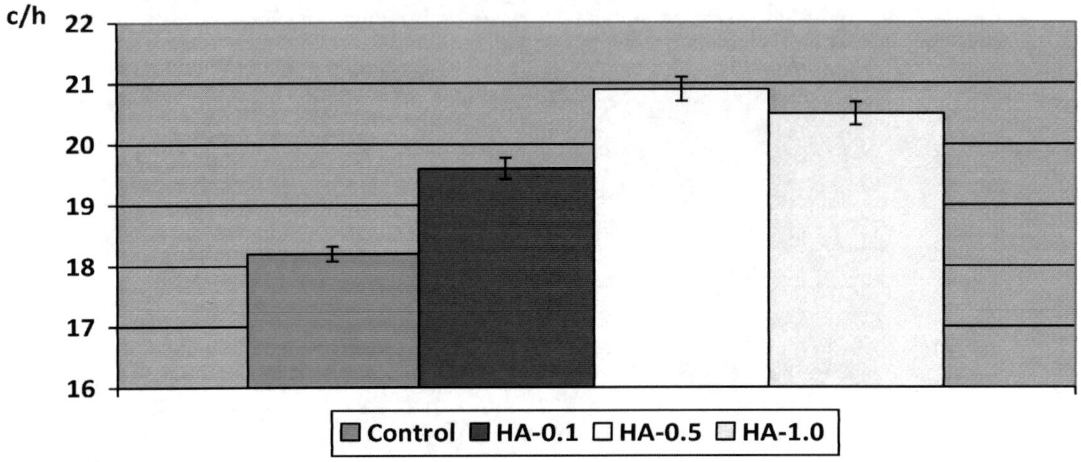

Figure 3. Yield of corn (c/ha).

The usage of humic acids increased the yield of corn seeds at concentration 0.1 g/h.s.r. by 7.7%, at a concentration of 0.5 g/h.s.r. by 14.8%, with 1.0 g/h.s.r. by 12.6% to control (figure 3). At the end of the growth season, it was observed that humic acids used with presowing treatment of corn seeds stimulated processes of growth and development of corn plants. Table 5 shows the chemical composition of the seeds of the test plants.

Table 5. Chemical composition of corn seeds

	Oil		Total ash		Crude protein		Humidity		Dry matter	
Variant	%	Rate to control	%	Rate to control.	%	Rate to control.	%	Rate to control	%	Rate to control.
Control	4.49	-	1.30	-	11.01	-	11.5	-	88.5	-
HA - 0.1	4.87	+0.38	1.26	-0.04	11.26	+0.25	10.3	-1.2	89.7	+1.2
HA - 0.5	5.02	+ 0.53	1.32	+0.02	11.12	+0.11	9.8	-1.7	91.2	+1.7
HA - 1.0	4.98	+0.49	1.28	-0.02	11.07	+0.06	10.5	-1.0	89.5	+1.0

Based on the presented data in Table 5, they have concluded the qualitative changes in the composition of crop production. The application of humic acids in the ultrafine state promoted qualitative changes in the chemical composition of the seeds of the test plants. The oil content at concentration of ultrafine humic acids 0.5 g/h.s.r. exceeded the control by 0.53%. At a concentration of 0.1 g/h.s.r. it was observed the maximum increase of crude protein 0.25% above control. On average, the amount of protein of all variants of experienced seeds exceeded the control. The increase of oil content, crude protein and ash indicate the activation of biochemical processes associated with the metabolism and syntheses of biologically active compounds. The activity of redox enzymes (peroxidase, superoxide dismutase) was investigated to confirm the safety use of humic acid as a growth promoter and the stimulator of the activity. It is known enzymes determine the biochemical status of the organism [6].

Table 6. Peroxidase activity (unit of optical density / g fresh matter · sec) and superoxide dismutase activity (в standard unit of activity/g fresh matter) in leafs and roots of corn plants treated with ultrafine humic acids

Variant	Peroxidase				Superoxide dismutase			
	Roots		Leafs		Roots		Leafs	
	absolute value	Rate to control	absolute value	Rate to control	absolute value	Rate to control	absolute value	Rate to control
Control	6.00±0.56	-	3.50±0.01		47.30±0.04	-	109.10±0.02	-
HA - 0,1	6.90±0.05	+15.0	3.60±0.02	+2.9	53.80±0.05	+13.7	111.50±0.04	+2.2
HA - 0,5	7.20±0.05	+20.0	3.70±0.01	+5.7	49.10±0.06	+3.8	109.90±0.03	+0.7
HA - 1,0	7.10±0.06	+18.3	4.0±0.24	+14.28	44.60±0.07	-5.7	102.40±0.08	-6.1

Humic acids in ultrafine state an average of 17.7% increase in peroxidase activity in the roots of maize plants. Activity of the enzyme was higher in sprouts control at a concentration of 0.1 g/h.s.r. 15.0% at a concentration of 1.0 g/h.s.r. – 18.3%. Peroxidase activity in roots of maize in all variants was above control (Table 6) and does not exceed the critical difference from the control of ± 30%, which is regarded as the limit for the normal development of plants (MU 1.2.2635-10). Therefore the plant organism in the concentration of humic acid 0.1 - 1.0 g/h.s.r. not experiencing a stress state which gives the electron transfer processes and neutralizing free radicals. Superoxide dismutase activity in the roots of the test plants was higher than the control by 8.9%. Superoxide dismutase activity in the roots at concentration of 0.5 g/h.s.r. was greater than control at 3.7%. It did not differ from controls in sprouts. With increasing concentration of the active substance of superoxide dismutase activity decreased. Probably contributed to increasing peroxidase activity reduce the overall oxidative stress plants because of the diminished superoxide. It is known that superoxide formed by the passage of oxygen atoms through a chain of electron carriers, in the case where the activity is reduced in the photochemical reaction centers photosystem 1 and photosystem 2 [7,8]. We can observe the formation of superoxide radicals particularly intensive in chloroplasts as a result of an imbalance between the rate of electron transfer and the rate of carbon fixation gas [9,10]. Presumably, the properties of nanoparticles, as the electron donor prevent the formation superoxide radicals and stabilize the flow of electrons in the mitochondrial respiratory chain, especially when it's passing in the intermembrane space.

3. Conclusions.

Humic acids in the ultrafine state stimulate the germination of corn seeds *in vitro* (by 3.4%) and in field experiments (up to + 10.2% compared to control). In this case, at the initial stages of germination the growth of the root system increased: the mass of roots exceeded the control in 3 times. It was activation of growth and development of corn plants during the growing season. The height and weight of the experimental plant exceeded the control one to 11.5% and 10.6%, respectively. Enhancement of photosynthesis processes resulted in increasing the leaf area and photosynthetic productivity to 5.1% and 13.4% in comparison with the control. Amplification processes contributed to increasing of the content of the synthesis of biologically active compounds in the seeds of the test plants: oil content by 0.53%, crude protein by 0.25%. Preliminary treatment of plants with humic acid has not experienced oxidative stress, which is confirmed by the study of the activity of redox enzymes.

References

[1] Polischuk S D, Churilov G I, Churilov D G and Chau N H 2013 *Proceedings of the 4th International Workshop on Nanotechnology and Application IWNA-2013.* 14-16 November 2013 6403.
[2] Polishuk S D, Nazarova A A, Churilov G I 2013 *Nanotechics* **3 (35)** 67.
[3] Polishuk S D, Nazarova A A and Churilov G I 2014 *J. of Mater., Sc. and Eng.* **2** 46
[4] Polischuk S D, Churilov G I, Churilov D G, Nazarova A A and Chau N H 2014 *Adv. Nat. Sci.: Nanosci. Nanotechnol.* 6.
[5] Crop seed. Methods of determination of germinating power. 1986 ***GOST RF***
[6] Nikolaeva M K, Maevskaja S N, Shugaev A V and Bukhov N G 2010 *Phytophisiology* **57** 94
[7] Bouvier F, Backhaus R A and Camara B J 1998 *Biol. Chem.* **273** 30651
[8] Stiborova M, Ditrichova M and Brezinova A 1988 *Photosynthetica* **22** 161
[9] Shioi Y, Tamii H, Sasa T 1978 *Plant Cell Physiol.* **19a** 203
[10] Li Y and Trash M A 1993 *Carcinogenesis* **14** 1303

Obtaining graphene nanoplatelets from various graphite intercalation compounds

A Melezhyk, E Galunin, N Memetov
Tambov State Technical University, 106, Sovetskaya Street, Tambov, 392000, Russian Federation

E-mail: evgeny.galunin@gmail.com

Abstract. The work compares the exfoliation ability of different graphite materials (expanded graphite intercalation compound, thermally expanded and oxidatively intercalated graphites) and describes the properties of graphene nanoplatelets (GNPs) obtained dependently on intercalation/deintercalation conditions and reagents. Among the studied materials, the graphite intercalated with ammonium persulfate in sulfuric acid and expanded at 40 ^{O}C possesses the maximum ability for ultrasonic exfoliation in the presence of a surfactant. The exfoliation efficiency strongly depends on the content of water in sulfuric acid during the intercalation. The highest efficiency was achieved for the expanded graphite intercalation compound (EGIC) prepared in sulfuric acid containing diluted oleum, which may be explained by increased acidity of the medium and, correspondingly, redox potential of the persulfate compound. This is also related to increased amounts of oxygen groups in the GNPs obtained from the EGIC synthesized in 100% sulfuric acid and diluted oleum. Besides, the nature of surface groups on the GNPs strongly depends on the nature of a deintercalating reagent. Thus, the treatment of the EGIC with different nucleophilic molecules (such as water, ammonia, carbamide, hexamethylenetetramine, organic amines, etc.) can yield GNPs with various surface groups. The interaction between the EGIC and nucleophilic molecules does not only include the substitution of sulfate groups, but also redox reactions with participation of graphene layers. Depending on the nature of the nucleophile, those reactions can lead to the formation of different groups attached to the graphene surface. GNPs with almost pure surface were obtained when using ammonia and carbamide.

1. Introduction

Nowadays, graphene-based nanostructures are of great interest for many technological applications (e.g., chemical current sources, adsorbents, sensors, polymer composites, etc.) due to their unique thermal, electrical, optical and mechanical properties [1-3]. They are usually produced as multi- or, especially, few-layered graphene nanoplatelets (GNPs) comprised of thin stacks of platelet-shaped graphene sheets. These nanoparticles have a wide range of thickness (0.34-20 nm) and may consist of up to 30-50 (or even more) monolayers.

In this regard, surfactant-assisted ultrasonic exfoliation of well-ordered crystalline graphite and its subsequent dispersion in water or organic solvents has recently been widely used to mass-fabricate GNPs [4-5]. Usually, surfactants and solvents efficient in dispersing carbon nanotubes (CNTs) can

Content from this work may be used under the terms of the Creative Commons Attribution 3.0 licence. Any further distribution of this work must maintain attribution to the author(s) and the title of the work, journal citation and DOI.
Published under licence by IOP Publishing Ltd

also be employed to obtain GNPs [6-7]. Therefore, previous studies performed on CNTs can provide useful information when working with GNPs.

However, the direct exfoliation of native graphite is not actually sufficiently effective. This process is slow, and only a small fraction of pristine graphite is transformed into thin (few-layered) GNPs. Separation of this few-layered fraction may become a serious problem in the large-scale fabrication of GNPs. Some promising approaches for the exfoliation of native graphite are under development nowadays; for instance, one of them is based on using supercritical carbon dioxide optionally combined with surfactants and ultrasonic irradiation [8].

To increase the efficiency of the exfoliation process, a certain kind of energy should be applied to a graphite material, either ultrasonically treated or not. In the case of applying chemical energy, some chemical transformations with large jumps in free energy take place. For instance, one treats starting graphite with a reagent and then with another one opposite in chemical nature, or after treating graphite with a reagent, the obtained graphite material is further processed under conditions, under which the previously treated material is highly unstable and decomposes quickly. It may be supposed that the free energy evolved during such chemical reactions can be transformed into the desired structural transformation. Sometimes, it is possible to find an effective combination of chemical reactions and physical treatments. The thermal expansion of graphite usually performed via oxidative intercalation of graphite and thermal shock resulting in chemical decomposition of graphite intercalation compounds accompanied by partial exfoliation appears to be a well-known example. However, the preparation of GNPs by ultrasonic exfoliation of common expanded graphite is not sufficiently effective, because rather thick platelets can only be obtained [9]. Much more interesting graphene materials were synthesized through exfoliation of so-called "highly thermally exfoliated graphite", which, in its turn, was obtained by thermal treatment of ClF_3-intercalated graphite [10]. Few-layered graphene products so fabricated contain fluorine and are hydrophobic, which makes them attractive as additives to various non-polar matrix composite materials. However, using such highly aggressive and toxic intercalating reagents demands special precautions and is potentially dangerous. Besides, to obtain high-yield GNPs, graphite is intercalated with potassium, followed by treatment with a reagent (water or alcohol) that can vigorously react with that metal, as well as with microwaves and ultrasound [11]. Such a combination of chemical and physical actions results in great jumps in the free energy of the system at several stages and, theoretically, can promote structural transformations. However, GNPs obtained according to this technique are rather thick. Treatment with such aggressive reagent as oleum is (sulfuric acid containing free sulfur trioxide) was also tried for exfoliation of graphite to GNPs [12].

One promising approach to the exfoliation of graphite materials is the electrochemical exfoliation of graphite [13-15]. In this case, the intercalation is followed by spontaneous or ultrasonic exfoliation. The electrochemical methods basically allow for more flexible tuning of intercalation and exfoliation processes, because the oxidation potential, particle nature and intercalation medium can be varied independently (within the limits of their compatibility). However, these techniques are sophisticated for laboratory experiments and, moreover, for industrial production because of arising numerous difficulties (e.g., more complex equipment, problems with stability of electrode materials, and limitations of ion/molecule diffusion and mass transfer). Considering this fact, purely chemical and physical methods of graphite exfoliation seem to be more attractive for the large-scale production of GNPs. The problem is to find an efficient combination of chemical reactions and physical treatments ensuring the desired structural transformations.

Previously, we have proposed a new method for producing GNPs that consists in cold expansion of peroxodisulfuric acid-intercalated graphite, spontaneous pre-exfoliation with the formation of an expanded graphite intercalation compound (EGIC) having an apparent volume of 300-320 cm3g-1 of the starting material, and its ultrasonic exfoliation into GNPs [16].

It should be noted that the idea of better exfoliation of the EGIC compared to the other intercalated materials (e.g., thermally expanded (TEG) and oxidatively intercalated (OIG) graphites) has come from the fact that the former consists of "live" GNPs that are not subjected to "hard" treatment like thermal shock. In this "live" expanded compound, the nanoplatelets are not only separated by vacant spaces, but also these spaces are filled with molecules of a reaction medium (acid or, after hydrolysis, water). When, for example, thermally expanded graphite is obtained after the thermal decomposition of intercalated graphite, the nanoplatelets have sintered edges which hinder their ultrasonic exfoliation.

Taking into account the aforementioned, the present work aims to compare the exfoliation ability of different graphite materials and study the properties of GNPs obtained under various intercalation/deintercalation conditions and reagents.

2. Research target and methods

Starting materials. Graphite GSM-1 (purity 99.9 %) was acquired from "Resurs-S" Ltd. (Ekaterinburg, Russia) and used as an initial material for obtaining GNPs. Ammonium persulfate (APS, analytical reagent grade) was purchased from "Reakhim" Ltd. (Moscow, Russia); it was used as an intercalation medium. To prevent aggregation of the synthesized GNPs in the system, surfactant NF (a sodium salt of naphthalenesulfonic acid/formaldegyde polycondensate) prepared according to the Russian National Standard GOST 6848-79 at "Pigment" Ltd. (Tambov, Russia) was employed.

Preparation of graphite intercalation compounds. The EGIC was synthesized through cold expansion of the graphite intercalated with an APS solution in 95 % and 100 % sulfuric acid, as well as in 100 % sulfuric acid containing 1% of free SO3, and used for ultrasonic exfoliation. The detailed description of the technique for obtaining the EGIC is given in our previous work [16].

As for the TEG and OIG, their production included oxidative intercalation of natural graphite in an acidic medium. In the case of the TEG, it was followed by rapid heating to high temperatures (thermal shock). Sulfuric acid combined with an oxidant (nitric acid, hydrogen peroxide, ammonium persulfate, etc.) is frequently used as intercalation medium. In some cases, other acids or oxidants can be used if the sulfur content in the TEG is undesirable [17-18].

Under laboratory conditions, the heat treatment of the OIG can be carried out by placing a sample into a hot muffle (900-1000 °C) for a short time. In the industrial production of the TEG, the thermal shock is performed by continuous blowing up the dust of dry OIG into a furnace heated by electricity or combustion of natural gas.

Sonication procedure and optical measurements. The sonication procedure was performed on an IL-10 device (telectric power of the emitter 2 kW) under cooling a glass with a processed solution in a cold water bath. The optical density (absorbance) was measured with a KFK-3 photoelectric colorimeter (Russia) in a 1 cm cuvette at a wavelength of 500 nm.

In these systems, the Beer-Lambert law was obeyed, and thus, the light absorption coefficient (K, dm^3/g·cm) can be estimated according to Eq. (1).

$$K=\frac{D}{CL} \tag{1}$$

where D is the optical density (dimensionless), C is the graphite concentration (g/dm^3), and L is the path length of the cuvette (cm).

To eliminate any uncertainties caused by the weight of oxygen groups, all K values were calculated considering the weight of the initial graphite, unless other stated. It was found that in experiments conducted under the similar conditions, the K value correlates with the dispersion/exfoliation degree– the smaller the thickness of GNPs and the lower their aggregation, the higher the K value. Thus, these parameters were evaluated through the K. In our previous experiments, the K values lied in the range of 19-33 dm^3/g.cm for the multi-layered GNPs (15-30 layers), and 50-63 dm^3/g.cm for the few-layered GNPs (2-5 layered). Besides, we calculated, with some assumptions, that the maximum theoretically possible K value for the dispersion of randomly oriented flat, non-aggregated graphene monolayers is 65.83 dm^3/g cm [16].

IOP Conf. Series: Materials Science and Engineering **98** (2015) 012041 doi:10.1088/1757-899X/98/1/012041

3. Results and discussion

In the Figure 1, there is compared efficiency of ultrasonic exfoliation of different graphite materials in aqueous solutions of the surfactant NF. To estimate the acoustic power, we assumed that (according to the manufacturer information) the efficiency of electric-to-acoustic energy transformation for the ultrasonic installation used was about 50%. We related the acoustic power to the volume of the solution processed (*i.e.*, in kWt-h/dm^3). It is a very approximate evaluation, since numerous technological factors affecting the exfoliation efficiency can hardly be considered correctly. From the Figure 1, it follows that the lowest exfoliation takes place for the TEG (which is known to be exfoliated better than common crystalline graphite).The OIG in sulfuric acid (after hydrolysis and washing with water) is exfoliated better than the TEG. There are data in the literature that the OIG so obtained (which is a common starting substance for producing the TEG) consists of pre-exfoliated crystals. Probably, this facilitates the ultrasonic exfoliation of this material compared to the common crystalline graphite. The EGICs in the system "graphite – sulfuric acid – ammonium persulfate (APS)" are exfoliated much better. This is natural, because the cold expansion process observed in these systems is extensive pre-exfoliation. The mechanism of the cold expansion was considered in our previous research [16]. Worm-like particles formed during this process consist of GNPs bound by edges, and it remains only to split these nanoparticles apart.

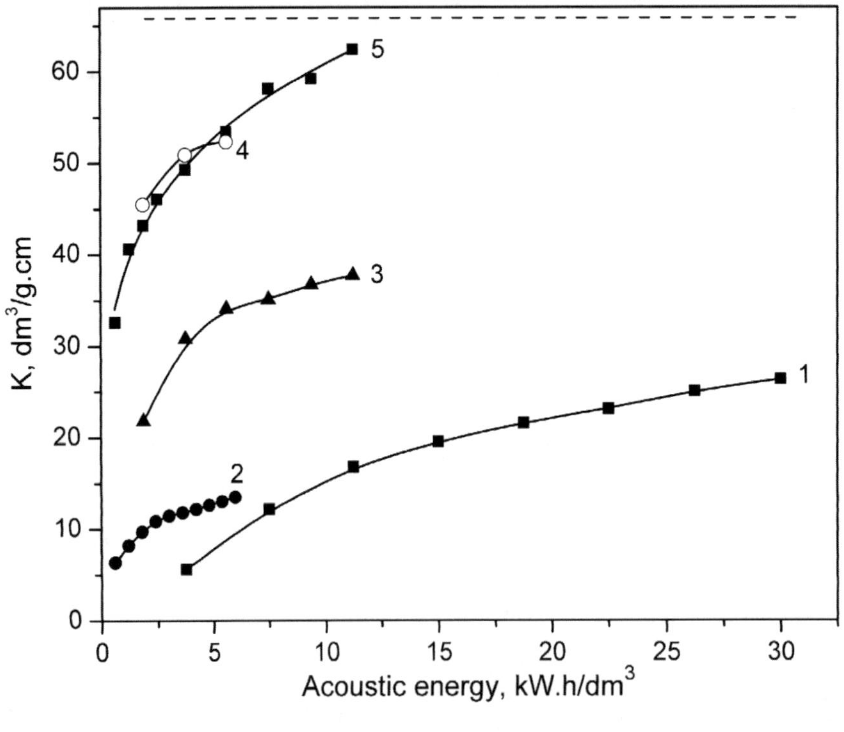

Figure 1. The ultrasonic exfoliation efficiency o different graphite intercalation compounds in aqueous solutions of the surfactant NF: 1 – TEG; 2 – OIG in 100 % sulfuric acid; 3 – EGIC in 95% sulfuric acid; 4 – EGIC in 100% sulfuric acid; 5 – EGIC in 100 % sulfuric acid containing 1 % of free SO_3. A dashed line indicates a theoretical limit for obtaining flat, non-aggregated graphene monolayers (K = 65.83 dm3/g.cm)

As seen on the Figure 1, the efficiency of the ultrasonic exfoliation of the EGICs strongly depends on the contents of water in sulfuric acid used as a medium for oxidative intercalation with APS.

Previously, we assumed that in those admixtures, water influences the acidity of sulfuric acid and, in its turn, the oxidative potential of peroxosulfate species [16]. The higher the oxidative potential, the more efficient the oxidation of the graphene framework and the cold expansion is. This also correlates with the increasing mass content of oxygen groups in the expanded graphite materials obtained after hydrolysis and washing (nearly 9 wt% for 100% sulfuric acid, 13 wt% for sulfuric acid containing 1.5% of free SO_3). We have studied in detail the effect of the water content in sulfuric acid used as a medium and solvent for APS on the K values obtained for GNPs. GNPs were synthesized through ultrasonic exfoliation of expanded graphite compounds in aqueous dispersion (3 g/dm^3) without surfactant. Then, to determine the K values, samples of the multi-layered GNPs so obtained were suspended in the solution of the surfactant NF (concentration of graphene material 0.025-0.035 g/dm^3), and additionally treated by ultrasound. The results are shown in the Figure 2. In this figure, the concentration of sulfuric acid below 100 % means simply "100% - content of water". To plot in the same scale concentration of sulfuric acid above 100%, we calculated the "negative" content of water by dividing the content of free SO_3 by its molecular weight (80) and by multiplying the molecular weight of water (18). Thus, 101% sulfuric acid corresponds to the content of 1%·80/18=4.44% of free SO3 (Figure 2).

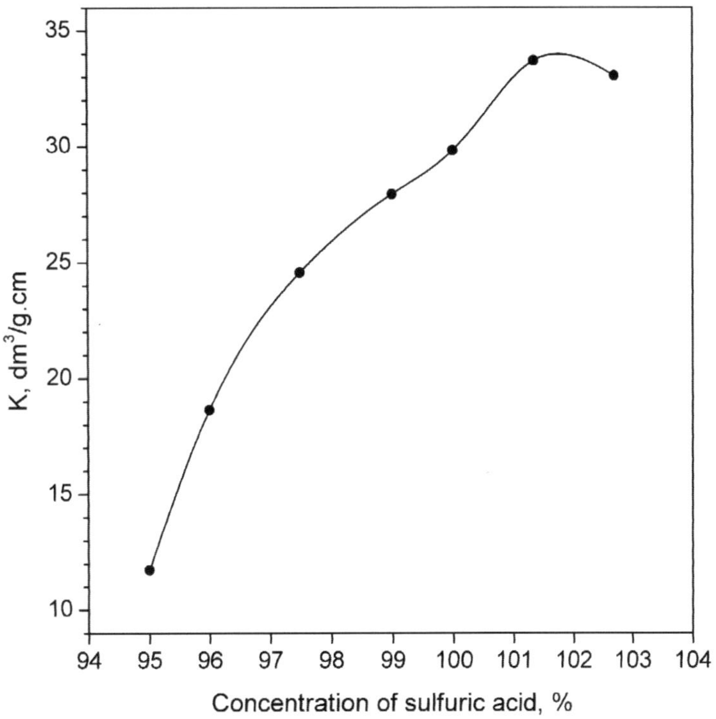

Figure 2. Effective K values for GNPs samples obtained from graphite intercalated by APS in sulfuric acid with different content of water or free SO_3. The K values calculated with respect to dry weight of GNPs including oxygen groups. An aqueous solution of surfactant NF was used in the process of ultrasonic exfoliation

The Figure 2 demonstrates the pronounced effect of the concentration of sulfuric acid on ability of the EGICs for the ultrasonic exfoliation. Even little admixtures of water in sulfuric acid substantially retard ability to exfoliation. On the other hand, a small content of free SO3 facilitates exfoliation. These data prove the interpretation made in our previous work [16].

The other essential factor determining the exfoliation effectiveness of the EGICs under the ultrasonic treatment, is the absence or presence of a surfactant. Generally saying, it is practically impossible to exfoliate common graphite materials in water without adding a surfactant, because of bad wettability of graphene planes with water. On the other hand, graphite oxide (GO) is spontaneously exfoliated in water due to large amounts of oxygen groups attached to the graphene skeleton. The expanded graphite materials obtained by the cold expansion of graphite intercalated with APS in sulfuric acid after hydrolysis contain 9-13 wt% oxygen groups. This in insufficient for spontaneous exfoliation of the GO in water, but greatly facilitates the ability for the ultrasonic exfoliation in pure water and aqueous solutions of surfactants compared to common graphite materials. For instance, the morphology of the EGICs is very similar to that of the TEG (worm-like particles, length up to 1-2 cm). However, as it was shown in the Figure 1, the TEG is much less exfoliated than the EGICs, which can be explained by the absence of oxygen groups in the TEG (they are eliminated after the high-temperature expansion process).

Principally, there are different methods of post-treatment of EGICs prior to ultrasonic exfoliation. The simplest one is hydrolysis and washing with water. Under these conditions, molecules of water behave as nucleophile and react with the positively charged graphene skeleton. As a result, the hydrolyzed product contains the oxygen groups. However, oxidized graphene layers are very reactive and easily interact with a lot of substances, even with inert hydrocarbons. In all the cases, the initial yellow-to-khaki color of the EGICs changes to dark-gray. One could think that different graphene substances can be obtained after post-treating the EGICs by varying reagents. However, our experiments indicate that in most cases it is not true. It seems that the treatment of the EGICs with water, alcohols and a large variety of organic compounds essentially result in similar, if not the same, product, namely, graphene containing oxygen groups. The reason of this is that positively charged graphene layers in the EGICs are very strong oxidizers, and the redox process is the first one occurring after the interaction with the organic compound. Usually, almost all organic compounds (with rare exceptions) contain hydrogen atoms connected to carbon. Thus, any oxidation process in an oxygen-containing medium (sulfuric acid) generates water, and, finally, almost the same oxygen-containing graphene product as after the direct hydrolysis with water can be obtained.

If one wants to obtain graphene products with different nature, the EGIC must be treated with nucleophiles which cannot be easily oxidized, or the oxidation reaction must not be the first one. There is a very restricted assortment of such reagents. We used, with some success, ammonia, urea, and hexamethylenetetramine. In the case of ammonia, the first stage of its interaction with the EGIC is protonation by sulfuric acid. The final graphene material obtained after the treatment of the EGIC with dry gaseous ammonia does not contain oxygen groups. It is obvious that some redox processes should occur for the reduction of the oxidized graphene skeleton, but now it is not possible to identify the nature of the processes taking place. Principally, the similar process occurs when the EGIC reacts with urea. In this case, carbon dioxide evolves, and amide groups react as base, substituting the sulfate functionality. However, contrary to the hydrolysis of the EGIC, its ammonolysis with gaseous ammonia or urea results in obtaining hydrophobic GNPs, the weight of which is precisely equal to the weight of the initial graphite. This indicates that no functional groups are attached to the surface of the GNPs so obtained. It may be that the positive charge of oxidized graphene layers is neutralized due to the redox reaction with ammonia or amide groups.

In the Figure 3, the plots of K vs. time of ultrasonic treatment for the samples of the EGICs deintercalated with ammonia (plot 3) and urea (plot 2) are demonstrated and compared to the original graphite (plot 1). In these experiments, the original graphite was GSM-2, and dimethylacetamide was used as a solvent instead of an aqueous surfactant solution.

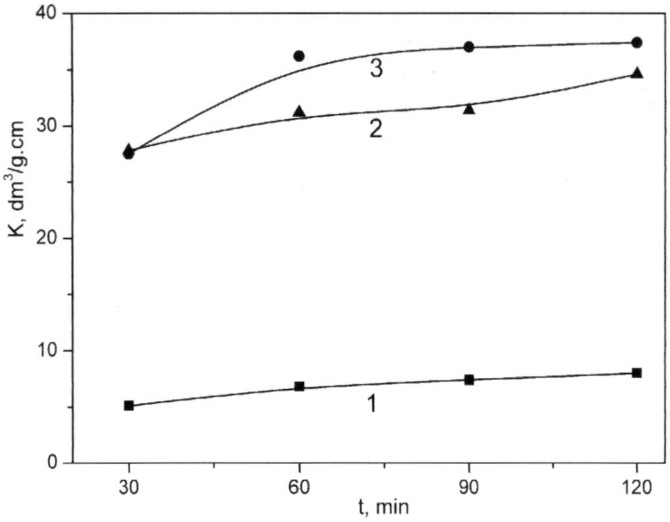

Figure 3. K vs. ultrasonic treatment time dependencies obtained for the EGIC samples deintercalated with ammonia (plot 3) and urea (plot 2) compared to the original graphite (plot 1). The concentration of the graphite materials in dimethylacetamide is 0.5 g/dm3

One can see that after the cycle "intercalation→expansion→deintercalation", the ability of the graphite material for the exfoliation considerably increases, which can be explained by the pre-exfoliation of graphite after that cycle. Multi-layered GNPs were obtained in these experiments, as indicated by K values.

However, the variants of deintercalation of EGICs with ammonia or urea, resulting in obtaining hydrophobic GNPs, are more complicated for scaling because of the evolution of heat and multiple technological stages. Up today, the hydrolysis remains the most simple and practical method for post-treating EGICs to synthesise GNPs, if the presence of surface oxide groups is allowable or favorable. Dependently on the absence or presence of surfactants, multi- or few-layered GNPs can be obtained by ultrasonic exfoliation. The typical SEM and TEM images of different kinds of GNPs so obtained are shown in the Figures 4-7.

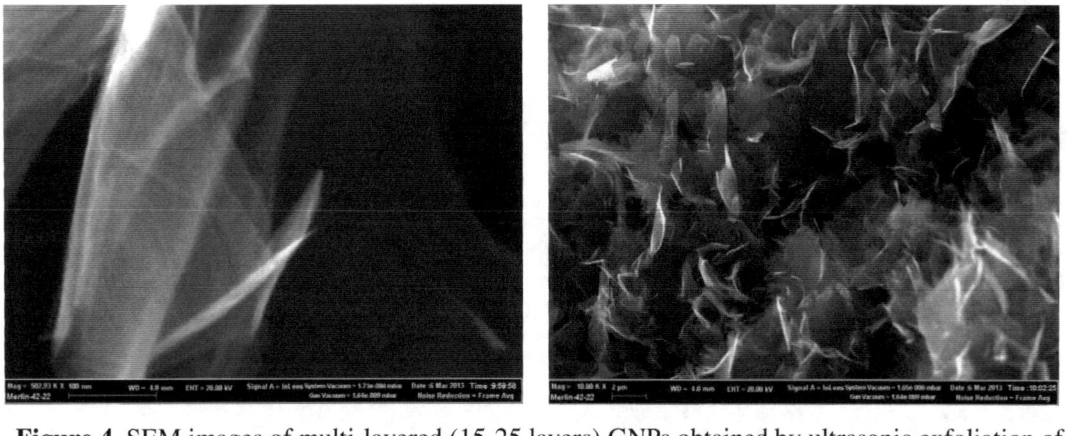

Figure 4. SEM images of multi-layered (15-25 layers) GNPs obtained by ultrasonic exfoliation of the hydrolyzed EGIC in water without surfactant

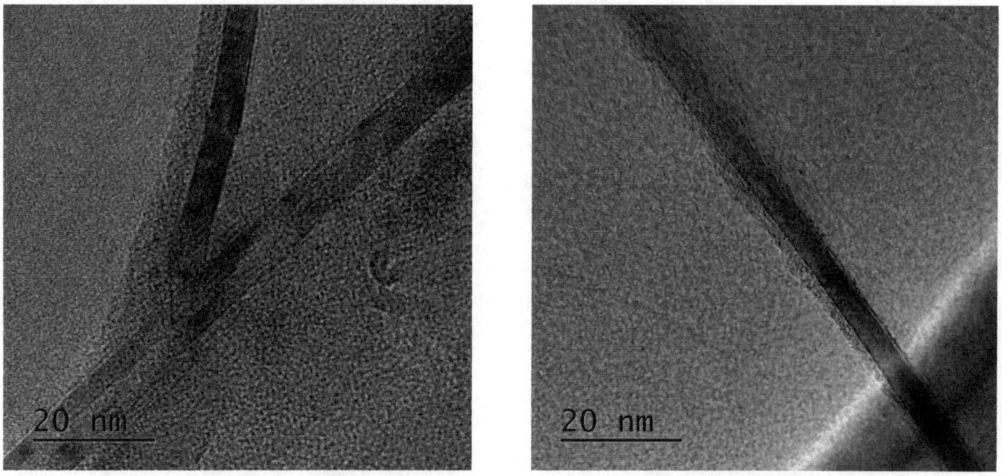

Figure 5. TEM images of multi-layered (15-25 layers) GNPs obtained by ultrasonic exfoliation of the hydrolyzed EGIC in water without surfactant

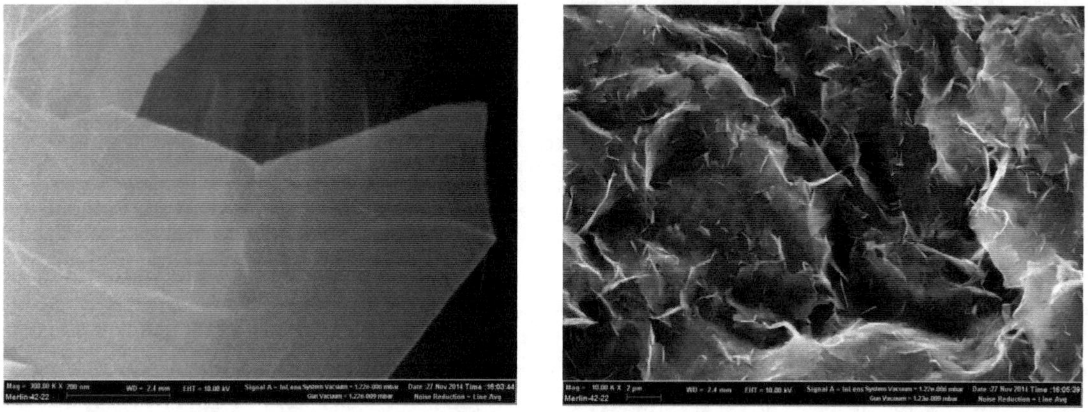

Figure 6. SEM images of few-layered (3-5 layers) GNPs obtained by ultrasonic exfoliation of the hydrolyzed EGIC in water with the surfactant

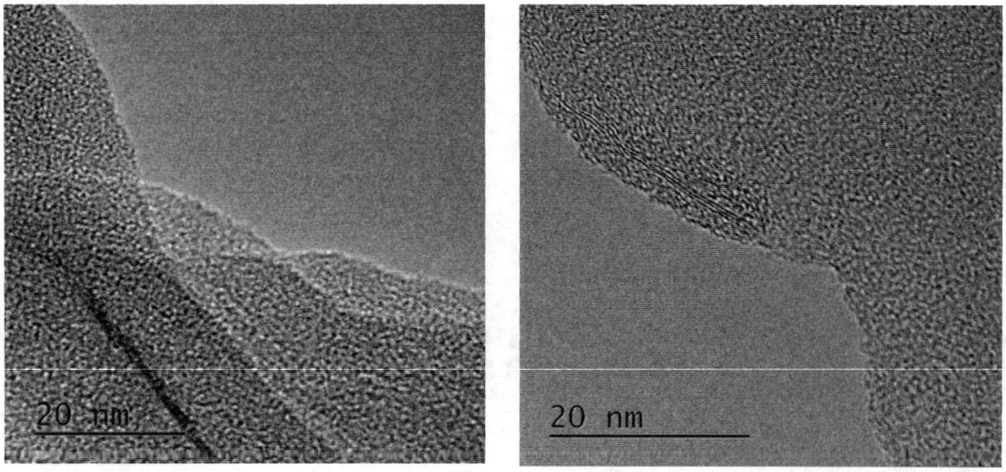

Figure 7. TEM images of few-layered (3-5 layers) GNPs obtained by ultrasonic exfoliation of the hydrolyzed CEGIC in water with the surfactant

4. Conclusions

In the process of cold expansion of the graphite using the APS, a covalent attachment of hydrosulfate groups to graphene layers probably takes place. Under the sonication, the cold EGICs are exfoliated into GNPs more easily than those ones obtained after hydrolysis or thermal expansion. The efficiency of the ultrasonic exfoliation of those EGICs strongly depends on the content of water in sulfuric acid (as oxidative intercalation medium): smaller amounts facilitate the process; this may be explained by increased acidity of the medium and redox potential of the APS compounds, as well as by larger amounts of oxygen groups available in the obtained GNPs.

The nature of groups available on the surface of the GNPs depends on the nature of a deintercalating reagent. This fact can be used regarding further applications of the synthesized GNPs (e.g., supercapacitor electrodes, composite materials, etc).

Acknowledgement

The research was performed within the framework of the Federal Target Program "Research and Development in Priority Areas of the Scientific and Technological Complex of Russia for 2014-2020" (State Contract No. 14.577.21.0091, Unique Identifier for Applied Scientific Research: RFMEFI57714X0091).

References

[1] Soldano C, Mahmood A and Dujardin E 2010 *Carbon* **48** 2127-2150
[2] Potts JR, Dreyer DR, Bielawski CW and Ruoff RS 2011 *Polymer* **52** 5-25
[3] Li A, Liu J and Feng S 2014 *Sci Adv Mater* **6** 209-234
[4] Hernandez Y, Nicolosi V, Lotya M, Blighe FM, Sun Z, De S, McGovern IT, Holland B, Byrne M, Gun'ko YK, Boland JJ, Niraj P, Duesberg G, Krishnamurthy S, Goodhue R, Hutchison J, Scardaci V, Ferrari AC and Coleman JN 2008 *Nat Nanotechnol* **3** 563-568
[5] Lotya M, Hernandez Y, King PJ, Smith RJ, Nicolosi V, Blighe FM, De S, Wang Z, McGovern IT, Duesberg GS and Coleman JN 2009 *J Amer Chem Soc* **131** 3611-3620
[6] Vaisman L, Wagner HD and Marom G 2006 *Adv Colloid Interfac* **128-130** 37-46
[7] Rastogi R, Kaushal R, Tripathi SK, Sharma Al, Kaur I and Bharadwaj LM 2008 *J Colloid Interf Sci* **328** 421-428
[8] Patent US 8696938 Zhamu A, Jang BZ (2014) Supercritical fluid process for producing nano graphene platelets.
[9] Chen G, Weng W, Wu D, Wu C, Lu J, Wang P and Chen X 2004 *Carbon* **42** (4) 753-759
[10] Lee JH, Shin DW, Makotchenko VG, Nazarov AS, Fedorov VE, Kim YH, Choi J-Y, Kim JM and Yoo J-B 2009 *Adv Mater* **21** 4383-4387
[11] Viculis LM, Mack JJ, Mayer OM, Hahn HT and Kaner RB *J Mater Chem* **15** 974-978
[12] Mukherjee A, Kang JH, Kuznetsov O, Sun Y, Thaner R and Bratt AS 2011 *Chem Mater* **23** 9-13
[13] Wei D, Grande L, Chundi V, White R, Bower C, Andrew P and Ryhanen T 2012 *Chem Commun* **9** 1239-1241
[14] Low CTJ, Walsh FC, Chakrabarti MH, Hashim MA and Hussain MA 2013 *Carbon* **54** 1-21
[15] Mahanandia A, Simon F, Heinrich G and Nanda KK 2014 *Chem Commun* **35** 4613-4615
[16] Melezhyk AV and Tkachev AG 2014 *Nanosyst Phys Chem Mat* **5** 294-306
[17] Celzard A, Mareche JF and Furdin G 2005 *Prog Mater Sci* **50** 93-179
[18] Zhao Q, Cheng X, Wub J and Yu X 2014 *J Ind Eng Chem* **20**(6) 4028-4032

IOP Conf. Series: Materials Science and Engineering **98** (2015) 012042 doi:10.1088/1757-899X/98/1/012042

Influence of Nano-Crystal Metals on Texture and Biological Properties of Water Soluble Polysaccharides of Medicinal Plants

G. Churilov[1], J. Ivanycheva[1], V. Kiryshin[1]

Division of General Chemistry with Course of Organic and Bio-Organic Chemistry, Ryazan State Medical University, Division of General Chemistry with Course of Organic and Bio-Organic Chemistry, Vysokovoltnaya Str., 9, Ryazan, 390026, Russia, tel./fax:+7(910)9013298

E-mail: genchurilov@yandex.ru

Abstract. When treating the plants seeds with nano-materials there are some quality and quantity changes of polysaccharides, the molecular mass increase and monosaccharides change that leads to the increase of physiological and pharmacological activity of carbohydrates got from medicinal plants. We have got water soluble polysaccharides and nano-metals combinations containing 0.000165- 0.000017 mg/dm^3 of the metal. In a case of induced anemia the blood composition has practically restored on the 10th day of the treatment with nano-composites. The use of pectin polysaccharides (that are attributed to modifiers of biological respond) to get nano-structured materials seems to be actual relative to their physiological activity (radio nuclides persorption, heavy metals ions, bacteria cells and their toxins; lipids metabolism normalization; bowels secreting and motor functions activation and modulation of the endocrine system.

1. Introduction

Most scientific publications nowadays are devoted to studying synthesis and mechanism of protein and nucleic acids action. Carbohydrates also perform important functions in micro-organisms, plants and animals [1, 2]. They are the most important sources of chemical energy transfer to biologically available energy accumulated in the form of human macro-energetic ATP links that later transfer to mechanic, electric, chemical and osmotic work. Carbohydrates have a plastic function, participate in bony tissue and cell formation, inhibit or activate the organism ferments [3]. Some carbohydrates and their derivatives are used as pharmaceuticals prohibiting radiation damages. Their ability to influence positively the inflammatory process, strengthening nervous and muscle tissues generation has led to the appearance of different native drugs of carbohydrate nature. Carbohydrates extracted from medicinal plants are non-toxic, have mild multidimensional influence on the organism. Their getting and use play a great role in medicine [4, 5]. The irregularity of pectin polysaccharides molecules structures and the block structure of macro-molecules determine a wide spectrum of pectins immune modulating effect based on their ability to influence the growth, generation and functional activity of immune competent cells [6]. Widening the range of pectin polysaccharides use in medicine is possible

Content from this work may be used under the terms of the Creative Commons Attribution 3.0 licence. Any further distribution of this work must maintain attribution to the author(s) and the title of the work, journal citation and DOI.
Published under licence by IOP Publishing Ltd

at the expense of determining the structural determinacy of their physiological activity and synthesis of polysaccharides with definite characteristics [7].

Nano-structured derivatives based on biocompatible natural polymers and products of their modification provoke great interest. Nano-composites based on plant polysaccharides (pectin polysaccharides, galactomannans) and chitosan are polymer systems used for transfer and controlled release of biologically active substances [8]. The use of pectin polysaccharides that are attributed to biological respond modifiers for getting nano-structured materials is actual in connection with their multi-dimensional physiological activity (radio nuclides persorption, heavy metals ions, bacteria cells and their toxins; lipid metabolism normalization; bowels secreting and motor functions activation; modulation of the endocrine system, etc.) [9, 10]. Using polysaccharides that differ by molecular mass, the degree of methyletherifying, galacturonic acid debris, neutral monosaccharides, line sections length and the branched domain molecule structure one can significantly change characteristics of nano-particles and nano-structures that in its turn can lead to significant change of their functional properties [11,12, 13].

2. Methods

We have studied knotgrass, bloodline buckwheat and vetch and bloodline beans. We have treated the seeds before planting with nano-powders of ferrum (Fe), cobalt (Co) and cuprum (Cu) produced at Moscow Institute of Steel and Alloys. The drugs have had particles of free form, high specific area up to 25 m^2/g and size of 20-50 nm.

When extracting and structuring water soluble polysaccharides we have used Sevaga's method (proteins deletion), Ceyzel's method (determining methoxy groups), chromatographic methods and dielectrolysis (estimating the polysaccharides homogeneity and monosaccharide composition) and Purdie, Hakomora & Coon's methods of methylation. We have used the infrared spectroscopy to investigate the methylation amplitude and control the polysaccharides discharge.

We have extracted proteins using the standard methods. We have had lectins extraction by Rigas & Osgud's method of divided ethapol fractioning and by the method of affine chromatography at sephadex G-150 (as matrix).

Having nano-composites as the object of study we have used water soluble polysaccharides of wild chamomile (Matricária chamomílla) and knotgrass (Polygonum aviculare) got by water extraction with further ethanol settling [14]. The extracted polysaccharides are pale pink powders when saluted in water form viscous solutions. In the infrared spectrum one can see the following absorption bands: 3600 cm^{-1} (OH-), 1739 cm^{-1} — valent variations – COOCH3, 1022 cm^{-1}, 1032 cm^{-1} and 1082 cm^{-1} — ring variations of the pyranoside cycle, band 840 cm^{-1} — characterizing α-configuration of the glygosidic linkage between hexose residues of galacturonic acid and band 890 cm^{-1} — characterizing β-linkage between sugar residues in the lateral chain. Fluctuations in band 1157 cm^{-1} are probably connected with the maximum degree of etherification.

To the extracted powders of polysaccharides we have added 1-10 % of nano-metals (Fe, Cu, Co) having characteristics APS 25 nm, purity 99.8 %, SSA 30-50 m^2/g, dissolved in water and processed in suprasonic disperser UZD-1-P for 8-10 minutes and deposited nano-composites with ethanol, washed with acetone and extracted in dry matter. Nano-composites are water soluble and 5-20 % solutions contain 0.000165-0.000017 mg/dm^3 of nano-metal. According to the infrared spectrography one can find in nano-composites containing the metal the weak line formation in range 1200 cm^{-1} together with the maximum growth in range ~1250 cm^{-1} that can be connected with type C-O-Me links formation in the probe.

3. Discussion

We have studied the increase of active substances in plants under the influence of ultra-fine ferrum, cobalt and cuprum micro-elements by the example of vetch seeds growth and development. When using low concentrations (0.01-0.08 g per hectare norm of seeds planting) there is activation of initial growing processes, and the yield increases by 25-32 %. At that we have seen protein and water soluble polysaccharides accumulation. According to the results of the vetch seeds chemical analysis it follows that the pre-plant treatment of seeds increases protein up to 40 % depending on the metal. We have seen the maximum protein when treating the seeds with nano-cobalt (0.03-0.06 g per hectare norm of seeds planting) that influences the nitrogen and protein metabolism promoting the protein synthesis. When treating with nano-cuprum and nano-cobalt there has been an increase of polysaccharides in the vetch seeds as compared with the control by 29 % and 32 % correspondingly. At that there has been a decline of lectin in total protein for nano-cuprum by 17 % and for nano-cobalt by 24 % that has increased the vetch seeds feed value. There has been an increase of polysaccharides probably connected with the decline of lectin that links carbohydrates. Water soluble polysaccharides extracted from vetch seeds have had the rotary power of $+120^0$….$+129^0$ and that of uronic acids up to 40 % that allows attributing them to glycoron glycans. The increase of polysaccharides has not influenced the quality of the extracted polysaccharides but changed their quantity correlations. The extracted polysaccharides have included D-galacturonic acid, galactose, glucose, mannose, arabinose aramnosa, xylose and their amount depends on the nano-metal. So treating the vetch with nano-cobalt has increased galactose by 71 % and decreased aramnose and arabinose by 43 % and 28 % correspondingly, the cuprum treatment has decreased aramnose by 8 %, arabinose by 6 %, xylose by 4 % (Figure 1). So it may be a condition for creating polysaccharides with definite properties.

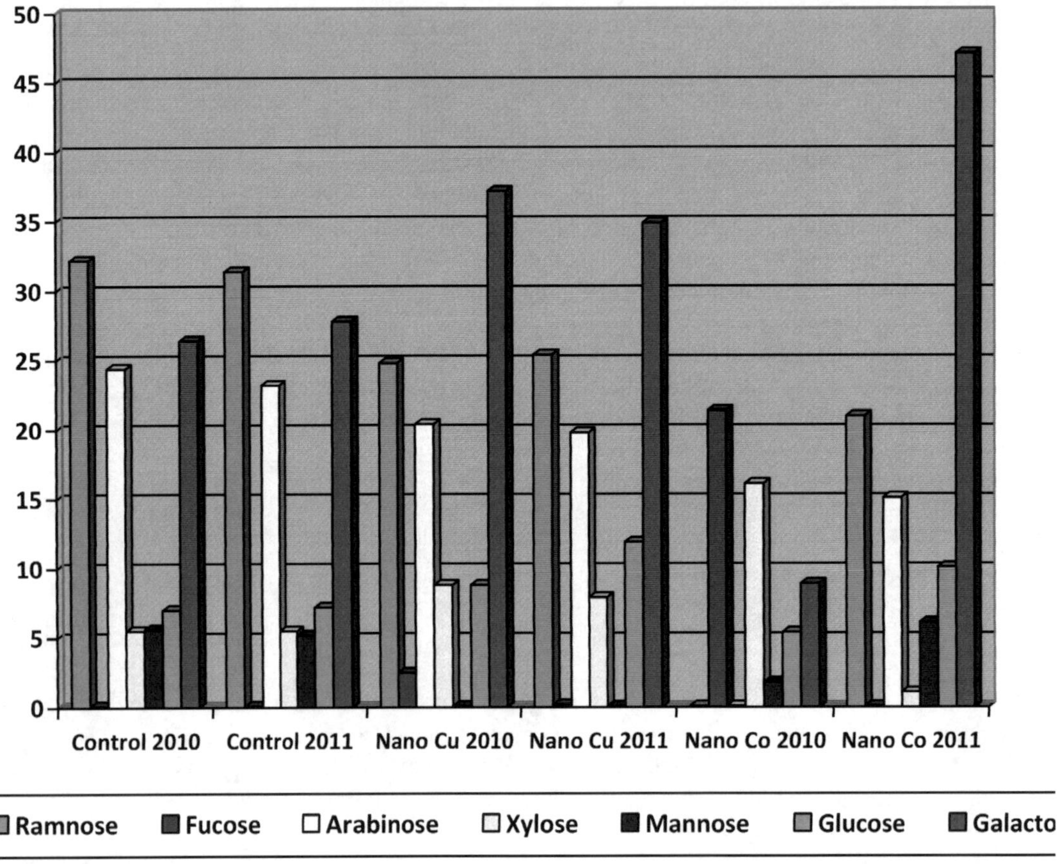

Figure 1. Percentage Composition of Monosaccharides in Vetch Seeds Polysaccharides

Polysaccharide composition of knotgrass. We have achieved the highest extraction of water soluble polysaccharides when treating the raw material in a boiling-water bath for 90 minutes in double replication. After fractioning from the aerial portions of knotgrass in the period of their maximum accumulation we have got homogenous fractions A, B, C, D differing in their monosaccharide composition and physical-chemical properties. In the extracted polysaccharides there prevailed sourer fractions A and B containing 87-92 % of galacturonic acid that allows attributing them to ramnogalacturons. Polysaccharide B has contained galactose besides ramnose that cuts D-galacturonic acid and the bio-polymer rotary power. Polysaccharides C and D are characterized by the small amount of uronic acids and low optical activities.

The pre-plant treatment of the knotgrass seeds has increased the polysaccharides extract on the average by 25 %. The comparison of the chromatography, infrared spectroscopy and ferment analysis results with the data of methylation and periodate oxidation has let determine the structural elements of homogenous A, B, C, D polysaccharides fractions extracted from the knotgrass without and after treating with nano-cobalt (Table 1). Polysaccharides A of both plants are close in their structure and composition. They consist of D-galacturonic acid blocks linked by α-D-1.4-glycoside link. The blocks are linked together with ramnose that is linked to the chain by 1.2-glycoside link. The amount of galacturonic acid after treating with nano-cobalt has increased in the chain. The correlation of galacturonic acid and ramnose residues in the polymer chin has been 9:1 for the knotgrass and 8:1 with nano-cobalt. The molecular masses have been 15000 and 16000 a.e.m. correspondingly.

Table 1. Physical-Chemical Characteristics of Extracted Polysaccharides

Plant Knotgrass	Polysaccharides Extraction	$[\alpha]^{20}_D$	Uronic Acids, %	Monosaccharides, mol					
				Galactose	Glucose	Mannose	Xylose	Arabinose	Ramnose
Control	Original	+116	46	14	1	+	+	28	4
Nano Co	Original	+126	42	29	3	2	1	2	22
Control	A	+269	92	—	—	—	—	—	+
Nano Co	A	+272	91	—	—	—	—	—	+
Control	B	+226	65	1	—	—	—	—	4
Nano Co	B	+236	72	3	—	—	—	—	1
Control	C	+73	+	2	—	—	—	1	9
Nano Co	C	+36	+	6	6	2	—	1	1
Control	D	+28	+	10	1	—	—	[illegible]	3
Nano Co	D	+21	+	9	10	1	—	3	9

Nano-cobalt has influenced changing the structure of the polysaccharide from the knotgrass. It has become branchier and got a higher degree of polymerization and as a result 6.7 % higher molecular mass than the control one. Nano-cobalt has not changed the structural elements. The main chain of polysaccharides B has consisted of D-galacturonic acid and ramnose. Ramnose has been linked with galacturonic acid by links 1, 2 in the main chain. At that it is the point of branching lining D-galacturonic acid in position 3. D-galacturonic acid also serves the branching point linking galactose in position 3.

Polysaccharides C and D being an inconsiderable part of the plants polysaccharide complex are branchy polymers that differ by the branchy degree and molecular masses. The assumed branchy points of polysaccharides C and D can be mannose, galactose and ramnose.

We have determined the monosaccharide composition of polysaccharides extracted in plants vegetation including galactose, glycose, xylose, arabinose, ramnose, mannose and fucosa (Figure 2).

Under nano-cobalt effect ramnose and galactose have been decreasing as the plant develops from 37.5 % to 18.0 % and from 50.1 to 21.3 % correspondingly. Glucose has changed from 12.5 to 7.8 % for 17 days (beginning of blossom), then there has been some increase of this monosaccharide from 12.5 % for 15 days (total blossom) and finally smooth down to 6.5 % for 45 days (end of blossom). Vice versa, arabinose increases as far as the plant vegetation up to 39 %. The results of the water extract analysis show the increase of potassium by 11.9 %, calcium by 10.5 % and phosphorus by 13 % as compared with the control that must influence the properties of the extracted polysaccharides increasing their water solubility.

Monosaccharides play different roles in plants development and the fact that under nano-metals during the vegetation there has been some change of the monosaccharide composition lets suppose different directivity of some physiological and synthetic processes necessary for the plant at the concrete moment. This allows supposing that nano-metals influence the activity of ferments responsible for oxidation-reduction processes and photosynthesis activity. At that there are some changes in many sides of the plant life activity including protein and carbohydrate metabolism that leads to accumulating biologically active compounds and changing their physiological activity.

Pharmacological activity of the knotgrass polysaccharide. A polysaccharide is a mixture of polysaccharide fractions: (I) State Registration Number - 6412486 and (II) State Registration Number - 6412585.

During the experiment with guinea pigs the polysaccharide complex has not had any allergenic properties not causing the allergized animals to have any fever, eosinophilia and leukogram change in a case of repeated administration. While studying the acute toxicity of the polysaccharide extracted from the knotgrass after the seeds' pre-plant treatment with ultra-fine Co powder it has been impossible to determine LD_{50} because of extremely low toxicity of the polysaccharide. Administrating its maximum dose [1000 mg/g] a day has not caused any changes of the animals' systemic condition and the composition of circumferential blood, therefore the ultra-fine metal powder do not cause any ghost effects of the macro-organism.

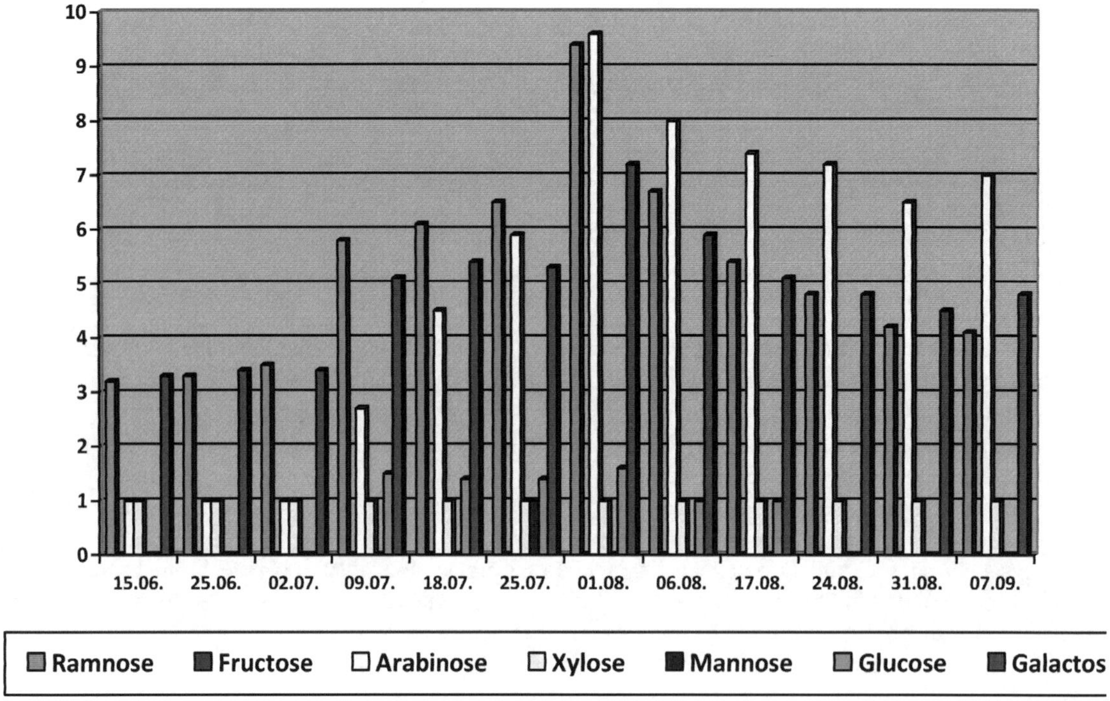

Figure 2. Percentage Composition of Monosaccharides in Knotgrass Polysaccharides in Progress

The amount of uronic acids, methoxy groups, ash content and rotary power of polysaccharides are presented in Table 2.

Table 2. Characteristics of Polysaccharides Extracted from Knotgrass Aerial Portions (average data, 2010-2011)

Date	Group	Yield, %	Ash, %	$[\alpha]^{20}{}_D$, degrees	% Content of Totally Dry Polysaccharide Mass: Uronic Acids	% Content of Totally Dry Polysaccharide Mass: Metoxy Groups
8-12.07	Control	6.2	18.4	+108	46.2	3.82
	Nano Co	6.9	22.3	+111	37.2	3.80
15-22.07	Control	7.4	19.9	+112	45.3	3.80
	Nano Co	12.2	24.2	+123	40.7	4.20
22-25.08	Control	12.8	18.2	+118	47.8	4.35
	Nano Co	14.2	21.1	+122	42.2	4.60
25.08-05.09	Control	5.8	18.3	+120	47.3	4.14
	Nano Co	11.6	22.7	+128	43.8	4.70

While using water soluble polysaccharides we have not discovered any significant difference of the blood cell composition. But there has been a tendency of increasing the number of erythrocytes and hemoglobin until the 15th day of the experiment and an authentic increase of leucocytes and later gradual decrease of these parameters until the 20th day of the experiment and initial parameters reconditioning after the drugs cancellation. Probably water soluble polysaccharides have stimulating effect on the state of the blood forming organs that activate them, where polysaccharides of the knotgrass pre-plant treated with ultra-fine cobalt powder have had the most stimulating effect.

Pharmacological activity of nano-composites. To have biological investigations of nano-composites we have chosen a group of 80 white rats aged 4-14 months and weighing 140-240 g. We have divided the group in the following way: 20 animals got the normal saline solution to compare the experiment results (control), 60 animals have got anemia by being administered subcutaneously with ethanoate lead every day during 5 days in the dose of 40 mg/g (experiment). The animals of the experiment group (that got ethanoate lead during 5 days) have shown the following dynamics of blood parameters by the 5th day: the number of erythrocytes - $2.41\pm0.6\cdot10^{12}$/l; leucocytes - $3.12\pm0.84\cdot10^{9}$/l; hemoglobin - 43.1 ± 4.2 g/l. The animals have become slack with considerable weight loss, fall-off, being aggressive, poor appetite and hemmoroidal syndrome.

To get rid of the experimental anemia the animals got 0.5ml of 5 % solution of ferrum nano-composite and Polygonumaviculare L. (experiment group 1) and 0.5 ml of 5 % solution of ferrum nano-composite and Matricariachamomilla L. (experiment group 2) every day (Table 3).

Table 3. Blood Basic Morphological Parameters of Rats when Treating the Experiment Anemia

	Blood Parameters	Control	Group 1 M±m	Group 1 %	Group 2 M±m	Group 2 %
Before Administ rating	Erythrocytes, $\cdot10^{12}$/l	6.20± 0.04	2.40± 0.06	38.7	2.20± 0.02	35.5
	Hemoglobin, g/l	110± 2	43.1± 0.02	39.3	44.0±0.3	40.0
	Leucocytes, $\cdot10^{9}$/l	10.50± 0.03	3.10± 0.04	29.5	3.20± 0.01	30.5

	3	Erythrocytes, $\cdot 10^{12}$/l	6.10±0.01	2.80± 0.05	45.9	2.80± 0.04	45.9
		Hemoglobin, g/l	112.0± 0.7	45.5± 0.06	40.6	49.60± 0.05	44.3
		Leucocytes, $\cdot 10^{9}$/l	11.0± 0.8	6.10± 0.07	55.5	7.50± 0.09	68.2
	5	Erythrocytes, $\cdot 10^{12}$/l	6.0±0.9	3.30± 0.03	55.0	4.70 ± 0.06	78.3
		Hemoglobin, g/l	115.0±0.3	59.2± 0.06	51.5	63.50± 0.02	55.2
		Leucocytes, $\cdot 10^{9}$/l	12.0± 0.5	6.8± 0.08	56.6	9.0± 0.7	75.0
DAYS	*7*	Erythrocytes, $\cdot 10^{12}$/l	6.0± 0.7	5.3± 0.05	83.3	5.70± 0.04	95.0
		Hemoglobin, g/l	117± 4	76.0± 0.5	65.0	86.0± 0.7	73.5
		Leucocytes, $\cdot 10^{9}$/l	12.0± 0.5	10.80± 0.06	90.1	10.0± 0.5	83.3
	10	Erythrocytes, $\cdot 10^{12}$/l	6.50±0.06	6.60± 0.07	101.5	6.90± 0.03	106.1
		Hemoglobin, g/l	110.2± 0.08	110.2± 0.06	100	110.3± 0.04	100.1
		Leucocytes, $\cdot 10^{9}$/l	10.60± 0.01	10.50± 0.08	99.0	10.20±0.05	97.2
	12	Erythrocytes, $\cdot 10^{12}$/l	6.50± 0.06	7.20± 0.09	110.8	6.70± 0.07	103.1
		Hemoglobin, g/l	115.0± 0.7	115.0± 0.4	100	115.6±0.01	100.1
		Leucocytes, $\cdot 10^{9}$/l	10.40± 0.06	10.40± 0.05	100	10.80± 0.06	103.8

During the experiment the control animals have got some increase of hemoglobin during the first 5 days. Later this parameter has come back to the norm that has not changed the results of the experiment. The animals of the experiment groups 1 and 2 have got some differences in dynamics of the registered parameters. So, for instance, hemoglobin in both experiment groups of animals have initially been pretty much the same – 43.1 ± 0.02 (group 1) and 44.0 ± 0.3 (group 2), but with nano-composites hemoglobin recovery of the animals from experiment group 2 has been faster than that of the 1st experiment group. So on the 3rd day the difference between the groups has been 4 % and on the 7th day - 8.4 %. On the 10th day of the experiment the animals of both groups have had the total recovery of hemoglobin (Figure 3).

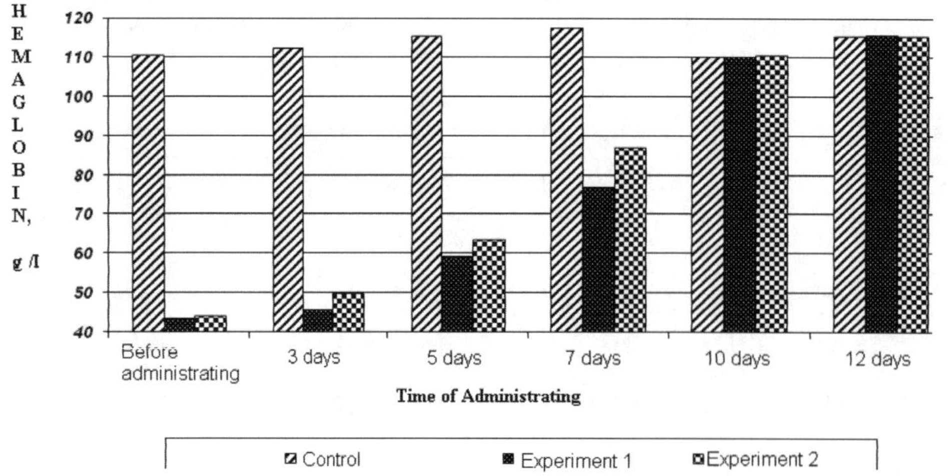

Figure 3. Dynamics of Hemoglobin Recover when Treating the Experiment Anemia

The dynamics of erythrocytes number recovery has also some differences between the groups. At the beginning of the treatment the 2nd experiment group animals have got the number of erythrocytes 3.2 % less than that of the 1st group, but erythrocytes recovery in it has been faster. So on the 3rd day the number of erythrocytes has been the same in the groups, on the 5th day the number of erythrocytes has been 11.7 % more than that of the1st experiment group; the erythrocytes number of the 2nd experiment group animals on the 7th day has reached 95.0 % of the control; on the 10th day it has exceeded the control by 6 %; and on the 12th day it has been only 3 % higher. The animals of the 1st experiment group have had erythrocytes recovery on the 10th day of treatment and on the 12th day it has exceeded the control one by 10.8 %.

The change of leucocytes happens fast enough in a case of giving nano-composites to animals. Polysaccharides that are their compounds give the plastic material for leucocytes cells building that has the carbohydrate part presented by some monosaccharides (mannose, fukose, galaktose and others), and nano-metal gives additional energy that promotes the leucocytes number recovery. This peculiarity is presented in Figure 4.

Though at the beginning of the treatment the leucocytes number of the 1st experiment group animals has been practically as that of the 2nd experiment group animals, its recovery has been more intensive. So on the 7th day of treatment the 1st experiment group animals have had leucocytes 6.8 % more than that of the 2nd experiment group. But the 2nd experiment group animals have had the leucocytes increase even on the 12th day of getting the nano-composite whereas the leucocytes number in the control and the 1st experiment groups has been declining but remained within the physiological norm.

So on the 10th day of treating the experiment anemia the blood picture has practically regenerated and on the 12th day the number of erythrocytes and leucocytes has even exceeded the control one.

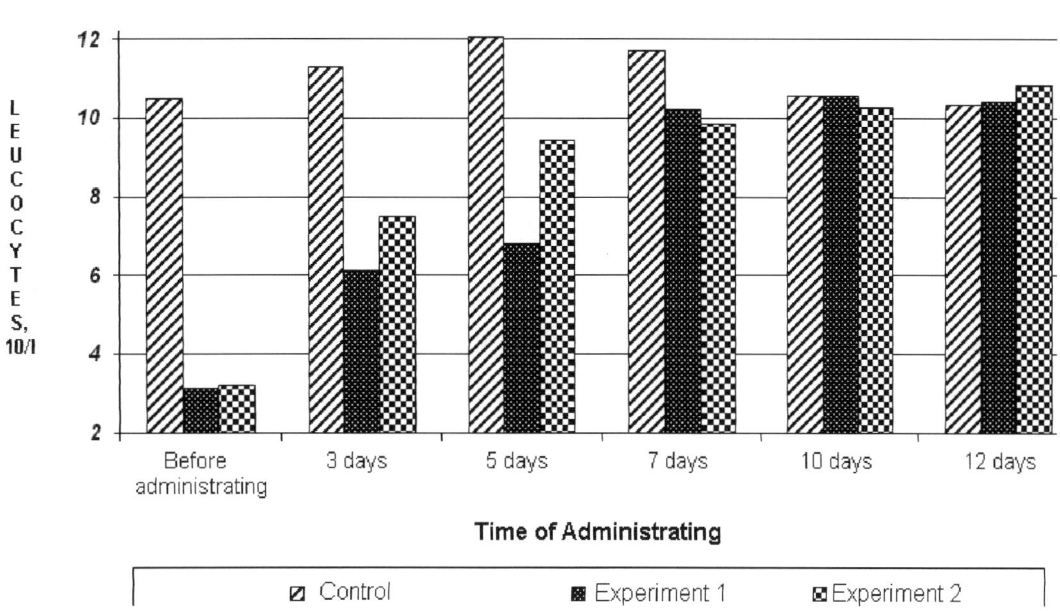

Figure 4. Dependence of Leucocytes on the Time of Administrating the Nano-Composites when Treating the Experiment Anemia

4. Conclusions

The seeds pre-plant treatment with the suspension of ferrum, cuprum and cobalt nano-powders increases accumulation of biologically active polymers by 30-40 %. At that there are changes of polysaccharides structural elements and quantitative composition of monosaccharides.

The polysaccharide complex extracted from Polygonumaviculare L after treating the seeds does not have any toxicity and has high pharmacological activity.

We have got water soluble nano-composites of plants polysaccharides Matricáriachamomílla and Polygonumaviculare containing a nano-crystal metal. We have proved that they are not a mechanical mixture. In a case of the experimental anemia ferrum and polysaccharides nano-composites of different plants are a stimulating factor for blood-forming function of the bone marrow and normalization of the blood state on 10-12th days of administrating the drug.

References

[1] Ovodov YuS 2009 Pectic Substances of plants in european north of russia *Russian Academy of Science* 243

[2] Zlobin AA, Martinson EA, Litvinets SG, Ovodova RG and Ovodov YuS 2014 Pectic polysaccharides of tylosis tissue of field ash stem *Bio-Organic Chemistry* **40** 211

[3] Golovchenko VV, Khramova DS, Ovodova RG, Ovodov YuS and Shashkov AS 2012 Structure of pectic polysaccharides isolated from onion *Allium Cepal.* using a simulated gastric medium and their effect on intestinal absorption *Food Chemistry* **134** 1813

[4] Golovchenko VV, Khramova DS, Ovodov YuS, Shashkov AS, Otgonbayar D and Chimidsogzo A 2012 Structural characterization of the polysaccharides from endemic mongolian desert plants and their effect on the intestinal absorption of ovalbumin *Carbohydrate Res.* **356** 265.

[5] Gorina YaV, Golovchenko VV, Ovodov YuS, Saprykina EV and Krasnov EA 2012 Elimination, characteristics and biological activity of polysaccharides of satinflower *Questions of Biol., Med. and Pharm. Chem.* **10** 9

[6] Markov PA Popov SV, Nikitina IR, Ovodova RG and Ovodov YuS 2011 Anti-Inflammatory activity of pectins and their galacturonan backbone *Rus. J. of Bioorg. Chem.* **37** 817

[7] Olennikov DN and Kashchenko NI 2014 Polysaccharides: modern state of study: experimental-scientometric research *Plant Mater. Chem.* **1** 5

[8] Pomogailo AD, Rozenberg AS and Uflyand IE 2000 Metals nano-particles in polymers *Chemistry* **13** 671

[9] Vainio U, Pirkkalainen K, Kisko K, Goerigk G and Serima R 2007 Copper and copper oxide nano-particles in a cellulose support studied using anomalous small-angle X-Ray scattering *European J. of Phys.* **42** 93

[10] Brusentsov NA, Brusentsova TN, Sergeev AV and Shumakov LI 2000 Ferri-magnetic liquids, fero-and ferri-magnetic suspensions for radio frequency induction hyperthermia of tumors *KhFZh* **34** 38

[11] Alekseev YuE, Garnovsky A D, Zhdanov Yu A 1988 Complexes of natural carbohydrates with metals cations *Chem. Success* **67** 723.

[12] Fengel D 1955 Structural changes of cellulose and their effects on the OH/CH2 valency vibration range in FTIR spectra cellulose and cellulose derivatives physical-chemical Aspects and industrial application *Woodhead Publishing Ltd.* 75-84

[13] Yakovlev AI, Churilov GI and Ginak AI 1985 Polysaccharides composition of p.avicularel *KhPS* **5** 619

IOP Conf. Series: Materials Science and Engineering **98** (2015) 012043 doi:10.1088/1757-899X/98/1/012043

Interaction of engineered nanoparticles with toxic and essential elements

AA Shumakova[1], IV Gmoshinski[1], SA Khotimchenko[1] and EN Trushina[1]
[1]Federal State Budgetary Science Institution «Institute of Nutrition», 2/14, Ust'inskiy proezd, Moscow, 109240 Russia

E-mail: Antonina_sh@list.ru

Abstract. Interaction of engineered nanoparticles with toxic and essential trace elements must be taken into consideration when estimating risks of NPs presented in the natural environment. The purpose of this work was to study the possible influence of silica, titanium dioxide (rutile) and fullerenol NPs on the toxicity of cadmium and to research the status of some trace elements and related indices of immune function in experiments on laboratory animals. Young male Wistar rats received cadmium salt (1 mg/kg b.w. Cd) orally for 28 days separately or in conjunction with the said kinds of NPs in different doses. A number of effects was observed as a result of combined action of Cd together with NPs, increase in bioaccumulation of this toxic trace element in the liver was most evident. The observed effects didn't show simple dose-dependence in respect to nanomaterials that should be taken into consideration when assessing the possible risks of joint action of nanoparticles and toxic elements existing in the environment in extremely low doses. Violation of microelement homeostasis caused by the combined action of Cd and NPs can have various adverse effects, such as inhibition of T-cell immunity induced by co-administration of Cd with rutile NPs.

1. Introduction

The assessment of potential risk of nanoparticles (NPs) and nanomaterials (NMs) is a difficult challenge, but it is of primary importance [1-3], considering rapid introduction of new products of nanotechnology (food, medicine, cosmetics, household chemicals, fuel components, construction materials and etc.) [4]. At the same time it's necessary to take into account that NPs are extremely seldom present in the environment and industry in the isolated form, instead they exist in complex with traditional components including contaminants, such as heavy metals, dioxins, etc [5].

NPs have physicochemical properties that are different from the same materials in the form of macroscopic dispersions or solids. One of them is the capability of NPs to penetrate through biological barriers and thereby facilitate trafficking of contaminants adsorbed on NPs surface or bond in their bulk phase. Thus, the question arises about possible risks, consisting in increased toxicity of chemical contaminants of food products and the environment as a whole in their conjunction with NPs. The possibility of such adverse effects was widely discussed [6-13].

Another source of NP's risk consists in their interaction with trace elements including essential ones (such as Fe, J, Se, Zn, Cu, Mn, Cr, Mo, Co) and other with less obvious biological function (such as Ag, V, B, Si, Br, F). Some of these elements are constituents of commercially manufactured NPs (such as Ag and Si), others may be adsorbed on NPs and penetrate with them through barriers of organism or, on the contrary, linger in the intestinal lumen due to binding to low-absorbable NPs.

Content from this work may be used under the terms of the Creative Commons Attribution 3.0 licence. Any further distribution of this work must maintain attribution to the author(s) and the title of the work, journal citation and DOI.
Published under licence by IOP Publishing Ltd

This may lead to changes in safety and bioavailability of some trace elements producing several systemic impacts, for example on immune system performance [14].

Therefore, the purpose of this research was to study the possible impact of some widespread NPs such as silica, titanium dioxide (rutile) and fullerenol NPs on the toxicity of cadmium together with status of some trace elements and related indices of immune function in experiments on laboratory animals.

2. Materials and methods

2.1. Nanoparticles and nanomaterials

Nanoparticles of titanium dioxide in the form of rutile (NPs TiO_2) were purchased from "Sigma-Aldrich" (Germany). The preparation was previously characterized by transmission electron microscopy (TEM) as consisting of partially agglomerated rod-shaped NPs with approximate size of 5×40 nm. Mean hydrodynamic diameter of particles equaled to 44.7 nm (51.5%) and 103.6 nm (48.5%), 90th percentile –129.3 nm according to laser dynamic light scattering (DLS). Fine dispersed amorphous silica "Orisil 300" was purchased in "Silica" LTD (Russia) and had BET specific surface 300 m^2/g according to the manufacturer. TEM characterized the sample as loose agglomerates of NPs sized from 5 to 100 nm with a very small number of free particles sized 5-20 nm. DLS-measured mean hydrodynamic diameter in sonicated water slurry was 56.6 ± 32.1 nm; 90th size percentile - 91.7 nm. Fullerenol ($C_{60}(OH)_{24}$) was purchased from "Fullerene-center" LTD (Russia) and had 99.5% purity in reverse-phase HPLC. DLS revealed marked aggregation of fullerenol molecules in water solution with formation of NPs (micelles) sized 2.89 nm (75.1%) and 1.05 nm (24.9 %) in diameter, 90th percentile – 4.68 nm.

2.2. Toxic heavy metal preparation

Cadmium chloride ($CdCl_2 \times 2.5\ H_2O$) was used as a source of toxic element with traditional level of dispersion. Preparation was analytical grade, purchased from «Khimmed», Russia.

2.3. Tested solutions preparation

Tested solutions of NPs were freshly prepared daily by dissolving nanomaterials and Cd salt in deionized water with further ultrasound treatment under the following conditions: frequency 44 kHz, power 2 W/ml, time 5 min, temperature 2-4^{0}C.

2.4. Animals

Experiment was performed on 128 adult male Wistar rats, weighing 67-109 g, kept in plastic cages in a temperature controlled (22°C) room, a 12 h light : dark cycle was maintained. Rats received standard full value semisynthetic rat diet prepared on the base of casein (coinciding with the composition to AIN76) and water *ad libitum* throughout the experiment. The animals were randomized into eight groups of 16 rats. The tested preparations were administered to animals once a day at a fixed time intragastrically by gavage in total volume of deionized water less than 2 ml. The experiment lasted for 28 days. The number of groups and doses of preparation tested are listed in Table 1.

2.5. Methods

Rats' body weight was registered daily throughout the experiment. Animals were withdrawn from the experiment by exsanguination under deep ether anesthesia, relative organ weights were determined. Hematological parameters of leucocytes were studied in hematology analyzer «Coulter AC TTM 5 diff OV» («Beckman Coulter», USA) with a standard set of reagents (manufactured by «Beckman Coulter», France). Expression of CD45RA, CD3, CD4, CD8, CD161a antigens was determined on peripheral blood lymphocytes (Ly) by direct immunofluorescence staining of whole blood with a panel of monoclonal antibodies conjugated with fluorescent dyes manufactured by «Beckman Coulter», USA. Analysis of the stained cells was performed on a flow cytometer "FC-500", manufactured by «Beckman Coulter» (USA) using program "Cytomics CXP Software". Ly populations were isolated by gating on parameters of small angle (FS) and side (SS) light scattering. The gating of CD3 + Ly populations via fluorescence channels FL1 and SS Lin was performed next. The results were recorded on a two-parameter histogram distribution of CD3+ (from gate B) using monoclonal antibodies against CD4 and CD8 detected in fluorescence channel FL5 and FL4, respectively. The expression of CD45RA and CD161a was determined in a separate test similarly. The total content of CD45RA + (B-Ly), CD3 + (T-Ly) and CD161a + (natural killer, NK) was expressed as per cent (%) of the total number of Ly analyzed considering at least 10^4 events per sample. The contents of CD3 + CD4 + (T-helper cells) and CD3 + CD8 + (cytotoxic T-Ly) were determined as % of their share in the total number of CD3 + cells. Dimensionless immunoregulatory index (IRI) was calculated as a ratio of CD4 + / CD8 + cells counts.

Table 1. List of groups of animals and studied preparation doses

Groups ##	Doses of preparation administered, mg/kg body weight			
	$CdCl_2 \times 2.5H_2O$ salt as Cd	Silica NPs as SiO_2	Rutile NPs as TiO_2	Fullerenol as $C_{60}(OH)_{24}$
1	0	0	0	0
2	1	0	0	0
3	1	1	0	0
4	1	100	0	0
5	1	0	1	0
6	1	0	100	0
7	1	0	0	1
8	1	0	0	10

The concentration of cadmium (Cd), lead (Pb), arsenic (As), silver (Ag), zinc (Zn), copper (Cu), chromium (Cr), manganese (Mg), aluminum (Al), nickel (Ni) and cobalt (Co) in organs (liver, kidney, brain) was determined by mass-spectrometry with inductively coupled plasma (ICP-MS) with octopole reaction system (ORS) by means of device "Agilent ICP-MS 7700x" ("Agilent Technologies", Japan). Sample preparation for ICP-MS measurements was carried out in microwave digestion system "TOP WAVE" ("Analytic Jena", Germany).

3. Results and discussion

Four rats died during the experiment in groups 1, 4 and 5; three – in groups 2 and 3, two – in group 7. A bilateral pneumonia was found on section in all the dead animals which was apparently a result of occasional aspiration of silica suspensions. The remaining animals had visibly normal behavior and performance, condition of hair and mucous surfaces and stool. At day 28 no pronounced and dose-dependent influences were observed on relative organ mass with exception of lung weight 16% ($p<0.05$) increase in group 8 (fullerenol, 10 mg/kg) compared to group 2 (Cd only).

As it's seen from **Table 2**, Cd intoxication didn't influence WBC indices in rats in the used doses. However, silica NPs (group 3) and rutile NPs (group 5) uniformly decreased Ly count and increased neutrophiles but this effect was not dose-dependent in respect to NPs dose.

Table 2. Mean (M ± m) indicators of leukocytes and platelets in rats from the treated groups

Groups ##	Number of rats	Total WBC count, 10^3 /μl	Contents of total WBC, %			Platelets 10^9 dm^{-3}
			Neutro-phyles	Lympho-cytes	Mono-cytes	
1	8	13.3±2.0	14.7±1.2	76.9±1.6	7.6±0.6	634.0±28.0

2	7	16.4±1.5	16.6±1.6	74.0±2.0	8.1±0.9	663.0±21.0
3	6	11.4±1. 9	19.7±1.9[1]	70.6±2.4[1]	8.7±1.2	624.0±38.0
4	5	12.2±1.2	18.2±1.9	71.7±3.4	8.8±1.9	689.0±55.0
5	6	12.7±2.1	19.4±1.1[1]	70.4±1.4[1]	9.1±0.8	706.0±34.0
6	8	18.9±2.0	15.5±2.4	76.2±2.6	7.2±0.5	685.0±16.0
7	6	14.6±2.6	17.6±1.2	73.4±1.2	8.1±0.9	674.0±27.0
8	9	14.3±1.0	16.0±1.5	74.6±2.0	9.0±0.7	661.0±27.0

[1]Difference with group 1 is significant, $p<0.05$

Study of cellular immune performance of animals (**Table 3**) revealed no influence of Cd on the studied parameters. NPs of all types had no effect on B-cells and NK counts. The most pronounced changes were the decrease in T-helper proportion and increase in T-cytotoxic cells with corresponding lowering of IRI in animals subjected to high dose of rutile NPs against the background of Cd intoxicaton. It's noticeable that such effect was absent in rats receiving this nanomaterial without Cd as it was shown in compatible conditions of experiment in [15].

ICP-MS studies of trace elements in the organs have clearly shown marked Cd accumulation in all tissues of the animals receiving this toxic element (**Table 4**). The most pronounced Cd rise was noticed in kidney (33.95%) and liver (45.14%) whereas in brain Cd concentration increased only by 55%. Cd accumulation in liver was strongly affected by silica NPs and (to lesser extent) fullerenol in small dose, whereas such effect from rutile NPs wasn't statistically significant. This result confirms the suggestion that Cd uptake in live organisms may be enhanced due to effect of its concomitant penetration through biological barriers together with some kinds of NPs [16]. Corresponding growth of kidney Cd accumulation in presence of NPs intake was far less pronounced and considered insignificant. Apparently no influence of Cd was noticed on NPs level in brain.

Table 3. Mean indices of cellular immunity performance in rats from the treated groups

Groups ##	Number of rats	Percentage of total lymphocytes, $M\pm m$			Percentage of total CD3+, $M\pm m$		CD4/CD8 (IRI)
		CD45RA+ (B-Ly)	CD3+ (T-Ly)	CD161a+ (NK-cells)	CD3+ CD4+ (Th-cells)	CD3+ CD8+ (T-cytotoxic)	
1	8	27.1±2.0	50.6±4.4	7.7±1.0	53.7±4.8	44.4±4.8	1.4±0.3
2	7	25.7±2.4	54.4±4.6	8.5±1.8	55.8±5.6	42.5±5.5	1.6±0.4
3	6	29.0±2.8	50.8±4.9	8.4±1.7	55.6±7.8	41.9±7.3	1.7±0.4
4	6	32.9±4.5	44.0±6.0	9.1±1.7	46.1±5.8	51.7±5.2	1.0±0.2
5	6	29.1±3.3	51.4±2.8	9.1±1.2	50.1±8.0	47.5±8.1	1.5±0.5
6	10	25.5±1.3	56.9±2.6	6.9±1.1	38.4±4.8[1,2]	60.1±4.8[1,2]	0.8±0.2[1,2]
7	6	29.1±3.2	50.0±3.6	9.5±1.4	49.6±7.0	48.6±7.2	1.3±0.3
8	10	27.2±1.9	53.1±3.3	6.9±0.6	50.4±5.1	47.9±5.0	1.3±0.3

[1,2]Superscript - group numbers, the difference with which was significant, $p<0.05$

Table 4. Cadmium accumulation in rat organs in presence of various NP types

Groups ##	Number of rats	Concentration μg/g tissue		
		Liver	Kidney	Brain
1	8	0.083±0.011	0.101±0.006	0.009±0.002

2	8	3.830±0.140[1]	3.530±0.180[1]	0.014±0.001[1]
3	8	5.200±0.490[1,2]	3.830±0.270[1]	0.016±0.002[1]
4	7	5.400±0.510[1,2]	3.430±0.260[1]	0.014±0.001[1]
5	8	4.210±0.420[1]	3.640±0.160[1]	0.013±0.001[1]
6	8	4.240±0.500[1]	3.680±0.150[1]	0.016±0.001[1]
7	7	4.740±0.250[1,2]	3.450±0.300[1]	0.014±0.001[1]
8	8	4.490±0.350[1]	3.720±0.200[1]	0.013±0.001[1]

[1,2]Superscript –see comment to table 3

Exposition of the animals to Cd demonstrated different impacts on levels of toxic, essential and conditionally essential trace elements (Tables 5, 6) that coincided with the data known from literature [17]. The effect of decrease in Pb accumulation was detected in liver, especially in rats additionally treated with low doses of silica (group 3), fullerenol (group 7) and high doses of rutile NPs (group 6). Unlike this Zn level significantly grew in following groups: 4 (high dose of silica), 5 (low dose of rutile) 7 and 8 (both doses of fullerenol). Consequently, this NPs markedly enhanced effect of Cd on Zn accumulation that was yet insignificant in group 2 (Cd only). Cu levels significantly grew in presence of fullerenol treatment (group 7) whereas Ag accumulation diminished and As increased in group 4 (high dose of silica NPs). Any differences in liver levels of Al, Ni and Cr weren't found in animals developed by both Cd and all kinds of NPs (data not shown).

Accumulation of Mn in liver was apparently not influenced by Cd itself but significantly enhanced by low dose of silica (group 5) and both doses of fullerenol (groups 7,8).

Table 5. Trace elements accumulation in rat liver in presence of cadmium and NPs

Groups ##	Number of rats	Concentration μg/g tissue						
		Pb	As	Ag	Zn	Cu	Mn	Co
1	8	0.064± 0.008	0.020± 0.002	4.14± 1.402	34.743 ±1.883	5.487± 0.318	2.164± 0.106	0.073± 0.004
2	8	0.046± 0.009	0.018± 0.001	2.306± 0.746	37.272 ±2.223	5.094± 0.434	2.212± 0.126	0.064± 0.005
3	8	0.038± 0.007[1]	0.018± 0.002	1.941± 0.747	39.277 ±1.432	5.201± 0.274	2.347± 0.115	0.071± 0.002
4	7	0.051± 0.009	0.024± 0.002[2]	2.272± 1.440	48.111 ±2.369[1]	6.597± 0.630	2.516± 0.073[1]	0.073± 0.005
5	8	0.042± 0.007	0.020± 0.001	1.465± 0.724	43.246 ±2.315[1]	6.522± 0.798	2.319± 0.122	0.083± 0.005[1]
6	8	0.031± 0.008[1]	0.020± 0.002	5.260± 1.585	38.748 ±1.566	5.007± 0.281	2.132± 0.080	0.067± 0.002
7	7	0.030± 0.006[1]	0.023± 0.002	1.001± 0.403	44.196 ±2.297[1]	6.320± 0.163[1,2]	2.612± 0.104[1,2]	0.073± 0.006
8	8	0.045± 0.009	0.020± 0.002	3.189± 0.825	42.694 ±1.791[1]	5.623± 0.221	2.466± 0.088[1]	0.077± 0.010

[1,2]Superscript –see comment to table 3

Table 6. Trace elements accumulation in rat kidneys in presence of cadmium and NPs

Groups ##	Number of rats	Concentration μg/g tissue					
		Pb	As	Zn	Cu	Mn	Co
1	8	0.066± 0.004	0.014± 0.001	21.722± 0.481	4.611± 0.064	0.915± 0.023	0.131± 0.003

2	8	0.060± 0.002	0.013± 0.001	23.952± 0.356[1]	5.287± 0.131[1]	0.979± 0.032	0.160± 0.006[1]
3	8	0.060± 0.002	0.013± 0.001	23.529± 0.750	5.241± 0.165[1]	0.873± 0.028[2]	0.160± 0.006[1]
4	8	0.052± 0.002[1,2]	0.014± 0.001	23.629± 0.741[1]	4.857± 0.246	0.954± 0.033	0.166± 0.004[1]
5	8	0.065± 0.004	0.014± 0.001	24.443± 0.493[1]	5.309± 0.087[1]	1.026± 0.021[1]	0.171± 0.007[1]
6	8	0.050± 0.002[1,2]	0.013± 0.001	25.342± 0.593[1]	4.865± 0.195	1.033± 0.047[1]	0.161± 0.012[1]
7	8	0.060± 0.003	0.016± 0.001[2]	23.391± 0.643	4.932± 0.187	0.950± 0.021	0.155± 0.005[1]
8	8	0.064± 0.001	0.016± 0.001[1,2]	23.945± 0.444[1]	5.184± 0.068[1]	0.957± 0.011	0.155± 0.008[1]

[1,2]Superscript –see comment to table 3

The landscape of trace elements distribution in kidneys was similar to some extent in regard to Pb, Zn and Mn, but the changes caused by NPs were less pronounced and could be explained in most cases by effect of Cd itself. In any way Cd strongly enhanced Co accumulation in kidney irrespective to NPs development. Effects on Al, Ni, Cr and Ag accumulation in all groups were insignificant (data not shown).

Most changes in trace elements levels in brain were small in dimension and insignificant that reflects high degree of brain-blood barrier impermeability to chemical factors and stability of brain media homeostasis (data no shown). The only exception was decrease in Co accumulation from 12.0±1.5 ng/g tissue (group 1) to 8.5±0.7 ng/g tissue (group 2) caused by Cd. Said effect was obvious also in combined treatment with all 3 types of NPs.

The results obtained confirm the presence of interactions of different types of NPs with toxic trace element Cd during its absorption and biodistribution in the bodies, which coincides with the data obtained earlier on alternative biological models [16,18,19]. Said changes of Cd accumulation may result in different manifestations of Cd toxicity, in particular its immunosuppressive action [20]. In this work it was demonstrated when joint treatment with Cd and rutile (TiO_2) NPs displayed pronounced decrease in Th mediated immunity in rats. The fact is noticeable that in many cases the effects of NPs appear to be dose independent i.e. pronounced at low rather than high doses. Today the data is insufficient to explain comprehensively this observation but one should keep in mind the instability in time and the variability of t properties of NPs depending on the biological environment [21]. It can be suggested in this case that NPs aggregation in gastrointestinal tract (that is more pronounced at high concentrations/doses in accordance with the law of mass action) is a factor that counteracts with NPs uptake and concomitant Cd trafficking. This may mean that effects potentiating toxic elements action may be more pronounced at extremely low doses of NPs that apparently take place in the real environmental conditions.

Numerous data on changes in the bioavailability of trace elements may be placed among the effects of "the interaction of trace elements" in animals treated with Cd [17]. Their explanation implies that trace elements compete for cellular sites responsible for uptake, transport, retention (fixation) and clearance both to influence of toxic metal (i.e. Cd) on activity of enzyme systems responsible for essential elements metabolism.

As shown by the data obtained in this work, various types of NPs, like silica, rutile and fullerenol are not neutral in respect to these processes, and may modify them to some extent, which is an additional source of risk of breaking the element homeostasis in the combined action of Cd and these NMs.

4. Conclusion

Thus, a number of effects were observed in combined action of Cd and NPs of practical importance including silica, titanium dioxide (rutile) and fullerenol on laboratory animals, the most significant effect was the strengthening of bioaccumulation of this toxic trace element in the liver. The effects observed do not show a simple dose-dependence towards NMs that should be considered when assessing the possible risks of joint action of NPs and toxic elements present in the environment in extremely low doses. Violation of microelement homeostasis caused by the combined action of Cd and NPs can have various adverse effects, such as inhibition of T-cell immunity induced by co-administration of cadmium with rutile NPs.

References

[1] Oberdörster G, Oberdorster E and Oberdorster J 2005*Environ. Health Perspect.* **113** 823
[2] Borm PJ, Robbins D, Haubold S, Kuhlbusch T, Fissan H, Donaldson K, Schins R, Stone V, Kreyling W, Lademann J, Krutmann J, Warheit D and Oberdorster E 2006 *Particle Fibre Toxicol.* **3** (11) 1–35
[3] Onischenko GG and Tutelyan VA 2007 *Voprosy pitaniya(Problems of nutrition)* **76** (6) 4 (in Russian)
[4] Wise K and Brasuel M 2011 *Nanotechnol Sci Appl* ***4*** 73
[5] Holsapple MP, Farland WH, Landry TD, Monteiro-Riviere NA, Carter JM, Walker NJ and Thomas KV2005 *Toxicol Sci.* **88** (1) 12-17
[6] Cui F, He C, Yin L, Qian F, He M, Tang C and Yin C2007*Biomacromolecules* **8**(9) 2845-2850
[7] El-Shabouri M.H. 2002 *Int. J. Pharm.* **249** (1-2) 101-108
[8] Thomas KV, Farkas J, Farmen E, Christian P, Langford K, Wu QandTollefsen KE 2011 *J Toxicol. Environ. Health A.* **74** (7-9) 466-477
[9] Cronholm P, Karlsson HL and Hedberg J2013 *Small.***9**(7)970-982
[10] Limbach LK, Wick P, Manser P, Grass RN, Bruinink A and Stark WJ2007 *Environ Sci Technol.***41**(11)4158-4163
[11] Ying Zhu, Yu Zhang, Jing Li, Xiaoyong Zhang, Chunhai Fan and Qing Huang 2014 *Material of «The 7th International Nanotoxicology Congress «NanoTox2014»*256
[12] Melnik EA, Buzulukov YP, Demin VF, Demin VA, Gmoshinski IV, Tyshko NV and Tutelyan VA2013 *ActaNaturae (Russian version)* **5** (3) 111
[13] Quadros ME and Marr LC2010 *J Air Waste Manag. Assoc.* **60** (7) 770-81
[14] Benetti F, Bregoli L, Olivato I and Sabbioni E 2014*Metallomics* **6** 729
[15] Arianova EA, Shumakova AA, Tananova ON, Trushina EN, Mustafina OK, Sharanova NE, Gmoshinski IV and Khotimchenko SA 2012 *Voprosy pitaniya (Problems of nutrition)* **84** (6) 47
[16] Zhang X, Sun H, Zhang Z, Niu Q, Chen Y, and Crittenden JC 2007*Chemosphere* **67** 160
[17] Avtsin AP, Zhavoronkov AA, Rish MA and Strochkova LS 1991 Human microelementoses *Moscow, "Meditsina"edition* 496 P.
[18] Sun H, Zhang X, Zhang Z, Chen Y and Crittenden JC 2009 *Environ Pollut***157** 1165
[19] Yang WW, Miao AJ and Yang LY2012 *PLoS One*7e32300
[20] Demenesku J, Mirkov I, Ninkov M, Popov-Aleksandrov A, Zolotarevski L, Kataranovski D and Kataranovski M 2014 *Toxicology* **326** 96
[21] Maynard RL 2012 *Emerg. Health Threats J* **5** 10